Deepen Your Mind

前言

Dart 是由 Google 公司推出的現代化程式語言，最初知道它是因為 Angular 框架推出了 Dart 版本。抱著好奇去看了 Dart 的官網，它以簡單、高效、可擴充為開發目標，將強大的新語言特性與熟悉的語言構造組合成清晰、可讀的語法，並提供很多語法糖來保證以更少的程式量完成指定功能。

Dart 不僅是一門語言，Dart 的各種開放原始碼專案和 Pub 套件管理工具幫助開發人員變得更有生產力。舉例來說，開發者可以使用 Pub 獲得與 JavaScript 互動操作的能力、Web UI 框架、單元測試庫、用於遊戲開發的庫及使用 Dart 語言開發的 Flutter 跨平台移動 UI 框架等。

本書主要內容：

第 1 章架設開發環境，安裝與設定編輯器。

第 2 章主要介紹內建類型，包括數字、字串、布林、List 集合、Set 集合、Map 集合及符文 (runes) 類型的定義及使用。

第 3 章講解函數的定義，主要包含可選參數、匿名函數、回呼函數及將函數作為物件傳遞。

第 4 章介紹運算子，包含算術運算子、關係運算子、設定運算子、邏輯運算子、位元運算符號、條件運算子及類型測試運算子。

第 5 章介紹流程控制敘述，包含分支敘述 if、switch；循環敘述 for、while、do-while；跳躍陳述式 break、continue。

第 6 章講解類，包含類的定義、屬性、建構函數、方法、介面、Mixin 及枚舉類。

第 7 章講解異常，包含異常的拋出、例外的捕捉、自訂異常。

第 8 章講解泛型，包含使用集合 List、Set、Map 提供的泛型介面，以及自訂泛型類和方法。

第 9 章介紹函數庫，包含函數庫的宣告、匯入、核心函數庫、數學函數庫、資料轉換函數庫、輸入輸出函數庫。

第 10 章介紹非同步，包含 Future、Stream 及生成器函數。

第 11 章介紹多執行緒實踐途徑 Isolate，包含 Isolate 的含義、事件循環、訊息傳遞及不同 Isolate 間相互通訊。

第 12 章是擴充閱讀，包含可呼叫類別的宣告、擴充方法、類型定義、中繼資料及註釋。

第 13 章介紹服務端開發，包含基礎的 HTTP 請求與回應、shelf 框架的使用及使用路由包定義服務的 API。

第 14 章介紹 Angular 框架的基礎知識，包含專案結構、資料綁定、內建指令、範本引用變數、服務、子元件及表單。 第 15 章介紹 Angular 框架的進階知識，包含屬性指令、元件樣式、依賴注入、生命週期掛鉤、管道、路由、結構指令、HTTP 連接及專案部署。

第 16 章介紹材質化元件庫 angular_components，該函數庫包含表單、業務流及版面配置中常用的元件。

第 17 章是專案實戰，介紹資料庫的安裝與連接，以及透過用於時間規劃的專案 Deadline 來溫習本書所學的基礎知識。

劉仕文

目錄

第一部分

01 簡介

02 變數和內建類型

03 函數

04 運算子

05 流程控制敘述

06 類別

07 異常

08 泛型

09 函數庫

10 非同步

11 Isolate

12 拓展閱讀

第二部分

13 服務端開發

第三部分

14 Angular 基礎

15 Angular 進階

16 材質化元件

第四部分

17 專案實戰 Deadline

Chapter 01

簡介

1.1 概述

Dart 是由 Google 開發的電腦程式語言，它的語法類似 C 語言。可以使用 Dart 編寫命令列指令稿、伺服器端應用、Web 應用和行動端應用。

Web 應用：常採用 Angular Dart 框架開發 Web 應用程式，它是 Angular 框架的 Dart 版本。Dart Web 採用編譯器將 Dart 程式編譯成 JavaScript，使得 Dart Web 應用程式可在瀏覽器中執行。在開發應用時採用 dartdevc 編譯器，部署應用時採用 dart2js 編譯器。

行動端應用：採用 Flutter 框架可以編寫 iOS 和 Android 應用程式。

1.2 環境安裝與設定

Dart SDK 支援 Windows、Mac OS、Linux 3 種作業系統，這裡提供資源站供 SDK 下載：

官網：dart.dev/get-dart

安裝套件分為 3 種類型：穩定版、測試版和開發版，推薦使用穩定版。

首先請根據對應平台下載檔案套件，預設為 zip 壓縮格式檔案。解壓 zip 檔案，解壓後的目錄裡包含所需的 dart-sdk 目錄。

1.2.1 Windows 使用者

第 1 步：複製 dart-sdk 目錄到 C:\Program Files 目錄下。

第 2 步：增加 dart-sdk\bin 路徑到 Path 環境變數，這樣在任何目錄下都可以執行 Dart SDK 提供的工具命令。

以 Windows 10 為例，按右鍵螢幕左下角的「開始」按鈕，在彈出的選單中選擇「系統」，彈出如圖 1-1 所示的設定對話方塊，點擊右側的系統資訊即可彈出如圖 1-2 所示的 Windows 系統對話方塊。

▲ 圖 1-1 設定對話方塊

▲ 圖 1-2 Windows 系統對話方塊

選擇左邊的「進階系統設定」，打開如圖 1-3 所示的系統內容對話方塊。

▲ 圖 1-3 系統內容對話方塊

在如圖 1-3 所示的系統內容對話方塊中，首先選擇「進階」，然後點擊「環境變數」按鈕打開環境變數對話方塊，如圖 1-4 所示。可以在使用者變數或系統變數中增加環境變數，一般情況下在使用者變數中增加，雙擊 Path 變數項目進入如圖 1-5 所示的編輯環境變數對話方塊。

▲ 圖 1-4 環境變數對話方塊

▲ 圖 1-5 編輯環境變數對話方塊

在圖 1-5 所示的編輯環境變數對話方塊中點擊右側的「新建」按鈕，輸入值 C:\Program Files\dart-sdk\bin，最後點擊「確定」按鈕即可完成編輯。

第 3 步：點擊螢幕左下角「開始」按鈕，在彈出選單中找到「Windows 系統」資料夾，點擊該資料夾下的「命令提示符號」選單。在命令提示符號中輸入 dart 命令。出現如圖 1-6 所示資訊表示環境設定成功。

```
:\Users\C01668>dart
Jsage: dart [<vm-flags>] <dart-script-file> [<script-arguments>]

Executes the Dart script <dart-script-file> with the given list of <script-argument
s>.

Common VM flags:
--enable-asserts
  Enable assert statements.
--help or -h
  Display this message (add -v or --verbose for information about
  all VM options).
--package-root=<path> or -p<path>
  Where to find packages, that is, "package:..." imports.
--packages=<path>
  Where to find a package spec file.
--observe[=<port>[/<bind-address>]]
  The observe flag is a convenience flag used to run a program with a
  set of options which are often useful for debugging under Observatory.
  These options are currently:
      --enable-vm-service[=<port>[/<bind-address>]]
      --pause-isolates-on-exit
      --pause-isolates-on-unhandled-exceptions
      --warn-on-pause-with-no-debugger
  This set is subject to change.
  Please see these options (--help --verbose) for further documentation.
--write-service-info=<file name>
```

▲ 圖 1-6 命令提示符號對話方塊

1.2.2 Mac 使用者

下載 Mac OS 版本的安裝套件到電腦，這裡預設放在下載目錄裡，如圖 1-7 所示。

▲ 圖 1-7 安裝套件存放目錄

雙擊檔案名稱即可解壓，得到 dart-sdk 資料夾。通常會將安裝目錄放在 /usr/local 目錄下，因此首先打開終端輸入命令 sudo mv，再將 dart-sdk 目錄拖到終端，這裡 mv 和路徑間有空格，如圖 1-8 所示。

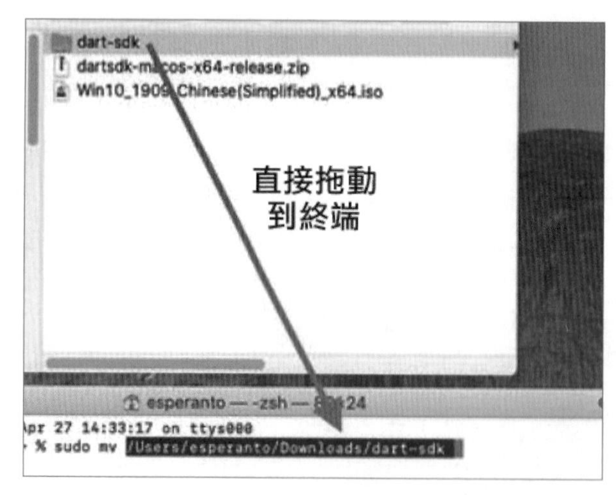

▲ 圖 1-8 行動資料夾

在終端中繼續輸入 /usr/local，這裡 /usr 前面需要空格，然後確認即可執行命令，如圖 1-9 所示。

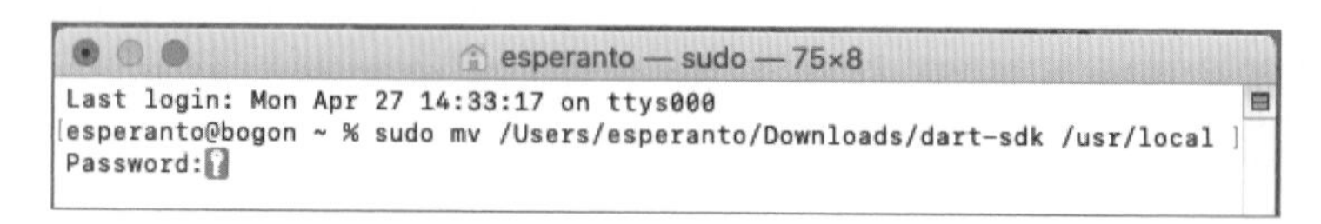

▲ 圖 1-9 執行移動目錄命令

接下來設定環境變數。

bash 終端設定程式如下：

```
echo 'export PATH=/usr/local/dart-sdk/bin:${PATH}' >> ~/.bash_profile
source ~/.bash_profile
```

zsh 終端設定程式如下：

```
echo 'export PATH=/usr/local/dart-sdk/bin:${PATH}' >> ~/.zshrc
source ~/.zshrc
```

在終端中輸入 dart 命令，出現如圖 1-10 所示資訊表示設定成功。

```
local — -zsh — 80×40
esperanto@bogon local % dart
Usage: dart [<vm-flags>] <dart-script-file> [<script-arguments>]

Executes the Dart script <dart-script-file> with the given list of <script-argum
ents>.

Common VM flags:
--enable-asserts
  Enable assert statements.
--help or -h
  Display this message (add -v or --verbose for information about
  all VM options).
--package-root=<path> or -p<path>
  Where to find packages, that is, "package:..." imports.
--packages=<path>
  Where to find a package spec file.
--observe[=<port>[/<bind-address>]]
  The observe flag is a convenience flag used to run a program with a
  set of options which are often useful for debugging under Observatory.
  These options are currently:
      --enable-vm-service[=<port>[/<bind-address>]]
      --pause-isolates-on-exit
      --pause-isolates-on-unhandled-exceptions
      --warn-on-pause-with-no-debugger
  This set is subject to change.
  Please see these options (--help --verbose) for further documentation.
--write-service-info=<file_name>
  Outputs information necessary to connect to the VM service to the
  specified file in JSON format. Useful for clients which are unable to
  listen to stdout for the Observatory listening message.
--snapshot-kind=<snapshot_kind>
--snapshot=<file_name>
  These snapshot options are used to generate a snapshot of the loaded
  Dart script:
    <snapshot-kind> controls the kind of snapshot, it could be
                    kernel(default) or app-jit
    <file_name> specifies the file into which the snapshot is written
--version
  Print the VM version.
esperanto@bogon local %
```

▲ 圖 1-10 執行 dart 命令

1.3 IntelliJ IDEA 的安裝與設定

在官網 https://www.jetbrains.com/idea/download 頁面有所需的下載連結，如圖 1-11 所示。IntelliJ IDEA 支援 Windows、Mac、Linux 3 種作業系統，並且提供 Ultimate 和 Community 兩種版本，選擇 Community 版就可以滿足需求。

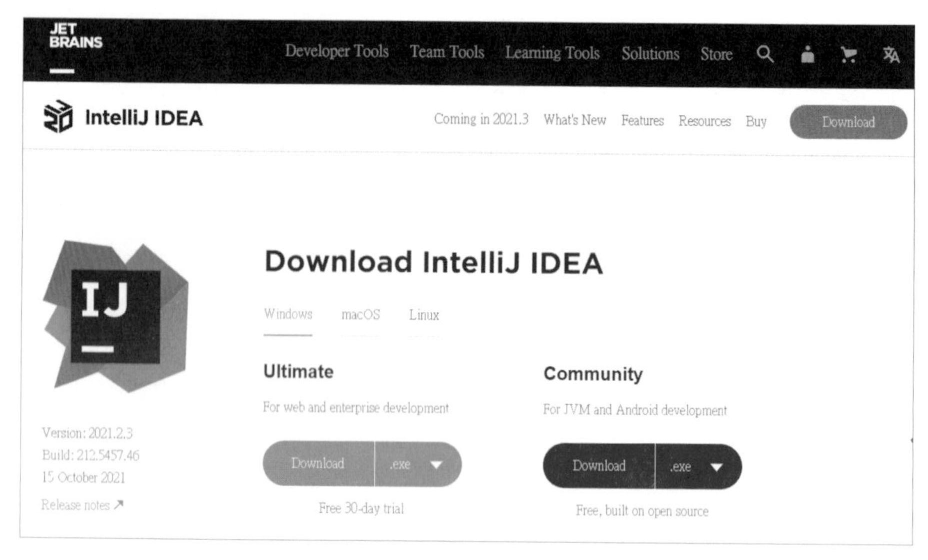

▲ 圖 1-11　編輯器 IDEA 下載頁面

1.3.1 Windows 使用者

雙擊安裝套件，進入安裝介面，如圖 1-12 所示。

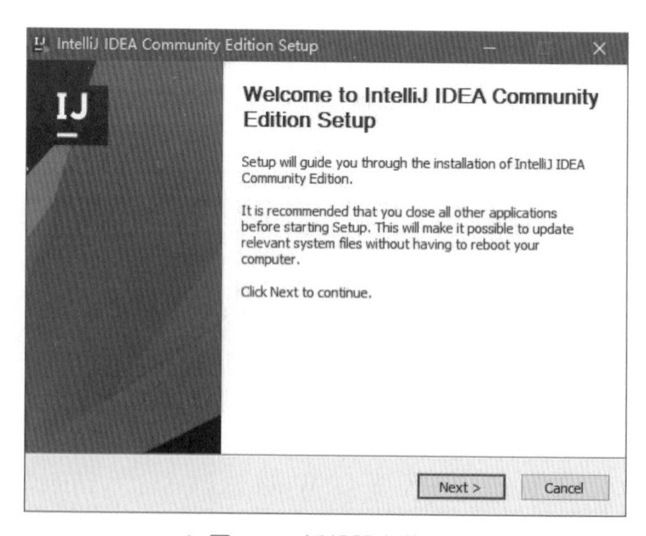

▲ 圖 1-12　編輯器安裝介面

點擊 Next 按鈕，進入如圖 1-13 所示安裝目錄對話方塊，通常採用預設安裝路徑。

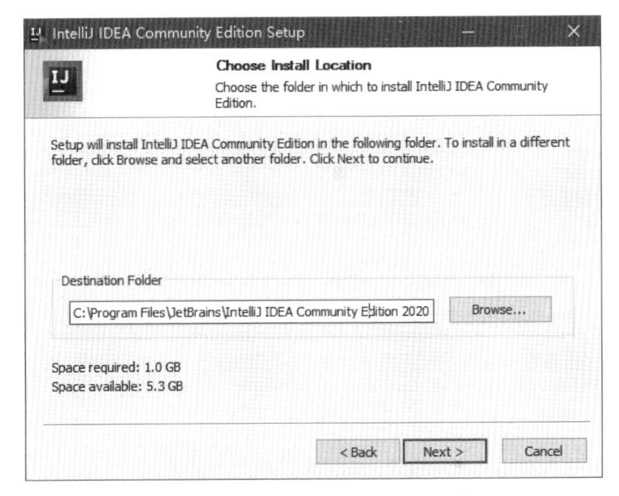

▲ 圖 1-13 安裝目錄對話方塊

點擊 Next 按鈕，進入如圖 1-14 所示安裝選項對話方塊。在 Create Desktop Shortcut 下選取啟動器，這裡選擇的是 64-bit launcher。在 Update PATH variable 下選取 Add launchers dir to the PATH。

▲ 圖 1-14 安裝選項對話方塊

點擊 Next 按鈕，進入如圖 1-15 所示的選擇啟動選單目錄對話方塊，這裡採用預設選單目錄即可，點擊 Install 按鈕，程式將自動安裝。

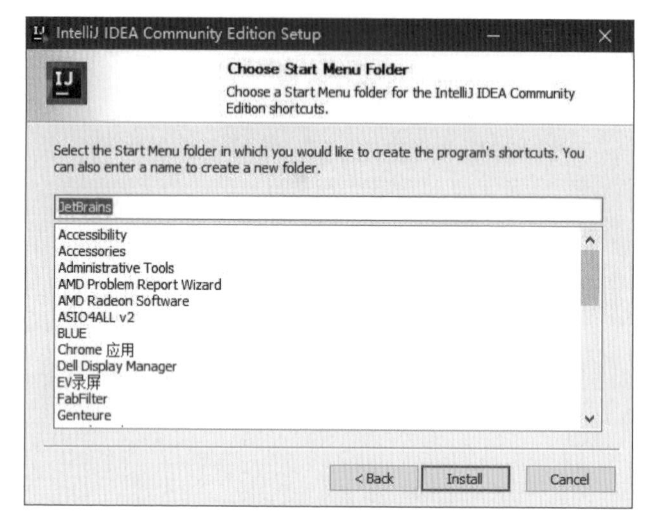

▲ 圖 1-15　選擇啟動選單目錄對話方塊

安裝完成後如圖 1-16 所示，選擇 Reboot now 選項後點擊 Finish 按鈕，系統將重新啟動。

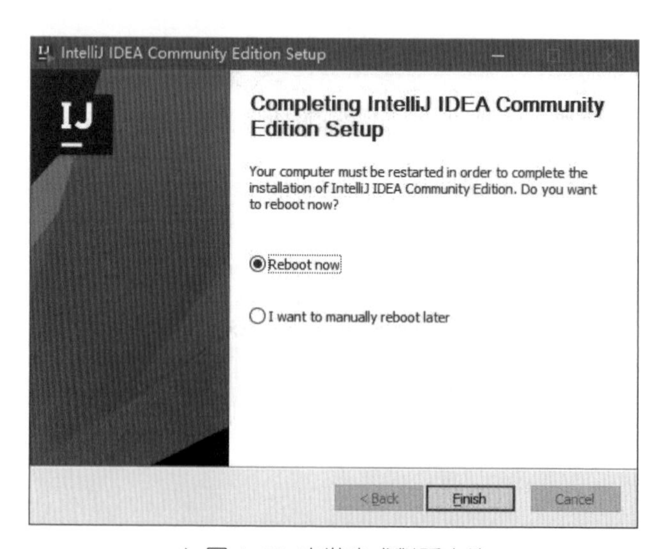

▲ 圖 1-16　安裝完成對話方塊

系統重新啟動後，點擊桌面上的 IntelliJ IDEA 編輯器啟動圖示，彈出如圖 1-17 所示的編輯器主題選擇對話方塊，根據習慣選擇對應主題，然後點擊對話方塊左下角的 Skip Remaining and Set Defaults 按鈕。

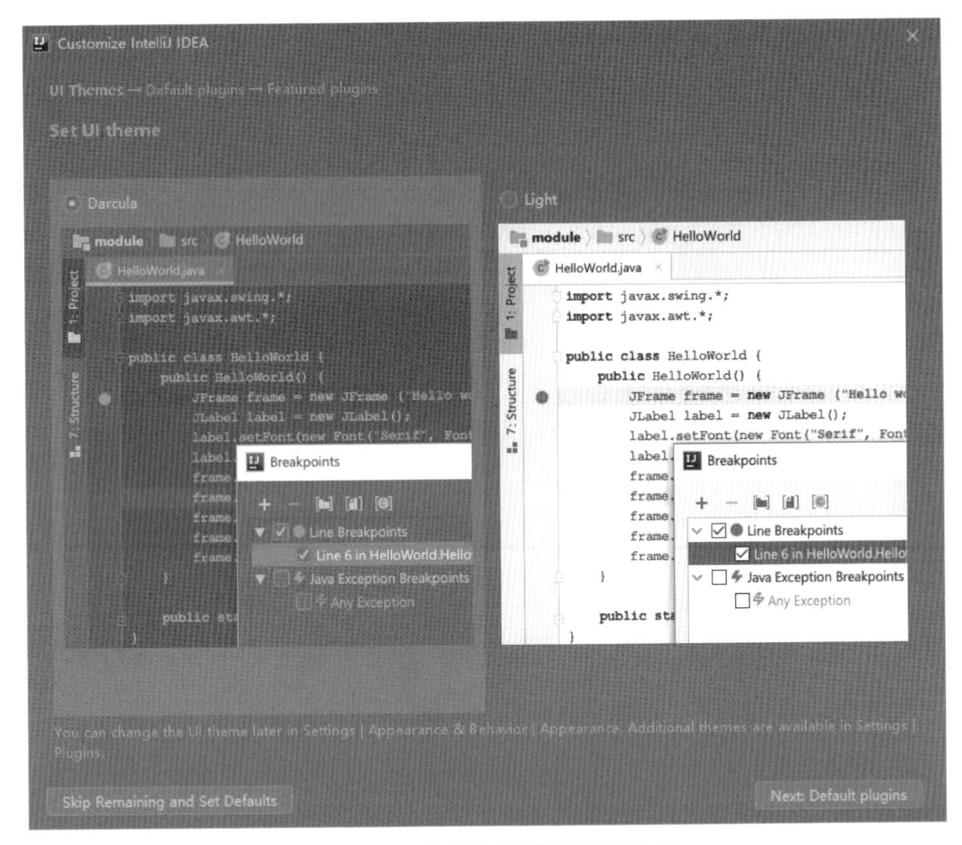

▲ 圖 1-17 編輯器主題選擇對話方塊

進入如圖 1-18 所示的編輯器歡迎介面對話方塊，選擇對話方塊右下角的 Configure 下拉式功能表，在選單中選擇 Plugins 選項，彈出外掛程式安裝對話方塊。

在外掛程式對話方塊中選擇 Marketplace 標籤，在搜索框中輸入 Dart，點擊 Dart 外掛程式右邊的 Install 按鈕安裝該外掛程式，如圖 1-19 所示。

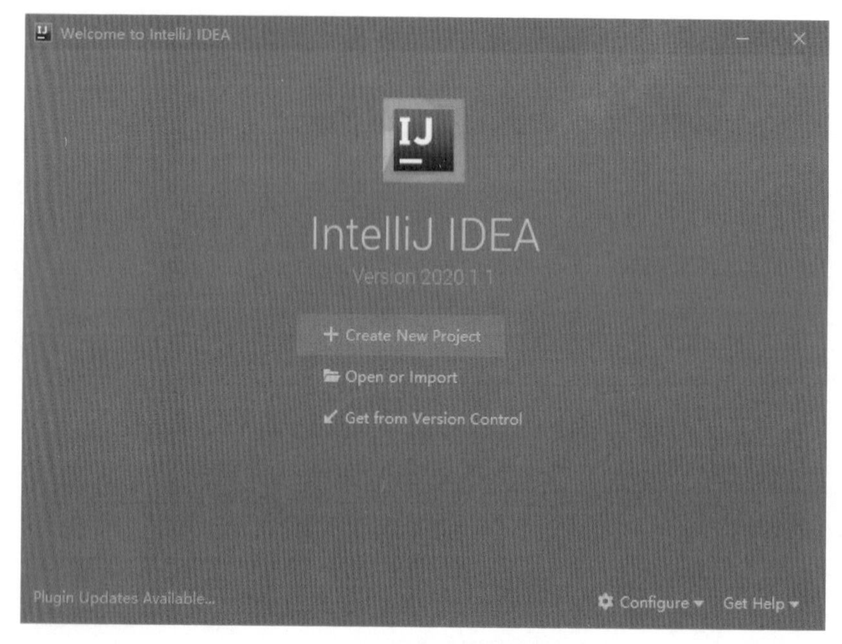

▲ 圖 1-18 編輯器歡迎介面對話方塊

▲ 圖 1-19 外掛程式安裝對話方塊

安裝完成後，重新啟動編輯器，在歡迎介面選擇 Create New Project 選項來創建新專案。彈出如圖 1-20 所示的專案創建對話方塊，點擊左側清單中的 Dart 選項，在右側標頭為 Dart SDK path 的路徑選擇框中選擇 Dart SDK 的安裝路徑，即 C:\Program Files\dart-sdk。在 Generate sample content 下會羅列一些 Dart 專案創建範本。

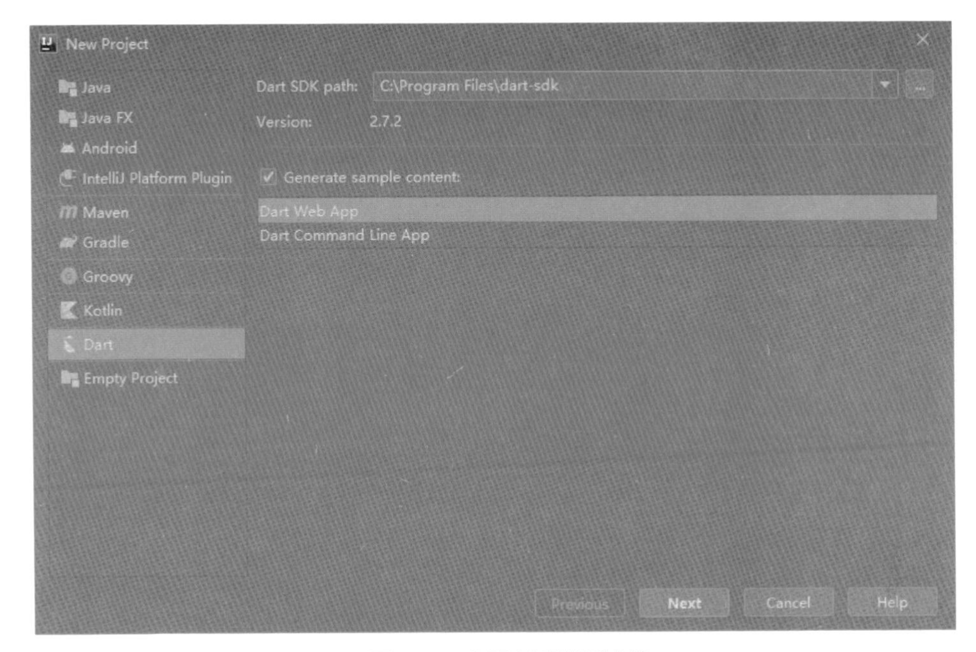

▲ 圖 1-20 專案創建對話方塊

實際上此時的範本並不夠用，因此還需要安裝第三方軟體套件 stagehand。在 Dart SDK 中包含 pub 工具，該工具可用於獲取 Dart 專案依賴的第三方軟體套件和工具。

pub 工具預設從 pub.dartlang.org 上獲取軟體套件，通常獲取軟體套件的速度較慢，有時會需要為 pub 工具更換映像檔來源。

打開環境變數對話方塊，打開步驟參見環境安裝與設定。在該對話方塊中點擊使用者變數下的「新建」按鈕，彈出「新建使用者變數」對話方塊。變數名稱填入值 PUB_HOSTED_URL，這裡選擇其它的映像檔來源，因此變數值填入 https://mirrors.tuna.tsinghua.edu.cn/dart-pub。

打開命令提示符號，輸入以下命令：

```
pub global activate stagehand
```

確認執行該命令，如果結果如圖 1-21 所示即表示 stagehand 軟體套件安裝成功。

```
C:\Users\C01668>pub global activate stagehand
Resolving dependencies...
+ args 1.6.0
+ charcode 1.1.3
+ collection 1.14.12
+ http 0.12.1
+ http_parser 3.1.4
+ meta 1.1.8
+ path 1.7.0
+ pedantic 1.9.0
+ source_span 1.7.0
+ stagehand 3.3.7 (3.3.9 available)
+ string_scanner 1.0.5
+ term_glyph 1.1.0
+ typed_data 1.1.6
+ usage 3.4.1
Downloading stagehand 3.3.7...
Precompiling executables...
Precompiled stagehand:bin\stagehand.
Activated stagehand 3.3.7.
```

▲ 圖 1-21 軟體套件安裝對話方塊

完成軟體套件安裝後，重新啟動編輯器，繼續創建專案，此時在專案生成範本清單下就會出現足夠的專案範本，如圖 1-22 所示。到此就完成了編輯器的安裝與設定。

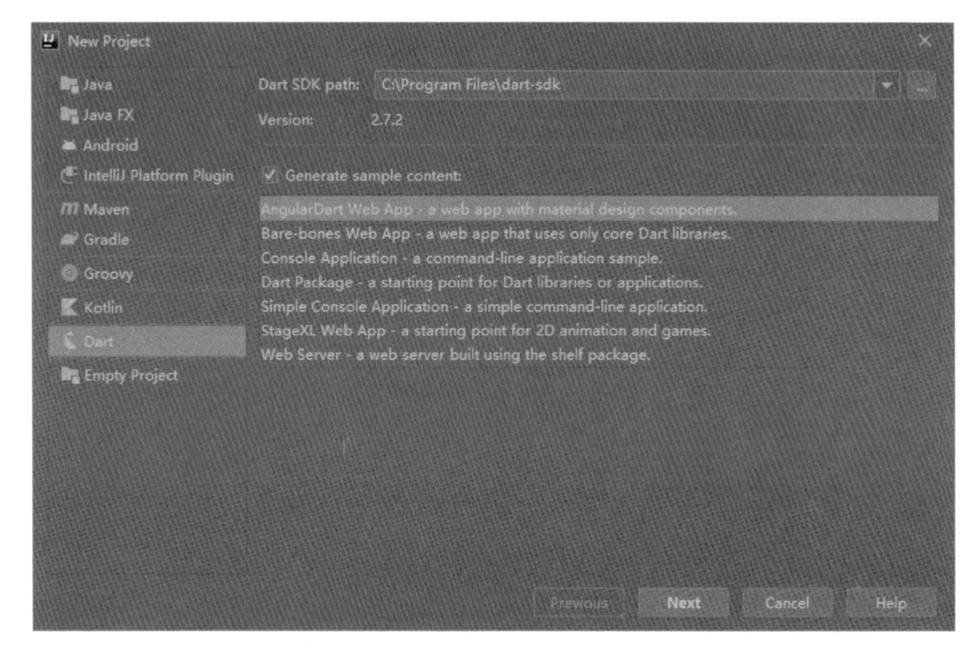

▲ 圖 1-22 使用專案範本

1.3.2 Mac 使用者

編輯器安裝與 Windows 使用者類似，這裡不再贅述。需要特別說明的是 pub 工具映像檔來源的更換，以某大學映像檔來源為例。

bash 終端設定程式如下：

```
echo 'export PUB_HOSTED_URL=https://mirrors.tuna.tsinghua.edu.cn/dart-pub' >>
~/.bash_profile
source ~/.bash_profile
```

zsh 終端設定程式如下：

```
echo 'export PUB_HOSTED_URL=https://mirrors.tuna.tsinghua.edu.cn/dart-pub' >>
~/.zshrc
source ~/.zshrc
```

Chapter 02

變數和內建類型

創建命令列應用程式，將專案命名為 chapter2。本章所有檔案都在該專案的 bin 資料夾中創建與執行。

2.1 識別符號

識別符號就是為變數、方法、列舉、類別、介面等指定的名字。在 Dart 語言中識別符號的命名風格有 3 種：小駝峰命名法、大駝峰命名法、底線命名法。

2.1.1 小駝峰命名法

小駝峰命名法的單字字首大寫，其餘字母小寫，但第一個單字除外。如：lowerCamelCase，適用於變數、參數、常數、列舉值及類別成員等。

2.1.2 大駝峰命名法

大駝峰命名法又稱帕斯卡命名法，單字字首大寫，其餘字母小寫。如：UpperCamelCase，適用於類別名稱、列舉類別型、類型定義、類型參數及副檔名等。

2.1.3 底線命名法

底線命名法的單字全部小寫，單字間使用底線連接。如：lowercase_with_underscores，適用於函數庫名稱、套件名、資料夾及原始檔案等。

如果對本節中出現的關鍵字不熟悉沒有關係，這些關鍵字將在後續一一呈現且遵循上述規則。

2.2 關鍵字

關鍵字是 Dart 中具有特殊用途的一系列詞，應當避免將它們用作識別符號。在這裡僅需了解，我們會在後面的章節中陸續使用這些關鍵字。

表 2-1 所示的是保留字，它們不能用作識別符號。

表 2-1 保留字

assert	break	case	catch	class	const	continue
default	do	else	enum	extends	false	final
finally	for	if	in	is	new	null
rethrow	return	super	switch	this	throw	true
try	var	void	while	with		

表 2-2 所示的是上下文關鍵字，它們僅在特定的位置具有含義，在所有地方都是有效識別符號。

表 2-2 上下文關鍵字

async	hide	on	show	sync		

表 2-3 所示的是內建識別符號，在大部分地方是有效識別符號，不能用作類別或類型名稱，也不能用作匯入字首。

表 2-3 內建識別符號

abstract	as	covariant	deferred	dynamic	export	external
factory	function	get	implements	import	interface	library
mixin	operator	part	set	static	typedef	

表 2-4 所示的是受限制的保留字，與非同步相關，在帶有 async、async* 或 sync* 標識的函數本體中不能使用 await 和 yield 作為識別符號。

表 2-4 受限保留字

await	ield					

2.3 變數

變數包含變數名稱和變數值，變數名稱屬於識別符號，故需遵守識別符號的命名規範。指定給變數的一切值都是物件，每個物件都是一個類別的實例。數字、函數、字串及 null 等都是物件，所有物件都繼承自 Object 類別。變數儲存的是物件的引用，當把物件指定給變數時，實際上是將物件的引用指定給變數。

變數的宣告格式如下，中括號中的內容代表可選。

```
資料類型變數名稱 [= 初值 ];
```

定義一個類型為 String 的變數 catName，且為其設定值。範例程式如下：

```
//chapter2/bin/example_01.dart
void main(){
  String catName = '千歲歲';
  print('變數 catName 執行時期類型為 ${catName.runtimeType}');
}
```

名為 catName 的變數儲存著 String 類型的物件的引用，該物件的值為千歲歲。

在 Dart 中可以使用 dynamic 或 var 宣告變數的類型，它們都會根據賦數值類型推斷該變數的類型。其中 var 常用於宣告區域變數，而不使用確定的類型，如上例中使用了 String 這一確定類型。範例程式如下：

```
//chapter2/bin/example_02.dart
void main(){
  dynamic dogName = '啵啵樂';
  print('變數 dogName 執行時期類型為 ${dogName.runtimeType}');
}
```

如上例，名為 dogName 的變數將被推斷為 String 類型。

2.3.1 預設值

在 Dart 中無論宣告類型是什麼，未初始化的變數都擁有一個預設值 null，即使其類型為數字。範例程式如下：

```
//chapter2/bin/example_03.dart
void main(){
  int sum;
  print(sum == null);
}
```

上例中運算子 "==" 用於判斷兩個物件的值是否相等，它的返回值是布林類型，即 true 或 false。

2.3.2 const 和 final

如果在設定值後不需要改變變數的值，那麼可以使用 const 或 final 來修飾變數，const 和 final 都可以代替 var，被修飾的變數的類型由設定值物件的類型決定，也可以放在確定的類型前共同修飾變數。

final 修飾變數的格式如下：

```
final [ 資料類型 ] 變數名稱 = 初值 ;
```

const 修飾變數的格式如下：

```
const [ 資料類型 ] 變數名稱 = 初值 ;
```

final 修飾的變數只能設定一次值，除了在宣告處設定值，還可以透過類別的建構函數為 final 修飾的變數設定值，這一點在後面會介紹。範例程式如下：

```
//chapter2/bin/example_04.dart
void main(){
// 不指定類型，設定值的類型作為推斷類型的依據，這裡為 String 類型
   final name = ' 貓貓村長 ';
// 可放在確定的類型前，共同修飾變數
   final String nickName = ' 貓貓 ';
}
```

不能修改由 final 修飾的變數的值。

```
// 錯誤，final 修飾的變數不能被修改
name = ' 貓貓 ';
```

const 修飾的變數應在宣告處設定值，且值必須為編譯時常數。用 const 修飾變數時可以直接使用常數為其設定值，也可以使用由 const 修飾的其他變數組成的運算式來設定值。範例程式如下：

```
//chapter2/bin/example_05.dart
void main(){
   // 創建常數
```

```
    const pi = 3.14;
    const r = 6;
    // 常數的運算式可由其他常數組成
    const l = 2*pi*r;
}
```

const 不僅用於定義常數，也可以用來創建常數值，範例程式如下：

```
var ls = const [];
```

沒有使用 final 或 const 修飾的變數，其值可以被修改，即使變數引用過 const 修飾的物件，範例程式如下：

```
ls = [1,2,3];
```

2.4 數字

數字在 Dart 中對應的類別為 num，num 類別中定義了 +、-、*、/ 等基本運算子，num 類別有兩個子類別：

(1) int：常被稱為整數，整數值長度不超過 64 位元，這依賴於平台，在 Dart VM 中，其設定值可從 -2^{63} 到 $2^{63}-1$。當將 Dart 程式編譯成 JavaScript 時將使用 JavaScript 中的數值範圍，其設定值可從 -2^{53} 到 $2^{53}-1$。

(2) double：64 位元雙精度浮點數，符合 IEEE 754 標準。

整數是不帶小數點的數字，定義時可以使用 int 或 num 修飾變數。範例程式如下：

```
//chapter2/bin/bin/example_06.dart
void main(){
    // 定義 int 型數字
    int i = 1;
    var j = 332131;
    num k = 1;
```

```
    print('k 的執行時期類型：${k.runtimeType}');
}
```

如果數字中包含小數點，那麼它就是浮點數。定義時可以使用 double 或 num 修飾變數。範例程式如下：

```
// 定義 double 型數字
double pi = 3.14;
var r = 6.7;
num w = 1.2;
print('w 的執行時期類型：${w.runtimeType}');
```

必要時，整數可以自動轉為浮點數。範例程式如下：

```
// 等於 double l = 3.0;
double l = 3;
print('l：$l');
```

2.5 字串

Dart 字串是 UTF-16 編碼的字元序列，對應的類型為 String，可以使用單引號或雙引號來創建字串。範例程式如下：

```
// 使用單引號創建字串
var s1 = 'Today is sunny';
// 使用雙引號創建的字串
var s2 = "Today is sunny";
```

也可以使用三重單引號或三重雙引號來創建多行字串。範例程式如下：

```
// 使用三重單引號創建多行字串
var s3 = '''Today
         is
         sunny''';
// 使用三重雙引號創建多行字串
var s4 = """Today
         is
         sunny""";
```

插值操作可以將運算式的值放入字串中，使用格式：${expr}，expr 代表的是運算式。如果運算式是識別符號，則可以省略 {}。使用格式：$Identifier，Identifier 代表的是識別符號。插值實際上是透過呼叫物件的 toString() 方法獲取該物件的字串形式。範例程式如下：

```
//chapter2/bin/example_07.dart
void main(){
   var pi = 3.14;
   var r = 6;
   var s = 'QianSuiSui is a cat';
// 插值運算式
   print(' 字串 s：$s');
   print(' 圓的周長 L：${2*pi*r}');
}
```

可以透過多個相鄰字串組成一個新的字串，也可以使用 "+" 運算子。範例程式如下：

```
// 將相鄰字串拼接在一起
   var s5 = 'Today'
            'is'
            'sunny';
// 透過使用 "+" 運算子將字串拼接在一起
   var s6 = 'Today' + 'is' + 'sunny';
```

2.6 布林

為了表示布林值，Dart 中有一個 bool 類型，只有兩個物件是 bool 類型：true 和 false。它們都是編譯時常數。範例程式如下：

```
bool isEmpty = true;
bool isNull = false;
```

不能為 bool 類型的變數設定值 true 和 false 之外的常數，也不能用運算結果不是 true 或 false 的運算式為其設定值，否則會發生編譯錯誤。範例程式如下：

```
// 下列程式將發生編譯錯誤
// 使用字串為 bool 變數設定值
bool isEmpty = 'AA';
// 使用數字為 bool 變數設定值
bool isNull = 1;
```

當 Dart 程式需要一個布林值時，只能向其提供 true 或 false。不能使用 if(非布林值) 或 assert(非布林值) 這樣的程式檢查布林值。應該顯性地檢查布林值，範例程式如下：

```
//chapter2/bin/example_08.dart
void main(){
// 檢查是否為空字串
   var name= '';
   print(name.isEmpty);

// 檢查是否小於或等於 0
   var topPoint = 0;
   print(topPoint <= 0);

// 檢查是否為 null
   var label;
   print(label == null);

// 檢查是否為 NaN
   var y= 0/0;
   print(y.isNaN);
}
```

2.7 List 集合

集合是程式設計中常用的資料類型，它們相比普通類型更加豐富、更具擴充性。

List 集合中的元素是有序的，並且可以重複出現相同元素。在其他語言中

會將陣列 (Array) 單獨拿出來作為一種類型，但在 Dart 中陣列 (Array) 由 List 集合來表示。

List 集合中可以存放任意類型的物件，也可以透過 "<>" 符號指定泛型，確保存入該集合的物件都是泛型所指定的單一類型。

Dart 中 List 字面量看起來與 JavaScript 中的陣列字面量一樣。範例程式如下：

```
var list = [1,2,3];
```

在上述程式中 Dart 推斷物件 list 的類型為 List<int>，因此只可以在該集合中增加 int 類型的物件，否則會顯示出錯。

上述程式也可以使用顯性類型宣告。範例程式如下：

```
Listlist = [1,2,3];
```

或顯性指定泛型為 int：

```
List<int> list = [1,2,3];
```

當使用字面量初始化 List 集合時，如果初始化值不是同一類型，則 List 集合中的元素為動態類型，即任意類型都可以存放進該集合。範例程式如下：

```
var list = [1,2,'abc'];
   //增加 int 類型的資料
   list.add(33);
   //增加字串類型的資料
   list.add('value');
   //增加布林類型的資料
   list.add(true);
```

此時物件 list 的推動類型為 List<Object>，這樣使用起來確實很方便。但是在處理集合中的物件時會增加難度，因此推薦使用泛型指定 List 集合中可存放的資料類型。

2.7.1 常用屬性

(1) first：返回 List 集合中的第一個元素。

(2) last：返回 List 集合中的最後一個元素。

(3) length：返回 List 集合中元素的數量，返回值為 int 型。

(4) isEmpty：判斷 List 集合中是否有元素，如果沒有元素則返回 true，否則返回 false。

(5) iterator：返回疊代器 (Iterator) 物件，疊代器物件用於遍歷集合，該屬性是從 Iterable 類別繼承而來。

使用範例程式如下：

```
//chapter2/example_09.dart
void main(){
   var list = [63, 70.9, 'abc', true];
// 列印第一個元素
   print(' 第一個元素：${list.first}');
// 列印最後一個元素
   print(' 最後一個元素：${list.last}');
// 列印集合的長度
   print('List 集合的長度：${list.length}');
// 列印執行時期類型
   print(' 執行時期類型：${list.runtimeType}');
// 列印疊代器的類型
   print(' 疊代器的類型：${list.iterator.runtimeType}');
}
```

執行結果如下：

```
第一個元素：63
最後一個元素：true
List 集合的長度：4
執行時期類型：List<Object>
疊代器的類型：ListIterator<Object>
```

2.7.2 常用方法

(1) 運算符號 [](int index)：返回 List 集合中指定索引處的物件。

(2) 運算符號 []=(int index,E value)：將 List 集合中指定索引處 index 的值替換為 value。

(3) add(E value)：將 value 增加到 List 集合的尾端，並使 List 集合長度增加 1。

(4) insert(int index,E element)：在 List 集合指定位置 index 插入元素 element。

(5) remove(Object value)：從 List 集合中刪除第一次出現 value 值的元素。

(6) removeAt(int index)：移除 List 集合指定位置的元素，並返回該元素。

(7) clear()：移除 List 集合中所有的元素，並將 List 集合的長度置為 0。

(8) sublist(int start,[int end])：返回 List 集合中位置 start 和 end 之間的元素的集合，包括 start 處的元素，但不包括 end 處的元素。

(9) indexOf(E element,[int start = 0])：按順序尋找 List 集合中的元素，返回第一次出現指定元素 element 的索引。如果集合中沒有該元素，則返回 -1。可選參數 start 表示從 List 集合中指定索引處開始尋找，一直尋找到集合的最後一個元素，其預設值為 0。

List 使用以 0 為基礎的索引，其中 0 是第一個元素的索引，list.length-1 是最後一個元素的索引。

使用範例程式如下：

```
//chapter2/bin/example_10.dart
void main() {
   var list = [63, 70.9, 'abc', true, 'abc'];
// 列印索引為 2 的元素，即集合中的第 3 個元素
   print(' 元素 list[2]：${list[2]}');
// 修改 list[1] 中的元素
   list[1]= false;
// 列印修改後 list[1] 中的元素
   print(' 元素 list[1]：${list[1]}');
```

```
// 在 list 增加元素
  list.add('new element');
  print(' 修改後的 list：$list');
// 向 list 索引 1 處插入元素
  list.insert(1, 'insert element');
  print(' 插入新元素後的 list：$list');
// 在 list 中尋找元素，返回其第一次出現的索引
  print('abc 在 list 中第一次出現的索引 :${list.indexOf('abc')}');
// 返回 list 索引從 2 到 5 的元素的集合
  print('list 中索引 2 到 5 的元素 :${list.sublist(2, 5)}');
// 在 list 中尋找元素，刪除第一次出現 abc 的元素
  print('list 中移除 abc 元素 :${list.remove('abc')}');
// 移除元素後的 list
  print(' 移除元素後的 list：$list');
// 清除 list 中的元素
  list.clear();
}
```

執行結果如下：

```
初始 list 中的元素：[63, 70.9, abc, true, abc]
元素 list[2]：abc
元素 list[1]：false
修改後的 list：[63, false, abc, true, abc, new element]
插入新元素後的 list：[63, insert element, false, abc, true, abc, new element]
abc 在 list 中第一次出現的索引 :3
list 中索引 2 到 5 的元素 :[false, abc, true]
list 中移除 abc 元素 :true
移除元素後的 list：[63, insert element, false, true, abc, new element]
```

若需要創建一個包含編譯時常數的 List，可以在 List 字面量前增加 const：

```
varconstantList = const [1,2,3];
```

需要創建固定長度的 List 集合，需使用建構函數指定：

```
varlist = List(6);
```

2.8 Set 集合

Set 集合中的元素無序且唯一，即不可重複。Set 的預設實現是 LinkedHashSet 類別，該子類別依據元素插入的順序進行疊代。Dart 中提供了 Set 字面量和 Set 類型兩種方式宣告 Set 集合。

採用 Set 字面量來創建 Set 集合的程式如下：

```
var halogens = {'fluorine','chlorine','bromine','iodine','astatine'};
```

Dart 推斷 halogens 變數是一個 Set<String> 類型的集合，只可以在該集合增加 String 類型的物件。

可以在 {} 前面加上泛型參數來創建一個空的 Set 集合，也可以將 {} 設定值給宣告類型為 Set 的變數。範例程式如下：

```
var cats = <String>{};
Set<String> dogs = {};
```

可以在 Set 字面量前增加 const 關鍵字創建一個 Set 類型的編譯時常數。範例程式如下：

```
final constantSet = const {
   'fluorine',
   'chlorine',
   'bromine',
   'iodine',
   'astatine',
};
```

2.8.1 常用屬性

(1) length：返回 Set 集合中元素的數量，返回值為 int 類型。
(2) iterator：返回疊代器 (Iterator) 物件，疊代器物件用於遍歷集合。
(3) isEmpty：判斷 Set 集合中是否有元素，如果有則返回 false，如果沒有則返回 true。

使用範例程式如下：

```
//chapter2/bin/example_11.dart
void main(){
   var set = {'abc',1,true};
   // 列印 Set 集合的長度
   print('Set 集合的長度：${set.length}');
   // 列印疊代器的類型
   print(set.iterator.runtimeType);
   // 判斷 Set 集合是否為空
   print('Set 集合是否為空：${set.isEmpty}');
}
```

執行結果如下：

```
Set 集合的長度：3
_CompactIterator<Object>
Set 集合是否為空：false
```

2.8.2 常用方法

(1) add(E value)：將元素增加到 Set 集合。

(2) clear()：清除 Set 集合中的所有元素。

(3) remove(Object value)：移除 Set 集合中的指定元素。

使用範例程式如下：

```
//chapter2/bin/example_12.dart
void main(){
   var set = {'abc',1,true};
   // 列印原 Set 集合
   print(' 原 Set 集合：$set');
   // 增加元素到 Set 集合
   set.add(99);
   print(' 增加元素後的 Set 集合：$set');
   // 移除 Set 集合中的元素
   set.remove(true);
   print(' 移除元素後的 Set 集合：$set');
   // 清除 Set 集合中的所有元素
```

```
   set.clear();
}
```

執行結果如下：

```
原 Set 集合：{abc, 1, true}
增加元素後的 Set 集合：{abc, 1, true, 99}
移除元素後的 Set 集合：{abc, 1, 99}
```

2.9 Map 集合

Map 集合是一種將鍵 (key) 和值 (value) 相連結的物件，key 和 value 都可以是任何物件。key 不可重複，但 value 可重複。Dart 中支援使用 Map 字面量和 Map 類型來建構 Map 物件。

採用 Map 字面量來創建 Map 集合的範例程式如下：

```
//chapter2/bin/example_13.dart
void main(){
// 鍵和值都是 String 類型
   var gifts = {
      //Key:Value
      'first':'partridge',
      'second':'turtledoves',
      'fifth':'golden rings'
   };
// 鍵是 int 型，值是字串型
   var nobleGases = {
      2:'helium',
      10:'neon',
      10:'argon',
   };
}
```

Dart 將名為 gifts 的變數的類型推斷為 Map<String,String>，將 nobleGases 推斷為 Map<int,String>。

也可以使用 Map 建構函數來創建 Map。範例程式如下：

```
//chapter2/bin/example_14.dart
void main(){
   // 用建構函數 Map() 建構 Map 物件
   var gifts1 = Map();
   gifts['first']= 'partridge';
   gifts['second']= 'turtledoves';
   gifts['fifth']= 'golden rings';

   var nobleGases1 = Map();
   nobleGases[2]= 'helium';
   nobleGases[10]= 'neon';
   nobleGases[18]= 'argon';
}
```

2.9.1 常用屬性

(1) length：返回 Map 集合中的鍵值對數量。

(2) keys：返回 Map 集合中的所有鍵 (key)，返回值是可疊代物件。

(3) values：返回 Map 集合中的所有值 (value)，返回值是可疊代物件。

使用範例程式如下：

```
//chapter2/bin/example_15.dart
void main(){
   var map = {
      1: 'partridge',
      'second': 'turtledoves',
      'age': 32
   };
   // 列印 Map 集合的鍵值對數量
   print('Map 集合的鍵值對數量：${map.length}');
   // 返回 Map 集合中的所有鍵
   print('Map 集合中的所有鍵：${map.keys}');
   // 返回 Map 集合中的所有值
   print('Map 集合中的所有值：${map.values}');
}
```

執行結果如下：

```
Map 集合的鍵值對數量：3
Map 集合中的所有鍵：(1, second, age)
Map 集合中的所有值：(partridge, turtledoves, 32)
```

2.9.2 常用方法

(1) 運算符號 [](Object key)：獲取指定 key 的值，如果該 key 不存在，則返回 null。
(2) 運算符號 []=(K key,V value)：將值 (value) 與指定鍵 (key) 相連結，如果該鍵存在於 Map 集合中，則將對應值修改為 value。如果對應鍵不存在，則將鍵值對增加到 Map 集合。
(3) containsKey(Object key)：判斷 Map 集合中是否包含指定鍵，如果包含則返回 true，否則返回 false。
(4) containsValue(Object value)：判斷 Map 集合是否包含指定值，如果包含則返回 true，否則返回 false。

使用範例程式如下：

```
//chapter2/bin/example_16.dart
void main(){
  var map = {
    1: 'partridge',
    'second': 'turtledoves',
    'age': 32
  };
  // 列印原 Map
  print(' 原 Map：$map');
  // 獲取 Map 集合中鍵為 age 的值
  print('Map 的鍵值對數量：${map['age']}');
  // 在 Map 集合中增加鍵值對
  map['name']= 'Bob';
  print(' 增加鍵值對後的 Map：$map');
  // 修改 Map 集合中鍵為 name 的值
```

```
    map['name']= 'Jobs';
    print(' 修改鍵的值後的 Map：$map');
    // 判斷 Map 集合中是否包含鍵 age
    print('Map 集合中是否包含鍵 age：${map.containsKey('age')}');
    // 判斷 Map 集合中是否包含值 32
    print('Map 集合中是否包含值 32：${map.containsValue(32)}');
}
```

執行結果如下：

```
原 Map：{1: partridge, second: turtledoves, age: 32}
Map 的鍵值對數量：32
增加鍵值對後的 Map：{1: partridge, second: turtledoves, age: 32, name: Bob}
修改鍵的值後的 Map：{1: partridge, second: turtledoves, age: 32, name: Jobs}
Map 集合中是否包含鍵 age：true
Map 集合中是否包含值 32：true
```

在 Map 字面量前增加 const 關鍵字可以創建 Map 類型的編譯時常數。範例程式如下：

```
final constantMap = const {
    2:'helium',
    10:'neon',
    18:'argon',
};
```

2.10 符文

在 Dart 中符文對應的類型為 Runes。Unicode 為世界上所有書寫系統中使用的每個字母、數字和符號定義了唯一的數值。由於 Dart 字串是 UTF-16 程式單元的序列，因此在字串中表示 Unicode 程式點需要特殊的語法。

表示 Unicode 程式點的常用形式是：\uXXXX，其中 XXXX 是 4 位十六進位數。例如：心臟字元是 \u2665。要指定多於或少於 4 個十六進位數字，需將值放在大括號中。例如：笑的表情符號是 \u{1f600}。

程式點 (code point)：程式點是指編碼字元集中，字元所對應的數字。有效範圍從 \u0000 到 \u10FFFF，其中 \u0000 到 \uFFFF 為基底字元，\u10000 到 \u10FFFF 為增補字元。

程式單元 (code unit)：程式單元對程式點進行編碼得到的 1 或 2 個 16 位元序列。其中基底字元的程式點直接用一個相同值的程式單元表示，增補字元的程式點用兩個程式單元進行編碼。

String 類別有幾個屬性可用於提取符文資訊，使用 codeUnits 屬性返回 16 位元程式單元，使用 runes 屬性獲取字串的符文。

為保證一致性，請在線上環境 DartPad(https://dartpad.dev) 中執行。範例程式如下：

```
//chapter2/bin/example_17.dart
void main() {
  // 創建符文字串
  var clapping = '\u{1f44f}';
  // 列印符文
  print(clapping);
  // 返回此字串的 UTF-16 程式單元
  print(clapping.codeUnits);
  // 返回此字串的 Unicode 程式點
  print(clapping.runes.toList());
   // 建構 Runes 物件
  Runes input = Runes(
    '\u2665 \u{1f605} \u{1f60e} \u{1f47b} \u{1f596} \u{1f44d}');
  // 將 Runes 物件轉為 String 並列印
  print(String.fromCharCodes(input));
}
```

函數

Dart 是物件導向的語言，即使函數也是物件，對應的類別為 Function。函數也是物件表示可以將函數設定值給變數，或將函數作為參數傳遞給另一個函數。函數有時又被稱為方法，它們是等同的，只是叫法上有差異。

宣告函數的格式如下：

```
[ 返回數值類型 ] 函數名稱 ([ 參數清單 ]){
   // 函數本體
   return expr;
}
```

每個函數都必須有返回值，如果不提供返回數值類型，則預設為 void。函數名稱是識別符號，應當遵循識別符號命名規範。參數清單中可能包含零個到多個參數，這取決於實際情況。函數本體由變數宣告、敘述組成。當函數存在返回值時，函數本體應當提供 return 敘述，例如：return expr;。如果函數返回值的類型為 void，則無須提供 return 敘述，等於 return null;。

定義函數的範例程式如下：

```
bool isEven(int x){
   return x%2 == 0 ? true : false;
}
```

該函數返回值的類型為 bool，接收一個 int 類型的參數，函數本體由一個 return 敘述組成，返回敘述的運算式是一個條件運算式，其返回值是布林值。對於只有一個運算式的函數可以採用簡寫形式，範例程式如下：

```
bool isEven(int x) => x%2 == 0 ? true : false;
```

=> expr; 是 {return expr;} 的簡寫形式，常稱為「=>」語法。「=>」(=>) 與分號 (;) 之間只能是運算式而不能是敘述，例如不能是 if 敘述，但可以是條件運算式 (expr1 ? expr2 : expr3)。

創建命令列應用程式，將專案命名為 chapter3。本章所有檔案都在該專案的 bin 資料夾下創建與執行。

函數包含兩種參數形式：必選參數和可選參數。參數清單先列出必選參數，再列出可選參數。

3.1 可選參數

可選參數分為具名引數和位置參數。在參數清單中只能選擇其中一種作為可選參數，不可同時出現。

3.1.1 具名引數

定義函數時使用大括號 ({}) 包裹可選參數清單，使用 {arg1,arg2,…} 的形式來定義具名引數，範例程式如下：

```
voidfindAll({int currentPage, int pageSize}){…}
```

呼叫函數時，採用 arg1:value1,arg2:value2 的形式來傳遞參數。即先提供參數名稱，接著是冒號 (:)，冒號後邊接著參數值：

```
findAll(currentPage: 1, pageSize: 10);
```

範例程式如下：

```
//chapter3/bin/example_01.dart
main(){
  // 以下定義了一個包含必選參數和可選具名引數的函數
  void message(String from, String content, {DateTime time, String device})
{
    // 呼叫該函數必須提供參數 from 和 content
    // 因此無須做任何判斷即可直接列印這兩個參數
    print(' 來自：$from, 正文：$content');
    // 如果參數 time 不為空，則列印 time
    if (time != null) {
    print(' 時間：$time');
    }
    // 如果參數 device 不為空，則列印 device
    if (device != null) {
      print(' 發送裝置：$device');
    }
  }
  // 呼叫 message 函數且提供兩個必選參數
  message('jobs', 'hello');
  // 呼叫 message 函數且提供可選參數 time
  message('jobs', 'hello', time: DateTime.now());
  // 呼叫 message 函數且提供可選參數 time 和 device
  message('jobs', 'hello', time: DateTime.now(), device: 'phone');
}
```

3.1.2 位置參數

使用中括號 ([]) 包裹可選參數清單來定義位置參數：

```
voidmessage(String from,String content,[DateTime time,String device]){...}
```

在呼叫包含可選位置參數的函數時應當按照位置參數清單的順序依次為參數設定值。範例程式如下：

```
//chapter3/bin/example_02.dart
void main(){
   // 定義帶可選位置參數 time 和 device 的函數
   void message(String from,String content,[DateTime time,String device]){
      print(' 來自：$from, 正文：$content');
      if(time != null){
         print(' 時間：$time');
      }
      if(device != null){
         print(' 發送裝置：$device ');
      }
   }
   // 呼叫函數時不指定可選參數
   message('jobs','hello');
   // 呼叫函數時指定可選位置參數 time
   message('jobs','hello',DateTime.now());
   // 呼叫函數時指定可選位置參數 time 和 device
   message('jobs','hello',DateTime.now(),'phone');
}
```

3.1.3 預設參數值

在宣告可選參數時可以使用等號運算子為可選參數指定預設值，預設值必須是編譯時常數，沒有指定預設值的參數將被設定值為 null。

為具名引數提供預設值的範例程式如下：

```
//chapter3/bin/example_03.dart
main(){
   // 為可選具名引數 device 提供預設值
   void message(String from,String content,{DateTime time,String
device='phone'}){
      print(' 來自：$from, 正文：$content');
      if(time != null){
         print(' 時間：$time');
```

```
        }
        // 參數 device 始終會被列印
        if(device != null){
            print(' 發送裝置：$device');
        }
    }
    // 呼叫函數時不指定可選參數
    message('jobs','hello');
    // 呼叫函數且指定可選具名引數 time
    message('jobs','hello',time: DateTime.now());
    // 呼叫函數且指定可選具名引數 time 和 device
    message('jobs','hello',time: DateTime.now(),device: 'pc');
}
```

為位置參數提供預設值的範例程式如下：

```
//chapter3/bin/example_04.dart
main(){
    // 為可選位置參數 device 提供預設值
    void message(String from,String content,[DateTime time,String
device='phone']){
        print(' 來自：$from, 正文：$content');
        if(time != null){
            print(' 時間：$time');
        }
        // 參數 device 始終會被列印
        if(device != null){
            print(' 發送裝置：$device');
        }
    }
    // 呼叫函數且不指定可選參數
    message('jobs','hello');
    // 呼叫函數且指定可選具名引數 time
    message('jobs','hello',DateTime.now());
    // 呼叫函數且指定可選具名引數 time 和 device
    message('jobs','hello',DateTime.now(),'pc');
}
```

3.2 main 函數

每個 Dart 程式都必須有一個 main 函數作為入口，main 函數的返回值為 void，它有一個 List<String> 類型的可選參數。

使用命令列存取帶有參數的 main 函數的範例程式如下：

```
//chapter3/bin/example_05.dart
// 帶有參數清單的 main 函數
void main(List<String> args){
   // 列印所有參數值
   print(args);
   var len = args.length;
   // 列印參數量
   print(' 參數量 :$len');
}
```

保存上述程式，檔案名稱為 example_05.dart。打開編輯器的命令列工具 Terminal，執行命令並攜帶參數：

```
dart bin/bin/example_05.dart first 2 dog
```

執行結果如下：

```
[first, 2, dog]
參數量 :3
```

也可以省略參數：

```
void main(){
// 函數本體
}
```

3.3 函數物件

可以將函數作為參數傳遞給另一個函數，下例中將函數 printElement 傳遞給 List 類型的物件 list 的 forEach 函數。範例程式如下：

```
//chapter3/bin/example_06.dart
void main(){
  // 定義一個帶有 int 類型參數的函數
  void printElement(int element) {
    // 列印傳入的參數值
    print(element);
  }
  // 定義一個 List 類型的物件並賦初值
  var list = [1, 2, 3];
  // 將 printElement 函數作為參數傳遞給 List 物件的 forEach 函數
  //forEach 函數會按疊代順序將函數 printElement 應用於 List 集合的每個元素
  list.forEach(printElement);
}
```

也可以將函數設定值給一個變數。範例程式如下：

```
//chapter3/bin/example_07.dart
void main() {
  void printElement(int element) {
    print(element);
  }
  // 將函數 printElement 設定值給變數 show
  var show = printElement;
  // 像執行函數一樣使用變數 show
  show(1);
}
```

3.4 匿名函數

大多數方法都是有名字的，例如 main 或 printElement。也可以創建一個沒有名字的函數，稱為匿名函數。匿名函數常以回呼函數或另一個函數的參數的形式出現。

匿名函數看起來與命名函數類似，在括號之間可以定義參數，參數之間用逗點分隔，後面大括號中的內容則為函數本體，也可以有返回值。

```
([ 類型 ][ 參數 ,…]){
// 函數本體 ;
}
```

匿名函數常用於集合疊代中，下面的程式定義了只有一個參數 item 且沒有參數類型的匿名函數。List 中的每個元素都會呼叫這個函數，函數本體列印元素在集合中的索引和值。範例程式如下：

```
//chapter3/bin/example_08.dart
void main(){
   var list = ['apples', 'bananas', 'oranges'];
   // 向 forEach 函數提供匿名函數
   list.forEach((item){
      print('${list.indexOf(item)}: $item');
   });
}
```

如果函數本體內只有一行敘述，則可以使用「=>」語法：

```
list.forEach((item) => print('${list.indexOf(item)}:$item'));
```

匿名函數也可以作為另一個函數的返回物件，返回敘述 return (num r) => 2*pi*r; 中的返回值就是匿名函數。範例程式如下：

```
//chapter3/bin/example_09.dart
void main(){
   // 定義一個返回數值類型為 Function 的函數
   Function perimeter(){
      var pi = 3.14;
      //return 關鍵字後面跟著的是一個匿名函數
      return (num r) => 2*pi*r;
   }
   // 將函數設定值給變數
   var per = perimeter();
   var r = 9;
   var l = per(r);
   print(' 圓的周長 :$l');
}
```

3.5 語法作用域

Dart 是有詞法作用域的語言，變數的作用域在寫程式的時候就確定了。大括號內定義的變數只能在大括號內被存取，在大括號內可以存取大括號外的變數，與 Java 類似。巢狀結構函數中變數在多個作用域中使用的範例程式如下：

```
//chapter3/bin/example_10.dart
// 定義一個頂層變數
String topLevel = 'top variable';
void main(){
   // 在 main 函數中定義變數
   var insideMain = 'insideMain variable';
   void myFunction(){
      // 在 myFunction 函數中定義變數
      var insideFunction = 'insideFunction variable';
      void nestedFunction(){
         // 在巢狀結構函數中定義變數
         var insideNestedFunction = 'insideNestedFunction variable';
         print('$topLevel');
         print('$insideMain');
         print('$insideFunction');
         print('$insideNestedFunction');
      }
   }
}
```

注意：nestedFunction() 函數可以存取包括頂層變數在內的所有變數。

3.6 語法閉包

語法閉包即一個函數物件，即使函數物件的呼叫在它原始作用域之外，依然能夠存取在它詞法作用域內的變數。

函數可以封閉定義到它作用域內的變數。下面的範例中，函數 makeAdder() 捕捉了變數 addBy。無論函數在什麼時候返回，它都可以使用捕捉的 addBy 變數。範例程式如下：

```
//chapter3/bin/example_11.dart
void main(){
  // 返回一個函數，該函數的參數將與 addBy 相加
  Function makeAdder(num addBy){
    return (num i) => addBy + i;
  }
  // 生成加 2 的函數
  var add2 = makeAdder(2);
  // 生成加 4 的函數
  var add4 = makeAdder(4);
  print(add2(3));
  print(add4(3));
}
```

3.7 函數相等性測試

函數相等性測試用來判斷兩個函數是否是同一個物件。

頂層函數、靜態方法和實例方法相等性的測試範例程式如下：

```
//chapter3/bin/example_12.dart
// 定義頂層函數
void foo(){}

// 定義一個類別
class A {
```

```
  // 定義靜態方法
  static void bar(){}
  // 定義實例方法
  void baz(){}
}

void main() {
  var x;
  x = foo;
  // 比較頂層函數是否相等
  print(foo == x);

  // 比較靜態方法是否相等
  x = A.bar;
  print(A.bar == x);

  //A 的實例 #1
  var v = A();
  //A 的實例 #2
  var w = A();
  var y = w;
  x = w.baz;
  // 這兩個閉包引用了相同的實例物件，因此它們相等
  print(y.baz == x);
  // 這兩個閉包引用了不同的實例物件，因此它們不相等
  print(v.baz != w.baz);
}
```

3.8 返回值

所有的函數都有返回值，返回值本質上是物件。返回數值類型可以是內建類型也可以是自訂類型，返回數值類型放在函數名稱前面。範例程式如下：

```
//chapter3/bin/example_13.dart
void main(){
```

```
    // 返回數值類型為 num
    num area(num pi,num r) {
        // 返回敘述
        return pi*r*r;
    }
    print(' 圓面積為 :${area(3.14,6)}');
}
```

當函數名稱前有類型修飾時，函數本體最後必須提供 return 敘述。當函數名稱前沒有類型修飾且未提供 return 敘述時，最後一行預設為執行 return null;。範例程式如下：

```
//chapter3/bin/example_14.dart
void main() {
    // 定義沒有提供返回數值類型的函數
    foo(){
        // 函數本體未提供 return 敘述
        var a;
    }
    // 判斷函數 foo 的返回值是否為 null
    print(foo() == null);
}
```

3.9 回呼函數

回呼函數是指作為參數傳遞的函數，回呼函數還會獲得原函數提供的值。

首先將函數作為參數：

```
void printProgress({Function(int) callback}){...}
```

有條件地在函數內執行回呼函數：

```
if(callback!=null) callback(progress);
```

接收原函數提供的參數：

```
printProgress(callback: (int progress){
  print('列印進度：$progress');
});
```

完整的範例程式如下：

```
//chapter3/bin/example_15.dart
void main(){
  // 定義 printProgress 函數，將回呼函數 callback 作為可選參數
  void printProgress({Function(int) callback}){
    for (int progress = 0; progress <= 10; progress++){
      // 判斷回呼函數是否為空
      if(callback!=null){
        // 若提供則呼叫回呼函數並傳遞循環變數 progress
        callback(progress);
      }
    }
  }
  // 呼叫 printProgress 函數並提供匿名回呼函數
  printProgress(callback: (int progress){
    print('列印進度：$progress%');
  });
}
```

Chapter 04 運算子

表 4-1 所示的是 Dart 語言定義的運算子，可以覆寫其中一部分運算子。

表 4-1 運算子

<table>
<tr><th>運算子名稱</th><th>運算符號</th></tr>
<tr><td>一元尾碼運算子</td><td>expr++ expr-- () [] . ?..</td></tr>
<tr><td>一元字首運算子</td><td>-expr !expr ~expr ++expr await expr</td></tr>
<tr><td>乘除</td><td>* / % ~/</td></tr>
<tr><td>加減</td><td>+ -</td></tr>
<tr><td>位元運算符號</td><td><< >> >>></td></tr>
<tr><td>二進位與</td><td>&</td></tr>
<tr><td>二進位互斥</td><td>^</td></tr>
<tr><td>二進位或</td><td>|</td></tr>
<tr><td>關係和類型測試</td><td>>= > <= < as is is!</td></tr>
<tr><td>相等判斷</td><td>== !=</td></tr>
<tr><td>邏輯與</td><td>&&</td></tr>
<tr><td>邏輯或</td><td>||</td></tr>
<tr><td>空判斷</td><td>??</td></tr>
<tr><td>條件運算式</td><td>expr1 ? expr2 : expr3</td></tr>
<tr><td>串聯</td><td>..</td></tr>
<tr><td>設定值</td><td>= *= /= += -= &= ^=</td></tr>
</table>

運算子存在優先順序，就像在數學中常說的先乘除後加減一樣。在運算子表中，在同一列中優先順序依次降低，即第一行優先順序最高，最後一行優先順序最低。在同一行中從左到右優先順序依次降低，即最左邊的優先順序最高，最右端的優先順序最低。

創建命令列應用程式，將專案命名為 chapter4。本章所有檔案都在該專案的 bin 資料夾下創建與執行。

4.1 算術運算子

表 4-2 所示的是 Dart 語言中的算術運算子，表中的 expr 代表的是運算式，var 代表的是變數。

表 4-2 算術運算子

運算子	說明
+	加
-	減
-expr	負號，也可以對運算式符號反轉
*	乘
/	除
~/	除，返回正整數部分
%	求模
++var	先自加再進行運算
var++	先運算再進行自加
--var	先自減再進行運算
var--	先運算再進行自減

常見算術運算子使用範例程式如下：

```
//chapter4/bin/example_01.dart
void main(){
```

```
    // 加號運算
    print('1+3=${1+3}');
    // 減號運算
    print('8-6=${8-6}');
    var a = 6;
    // 負號運算
    print('-a=${-a}');
    // 對運算式符號反轉
    print('-(-a)=${-(-a)}');
    // 乘號運算
    print('6*7=${6*7}');
    // 除號運算
    print('8/2=${8/2}');
    // 除並取整數運算
    print('9~/2=${9~/2}');
    // 除並取餘數運算
    print('9%2=${9%2}');
}
```

自動增加和自減需要特別注意。範例程式如下：

```
//chapter4/bin/example_02.dart
void main(){
    var a = 10;
    var b = 20;
    var c = 30;
    var d = 40;
    var e,f,g,h;
    //a 先自加再設定值給 e
    e = ++a;
    print('e:$e a:$a');
    //b 先設定值給 f 再自加
    f = b++;
    print('f:$f b:$b');
    //c 先自減再設定值給 g
    g = --c;
    print('g:$g c:$c');
    //d 先設定值給 h 再自減
    h = d--;
    print('h:$h d:$d');
}
```

4.2 關係運算子

表 4-3 中的關係運算子常用來比較兩個物件是否相等或大小關係。關聯運算式屬於布林運算式，所以它們的返回值均為布林值，即 true 或 false。

表 4-3 關係運算子

運算子	說明
==	相等
!=	不等
>	大於
<	小於
>=	大於或等於
<=	小於或等於

要測試兩個物件是否表示同樣的值，例如數值、布林值或字串值等，需使用 == 運算子。有時需要知道兩個物件是否是完全相同的物件，需改用 identical() 函數。== 運算子的工作方式如下：

(1) 判斷 x 或 y 是否為 null 的情況：如果兩個均為 null，則返回 true；如果只有一個為 null，則返回 false。

(2) 其餘情況下返回方法呼叫 x.==(y) 的結果。值得注意的是：== 之類的運算子是在第一個運算元上呼叫的方法。

範例程式如下：

```
//chapter4/bin/example_03.dart
void main(){
   var a = 10;
   var b = 10;
   var c = 30;
   var d = 40;
   // 定義 e 和 f 且未設定值，預設值為 null
   var e,f;
```

```
    // 相等運算
    print('a==b:${a==b}');
    print('e==f:${e==f}');
    print('a==e,${a==e}');
    // 不等運算
    print('a!=c,${a!=c}');
    // 大於運算
    print('d>c,${d>c}');
    print('d>c,${d>c}');
    // 小於運算
    print('b<d,${b<d}');
    // 大於或等於運算
    print('a>=b,${a>=b}');
    // 小於或等於運算
    print('b<=c,${b<=c}');
}
```

4.3 類型測試運算子

表 4-4 中的類型測試運算子可以在執行時期檢查類型。

表 4-4 類型測試運算子

運算子	說明
as	類型轉換
is	類型判斷，如果物件是指定物件則返回 true
is!	類型判斷，如果物件不是指定物件則返回 true

運算子 is 檢查的不是物件是否屬於某個類別或其子類別，而是檢查物件所屬的類別是否實現了某個類別的介面 (直接或間接)。如果物件 obj 實現了類別 T 定義的介面，obj is T 則返回 true。obj is Object 則始終返回 true，因為 Object 是所有類別的父類別。

運算子 as 也是用來檢查物件是否屬於某個類型或其子類別，區別在於，如果測試成功則返回被檢測的物件，如果測試失敗則會拋出一個

CastError。範例程式如下：

```
//chapter4/bin/example_04.dart
// 定義類 Car
class Car{
   String length;
   String color;
}
// 定義類 Taxi 繼承類別 Car
class Taxi extends Car{
   double fee;
}
void main(){
   var t = Taxi();
   (t as Car).color = 'RED';
   print(' 變數 t 的引用物件執行時期類型為 ${t.runtimeType}');
   print('t 的引用物件是類別 Car 的子類別 :${t is Car}');
   print('t as Car 執行時期類型為 ${(t as Car).runtimeType}');
}
```

使用 as 運算子，即類型轉換的工作邏輯大致可以透過以下程式表述：

```
var t = someObject;
t is T ? t : throw CastError();
```

運算子 as 的主要使用場景是資料驗證：

```
List ls = getTaxis() as List;
```

程式部分中的 getTaxis() 方法用於獲取一個泛型為 Taxi 的清單，如果返回值確實是清單，則繼續其他程式的執行，如果返回的不是清單，則立即拋出錯誤。

```
(t as Car).color = 'RED';
```

上述程式部分實際上是對 as 運算子的濫用，因為類型轉換是在執行時期才執行的，因此會造成運算資源消耗。這裡只是為了演示它的返回值是被檢測物件 t，因為已經明確知道物件 t 所屬的類別就是 Car 類別的子類別，在實際開發中應當避免類似操作。

4.4 設定運算子

表 4-5 中列出了常用設定運算子及含義，它們的作用是按對應規則為變數設定值。

表 4-5 常用設定運算子及含義

<table>
<tr><th>運算子</th><th>含義</th></tr>
<tr><td>=</td><td>a = b</td></tr>
<tr><td>??=</td><td>a ??= b 等值於 a 等於 null 時 a=b</td></tr>
<tr><td>-=</td><td>a -= b 等於 a = a-b</td></tr>
<tr><td>/=</td><td>a /= b 等於 a = a/b</td></tr>
<tr><td>%=</td><td>a %= b 等於 a = a%b</td></tr>
<tr><td>>>=</td><td>a >>= b 等於 a = a>>b</td></tr>
<tr><td>^=</td><td>a^= b 等於 a = a^b</td></tr>
<tr><td>+=</td><td>a += b 等於 a = a+b</td></tr>
<tr><td>*=</td><td>a *= b 等於 a = a*b</td></tr>
<tr><td>~/=</td><td>a ~/= b 等於 a = a~/b</td></tr>
<tr><td><<=</td><td>a <<= b 等於 a = a<<b</td></tr>
<tr><td>&=</td><td>a &= b 等於 a = a&b</td></tr>
<tr><td>|=</td><td>a |= b 等於 a = a|b</td></tr>
</table>

可以使用符號 = 為變數設定值，使用符號 ??= 為值為 null 的變數設定值。範例程式如下：

```
//chapter4/bin/example_05.dart
void main(){
   // 為變數 a 設定值
   var a = 12;
   print('a：$a');
   var b1=10,b2;
   //b1 已有初值所以將保持原值
   b1??=13;
```

```
  //b2 沒有初值所以將被設定值
  b2??=13;
  print('b1：$b1, b2：$b2');
}
```

複合設定運算子的使用範例程式如下：

```
//chapter4/bin/example_06.dart
void main(){
  var a = 12,b = 5;
  print('a%=b:${a%=b}');
  print('a~/=b:${a~/=b}');
}
```

4.5 邏輯運算子

表 4-6 中的邏輯運算子運算後的結果都為布林值。邏輯與和邏輯或都是先計算左運算元，再計算右運算元。如果邏輯與左運算元計算結果是假值，則返回假值，無須計算右運算元。如果邏輯或左運算元計算結果是真值，則返回真值，無須計算右運算元。

表 4-6 邏輯運算子

運算子	描述
!expr	對運算式結果反轉
||	邏輯或
&&	邏輯與

常見使用範例程式如下：

```
//chapter4/bin/example_07.dart
void main(){
  var c = true,d = false,e = true;
  // 非運算
  print('!c:${!c}');
```

```
  print('!d:${!d}');
  // 邏輯或運算
  print('c||d:${c||d}');
  print('c||e:${c||e}');
  // 邏輯與運算
  print('c&&d:${c&&d}');
  print('c&&e:${c&&e}');
}
```

4.6 位元運算符號

使用表 4-7 中的位元運算符號可以運算數字的各個二進位位元。一般來説可以對整數使用這些位元運算符號和移位元運算符號。

表 4-7 位元運算符號

運算子	描述
&	逐位元進行與操作
|	逐位元進行或操作
^	逐位元進行互斥操作
~expr	逐位元進行反轉操作
<<	逐位元進行左移操作
>>	逐位元進行右移操作

下例中使用了兩個十六進位數，它們實際上是先轉為二進位數字，然後進行位元運算。範例程式如下：

```
//chapter4/bin/example_08.dart
void main(){
  final value = 0x22;
  final bitmask = 0x0f;
  // 逐位元進行與操作
  print((value & bitmask) == 0x02);
  // 反轉後逐位元進行與操作
```

```
    print((value & ~bitmask) == 0x20);
    // 逐位元進行或操作
    print((value | bitmask) == 0x2f);
    // 逐位元進行互斥操作
    print((value ^ bitmask) == 0x2d);
    // 逐位元進行左移操作
    print((value << 4) == 0x220);
    // 逐位元進行右移操作
    print((value >> 4) == 0x02);
}
```

4.7 條件運算式

表 4-8 中列出的兩種條件運算式常用於建構一個設定陳述式，它們也用於替代簡單的 if-else 敘述。

表 4-8 條件運算式

運算子	描述
condition ? expr1 : expr2	如果條件為真則返回 expr1，否則返回 expr2
expr1 ?? expr2	如果 expr1 為非 null 則返回 expr1，否則返回 expr2

使用範例程式如下：

```
//chapter4/bin/example_09.dart
void main(){
   bool isPublic = true;
   // 三元條件運算式
   print('${isPublic ? 'public' : 'private'}');
   var x,y=10,z=20;
   // 二元條件運算式
   print('${x ?? y}');
   print('${y ?? z}');
}
```

4.8 其他運算子

表 4-9 中列出的其他運算子涉及後續的知識，這裡大概知道即可，可以在掌握相關知識後再瞭解。

表 4-9 其他運算子

運算子	描述
( )	代表呼叫的是函數
[ ]	存取 List 特定位置的元素
.	存取物件的成員
?.	有條件地存取物件的成員，若物件為 null 則返回 null

使用範例程式如下：

```
//chapter4/bin/example_10.dart
// 定義類別 Point
class Point{
   int x,y;
   Point(this.x,this.y);
}
// 定義頂層函數 printTime
void printTime() {
   print(DateTime.now());
}
void main() {
   // 透過使用 () 運算子標識呼叫的是函數
   printTime();

   var ls = [1, 2, 3];
   // 透過 [] 運算子存取 ls 陣列的第一個元素
   var a = ls[0];

   var p1 = Point(3,6);
   // 透過 . 運算子存取 p1 的成員 x
   print('${p1.x}');
```

```
    var p2;
    // 透過 ?. 運算子存取 p2 的成員 x
    // 因為 p2 沒有成員 x ，所以返回 null
    print('${p2?.x}');
}
```

流程控制敘述

在一個程式執行的過程中，各行敘述的執行順序對程式的結果是有直接影響的。控制敘述用實現對程式流程的選擇、迴圈、轉向和返回等進行控制。只有在清楚每行敘述的執行流程的前提下，才能透過控制敘述的執行順序實現要完成的任務。

Dart 中的控制敘述分為以下幾類：

(1) 分支敘述：if 和 switch。
(2) 迴圈敘述：while、do-while 和 for。
(3) 跳躍陳述式：break、continue、return 和 throw。

創建命令列應用程式，將專案命名為 chapter5。本章所有檔案都在該專案的 bin 資料夾下創建與執行。

5.1 分支敘述

分支敘述又稱條件陳述式，條件陳述式使得部分程式區塊根據運算式的值有選擇地執行。Dart 語言提供了 if 和 switch 兩種分支敘述。

5.1.1 if 敘述

具體來說分支敘述由選擇結構組成，if 敘述啟動的選擇結構包括 if 結構、if-else 結構和 else-if 結構。

1. if 結構

if 結構包含條件運算式和敘述區塊，當條件運算式為 true 時執行敘述區塊，否則執行 if 結構後面的敘述。通常來說敘述區塊由大括號 ({}) 包裹，如果 if 結構中的敘述區塊只有一行敘述，則可以省略大括號。

if 結構宣告語法如下：

```
if( 條件運算式 ){
   // 敘述區塊
}
```

範例程式如下：

```
//chapter5/bin/example_01.dart
void main(){
   var x = true,y = 10;
   if(x){
      // 此處 x 為真，敘述體得到執行
      print(' 條件 x 為 $x');
   }
   if(!x){
      // 此處 !x 為假，敘述體得不到執行
      print(' 條件 !x 為 $x，if 敘述體得到執行。');
   }
   // 此處省略大括號
   if(y<=10)
      // 條件 y<=10 為真，敘述體得到執行
      print(' 條件 y<=10 為 ${y<=10}');
}
```

2. if-else 結構

在 if 結構中，只有條件運算式為 true 時提供對應的處理敘述區塊。有時

無論條件運算式為 true 還是 false 都需進行對應處理，此時可以使用 if-else 結構。宣告格式如下：

```
if( 條件運算式 ){
   // 敘述區塊 1
}else{
   // 敘述區塊 2
}
```

當程式執行到 if 敘述時，先判斷條件運算式，如果值為 true，則執行敘述區塊 1，然後跳過 else 敘述區塊，繼續執行後續敘述。如果條件運算式值為 false，則忽略敘述區塊 1 直接執行敘述區塊 2，然後繼續執行後續敘述。

if-else 結構使用範例程式如下：

```
//chapter5/bin/example_02.dart
void main(){
   var x = true;
   if(!x){
      // 條件 !x 為真時，if 敘述區塊得到執行
      print('if !x：${!x}');
   }else{
      // 條件 !x 為假時，else 敘述區塊得到執行
      print('else !x：${!x}');
   }
}
```

3. else-if 結構

else-if 結構是 if-else 結構的多層巢狀結構形式，它會在多個分支中執行一個敘述區塊，其他分支將不會得到執行，所以這種結構常用於有多種判斷結果的分支中。

else-if 結構如下：

```
if( 條件運算式 1){
   // 敘述區塊 1
```

```
}else if( 條件運算式 2){
    // 敘述區塊 2
}
...
}else if( 條件運算式 n){
    // 敘述區塊 n
}else{
    // 敘述區塊 n+1
}
```

else-if 結構使用範例程式如下：

```
//chapter5/bin/example_03.dart
void main(){
    var x = true,y = 10;
    if(!x){
        // 條件 !x 為假，if 敘述區塊不會執行
        print(' 條件 x 為 $x');
    }else if(y<=10){
        // 如果條件 y<=10 為真，則 else if 敘述區塊得到執行
        print(' 條件 y<=10 為 ${y<=10}');
    }else{
        // 當所有條件都為假時，else 敘述區塊才會執行
        print('else !x：${!x}。');
    }
}
```

5.1.2 switch 敘述

可以使用 switch 敘述選擇要執行的多個程式區塊中的程式區塊，語法如下：

```
switch(expr){
    case 值 1:
        // 敘述區塊 1
        break;
    case 值 2:
        // 敘述區塊 2
        break;
```

```
      ...
   case 值 n:
      // 敘述區塊 n
      break;
   default:
   // 敘述區塊 n+1
}
```

首先設定運算式 expr，通常是一個變數。隨後運算式的值會與結構中的每個 case 的值做比較。如果存在匹配，則與該 case 子句連結的程式區塊會被執行。如果 case 子句中沒有與 expr 匹配的值，則會執行 default 子句的敘述區塊，然後跳出 switch 結構。

switch 敘述使用 == 來比較整數、字串或編譯時常數，比較的兩個物件必須是同一個類別的實例且沒有覆寫 == 運算子。

每個不可為空的 case 子句必須包含一個 break 敘述。當沒有匹配的子句時，可以使用 default 子句匹配這種情況。範例程式如下：

```
//chapter5/bin/example_04.dart
void main(){
   var color = 'Red';
   switch(color){
      case 'Green':
         print('color 值為 Green');
         break;
      case 'Orange':
         print('color 值為 Orange');
         break;
      case 'Red':
         print('color 值為 Red');
         break;
      default:
         print('color 值未匹配 ');
   }
}
```

下面的例子忽略了 break 子句，因此產生錯誤。

```
//chapter5/bin/example_05.dart
void main(){
   var color = 'Red';
   switch(color){
      case 'Green':
         print('color 值為 Green');
// 執行出錯，沒有 break 敘述
   case 'Orange':
      print('color 值為 Orange');
      break;
   case 'Red':
      print('color 值為 Red');
      break;
   default:
      print('color 值未匹配 ');
   }
}
```

Dart 語言支援 case 子句的內容為空，這種情況稱為貫穿。下例中值為 Orange 的 case 子句內容為空，它會造成貫穿。當傳入變數值為 Orange 時，程式會繼續執行與它相鄰的值為 Red 的 case 子句。範例程式如下：

```
//chapter5/bin/example_06.dart
void main(){
   var color = 'Orange';
   switch(color){
      case 'Green':
         print('color 值為 Green');
         break;
      case 'Orange':
         // 空敘述貫穿
      case 'Red':
         //color 值為 Orange 和 Red 都會被執行
         print('color 值為 Red');
         break;
      default:
         print('color 值未匹配 ');
   }
}
```

不可為空 case 子句實現貫穿可以透過 continue 敘述和標籤來完成。下例中匹配到值為 Green 的 case 子句時，不僅會執行該子句的程式區塊，還會執行值為 Red 的 case 子句中的程式區塊。範例程式如下：

```
//chapter5/bin/example_07.dart
void main(){
  var color = 'Green';
  switch(color){
    case 'Green':
      print('color 值為 Green');
      // 繼續執行標籤為 red 的 case 子句
      continue red;
    case 'Orange':
      print('color 值為 Orange');
      break;
    // 標籤 red，它代表了 case 值為 Red 的子句
    red:
    case 'Red':
      //color 值為 Green 和 Red 都會被執行
      print('color 值為 Red');
      break;
    default:
      print('color 值未匹配 ');
  }
}
```

每個 case 子句都可以有區域變數且僅在該 case 子句內有效。

5.2 迴圈敘述

迴圈敘述能夠根據迴圈條件使程式碼重複執行。Dart 支援 3 種迴圈結構：for、while 和 do-while。

5.2.1 for 敘述

for 敘述是一種應用廣泛、功能最強的迴圈敘述。宣告格式如下：

```
for( 運算式 1; 運算式 2; 運算式 3){
    // 迴圈敘述區塊
}
```

運算式 1 是初始化敘述，用於初始化迴圈變數和其他變數。運算式 2 是迴圈條件，運算式 3 用於更新迴圈變數，使迴圈變數趨近於某個值，直到使迴圈條件變為 false。

開始 for 迴圈時，會執行運算式 1 初始化迴圈變數，運算式 1 只會執行一次。然後執行運算式 2，判斷迴圈條件是否滿足，如果滿足，則繼續執行迴圈本體。執行完成後繼續執行運算式 3，更新迴圈變數，然後接著再判斷迴圈條件。如此反覆，直到迴圈條件不滿足時跳出迴圈。執行流程如圖 5-1 所示。

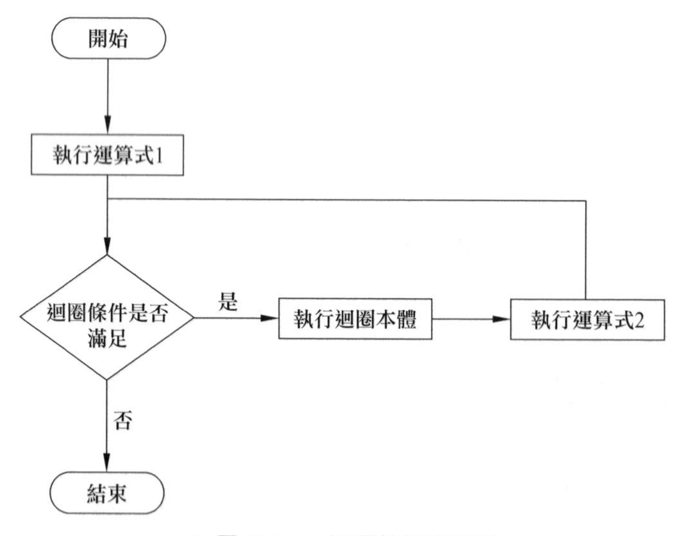

▲ 圖 5-1 for 迴圈執行流程圖

for 迴圈使用範例程式如下：

```
//chapter5/bin/example_08.dart
```

```
void main(){
  for (var i = 0;i < 5;i++){
    // 此處 i 的初值為 0，所以使用 i+1 保證其與實際迴圈次數一致
    print('第 ${i+1} 次迴圈。');
  }
}
```

迴圈開始時，給迴圈變數 i 設定值為 0，每次執行迴圈本體前都會判斷 i 的值是否小於 5，如果為 true，則執行迴圈本體，然後執行 i++，使得迴圈變數 i 的值加 1。然後繼續判斷迴圈條件，直到判斷結果為 false 時跳出迴圈。

對於實現了 iterable 介面的類別可以使用 forEach() 函數進行疊代。例如 List 和 Set 類別，它們還支持 for-in 形式的疊代。範例程式如下：

```
//chapter5/bin/example_09.dart
void main(){
  var collection = [0, 1, 2];
  // 這裡給 forEach 函數傳入了一個匿名函數
  collection.forEach((i)=>print(i));
  //for-in 形式的疊代
  for (var x in collection) {
    print(x);
  }
}
```

在 Dart 語言中，for 迴圈中的閉包會自動捕捉索引值，這在某些情況下很有用。範例程式如下：

```
//chapter5/bin/example_10.dart
void main(){
  var callbacks = [];
  for (var i = 0; i < 2; i++) {
    callbacks.add(() => print(i));
  }
  callbacks.forEach((c) => c());
}
```

5.2.2 while 敘述

while 迴圈在執行迴圈本體前先判斷迴圈條件，如果條件為真則執行迴圈本體，如果條件為假則結束迴圈。

宣告格式如下：

```
while( 迴圈條件 ){
   // 敘述區塊
   }
```

在 while 迴圈本體中一定要存在影響迴圈條件的敘述，其作用是使迴圈條件最終變為 false 以避免無限迴圈。

範例程式如下：

```
//chapter5/bin/example_11.dart
void main(){
   var i=0,x=10;
   while(i<x){
      print('第${i+1}次 while 迴圈。');
      //影響條件運算式並使其趨於假的敘述
      i++;
   }
}
```

範例中 i++ 是影響迴圈條件的因數，也是它使得迴圈能夠在有限次迴圈後結束。

5.2.3 do-while 敘述

do-while 迴圈與 while 迴圈類似。宣告格式如下：

```
do
   // 敘述區塊
}while( 迴圈條件 ){
```

不同之處在於 do-while 迴圈會先執行一次迴圈本體，再判斷迴圈條件，如果條件為真則執行迴圈本體，如果條件為假則結束迴圈。

範例程式如下：

```
//chapter5/bin/example_12.dart
void main(){
   var i=0,x=10;
   do{
      print('第${i+1}次do while迴圈。');
      //影響條件並使其趨於假的敘述
      i++;
   }while(i<x);
}
```

5.3 跳躍陳述式

跳躍陳述式可以改變程式的執行順序，從而可以實現跳躍。

5.3.1 break 敘述

break 敘述可用於 for、while 和 do-while 迴圈結構，它的作用是強行退出迴圈本體，且不再執行迴圈本體中剩餘的敘述。

在 while 迴圈中 break 敘述使用範例程式如下：

```
//chapter5/bin/example_13.dart
void main(){
   var i=0,x=10;
   while(i<x){
      print('第${i+1}次while迴圈。');
      if(i==5){
         //當迴圈變數i等於5時跳出for迴圈
         break;
      }
```

```
        // 影響條件運算式並使其趨於假的敘述
        i++;
    }
}
```

5.3.2 continue 敘述

continue 敘述的作用是結束本次迴圈，跳過迴圈本體中尚未執行的敘述，接著繼續判斷迴圈條件，以決定是否繼續迴圈。

在 for 迴圈中 continue 敘述使用範例程式如下：

```
//chapter5/bin/example_14.dart
void main(){
    for (var i = 0; i < 10; i++) {
        if(i==5) continue;
        // 此處 i 的初值為 0，所以使用 i+1 保證其與實際迴圈次數一致
        print(' 第 ${i+1} 次迴圈。');
    }
}
```

5.3.3 assert

assert 函數又稱斷言，它的第一個參數是布林運算式，在運算式的值為 false 時中斷程式執行，斷言在開發過程中用於檢測物件是否符合期待。範例程式如下：

```
//chapter5/bin/example_15.dart
void main(){
    var x,y=9;
    // 確保變數 x 的值不可為空
    assert(x != null);
    // 確保變數 x 的值小於 8
    assert(y < 8);
}
```

也可以向 assert 函數傳遞一個 String 類型的可選參數。範例程式如下：

```
//chapter5/bin/example_16.dart
void main(){
      var URLString = 'http://dartlang.org';
      // 確保網址應該以 https 開始
      assert(URLString.startsWith('https'));
}
```

斷言什麼時候起作用取決於使用的工具和框架：

(1) Flutter 在偵錯模式下啟用斷言。
(2) 預設情況下，僅開發工具 (例如 dartdevc) 通常啟用斷言。
(3) 某些工具 (例如 dart 和 dart2js) 透過命令列標示 ---enable-asserts 支持斷言。
(4) 在生產程式中，斷言將被忽略，並且不會評估斷言的參數。

類別

Dart 是一種物件導向的程式語言，它支援以 mixin 為基礎的繼承機制和擴充方法。Dart 中每個物件都是一個類別的實例，並且所有類別都來自 Object 類別。

使用 class 關鍵字宣告類別，類別由類別名稱和大括號包裹的類別本體組成，類別本體可以為空，格式如下：

```
class Point{}
```

可以透過以下格式創建該類別的新實例。

```
Pointp = Point();
```

當然也可以使用 var、dynamic、Object 代替變數 p 前的類型名稱。

```
varp = Point();
```

在這裡該類別不具有任何屬性和方法，單純地僅是一個符合規範的類別而已，它不具有任何實用價值。

創建命令列應用程式，將專案命名為 chapter6。本章所有檔案都在該專案的 bin 資料夾下創建與執行。

6.1 屬性

所謂的屬性，就是在類別中定義的變數。根據使用方式的不同，屬性分為實例屬性和類別屬性。

宣告屬性的格式如下：

```
[ 修飾符號 ] 資料類型屬性名稱 [= 值 ];
```

修飾符號是可選的，修飾符號可以是多個。資料類型可以是內建類型也可以是自訂類型，屬性的命名應當符合規範，屬性的設定值是可選的。

沒有 static 修飾的屬性就是實例屬性，也可以稱作執行個體變數。實例屬性只能透過類別的實例引用，宣告實例屬性的範例程式如下：

```
class Point{
   // 宣告實例屬性 x, 也可稱作執行個體變數 x，預設初值為 null
   num x;
   // 宣告實例屬性 y, 也可稱作執行個體變數 y，設定初值為 0
   num y = 0;
}
```

所有未初始化的屬性的值預設為 null，如果在宣告處就為屬性初始化，其值在實例創建之前就被設定，甚至在建構函數初始化清單執行之前。所有的屬性都會生成一個隱式的 getter 方法，非 final 修飾的屬性還會生成一個 setter 方法。

實例物件透過使用點號 (.) 來引用屬性。範例程式如下：

```
//chapter6/bin/example_01.dart
// 宣告 Point 類別
class Point{
   num x;
   num y;
}
void main(){
   // 創建 Point 類別的實例
```

```
    var point = Point();
    // 呼叫屬性 x 的 setter 方法
    point.x = 4;
    // 呼叫屬性 x 的 getter 方法
    print(point.x);
    // 屬性 y 的值預設為 null
    print(point.y);
}
```

為了避免物件 p 為 null 而引起異常，可以使用 ?. 代替 .。

```
void main(){
    // 如果物件 p 非 null, 設定 y 的值為 4
    p?.y = 4;
}
```

類別屬性是由關鍵字 static 修飾的屬性，也可以稱作類別變數。類別屬性對於表示類別範圍的狀態變數和常數很有用，類別屬性在第一次使用時會被初始化。類別屬性透過類別直接存取和修改值，類別屬性不能透過類別的實例存取和修改值，定義類屬性 z 的範例程式如下：

```
//chapter6/bin/example_02.dart
class Point{
    num x;
    num y;
    // 定義類屬性 z, 也可稱作類別變數 z
    static double z = 1;
}
void main(){
    // 直接透過類別存取類別屬性 z
    print(Point.z);
    // 呼叫類別屬性 z 的 setter 方法來修改值
    Point.z =10;
    // 查看類別屬性 z 修改後的值
    print(Point.z);
}
```

6.2 建構函數

透過創建與類別名稱相同的方法宣告建構函數，下例中 Point 方法就是 Point 類別的名稱相同建構函數。範例程式如下：

```
//chapter6/bin/example_03.dart
class Point{
   num x,y;
   Point(num x,num y){
      // 建構函數本體
      // 透過傳入的參數為屬性設定值
      this.x = x;
      this.y = y;
   }
}
```

this 關鍵字引用的是當前實例，將建構函數參數的值分配給實例屬性的模式非常普遍，Dart 擁有語法糖使其變得簡單。範例程式如下：

```
class Point{
   num x,y;
   // 使用語法糖設定 x 和 y, 在建構函數本體執行之前執行
   Point(this.x,this.y);
}
```

在 Dart 中與類別名稱相同的建構函數只能有一個，這與其他語言中可以擁有多個名稱相同建構函數不同，為了實現相同的功能，可以將所有與實例屬性相關的參數放在可選參數清單中。範例程式如下：

```
//chapter6/bin/example_04.dart
class Person{
   int age;
   double hight;
   String name;
   // 將所有參數置於可選參數清單中
   Person({this.age,this.name,this.hight});
}
```

```
void main(){
  // 創建只為 age 屬性設定值的 Person 類別的實例
  var p1 = Person(age:10);
  // 創建為 name 和 age 屬性設定值的 Person 類別的實例
  var p2 = Person(name:'jobs',age:33);
  // 創建為所有屬性設定值的 Person 類別的實例
  var p3 = Person(name:'jobs',age:33,hight:1.76);
  print(p1.age);
  print(p2.name);
  print(p3.hight);
}
```

如上例所示，向建構函數傳入不同數量或類型的參數來創建新實例以滿足實際需求，將非常實用。

6.2.1 預設建構函數

如果在定義一個類別時未宣告建構函數，將為此類提供一個預設建構函數，預設建構函數沒有參數，並會呼叫父類別中的無參數建構函數。

6.2.2 命名建構函數

透過附加的識別符號宣告命名建構函數，使用命名建構函數可為一個類別實現多個建構函數以供滿足明確的使用場景。定義命名建構函數 Point.origin 和 Point.fromJson 的範例程式如下：

```
//chapter6/bin/example_05.dart
class Point{
  num x, y;
  // 名稱相同建構函數
  Point(this.x,this.y);
  // 命名建構函數 origin
  Point.origin(){
    x = 0;
    y = 0;
  }
```

```
  // 命名建構函數 fromJson
  Point.fromJson(Map json){
    this.x = json['x'];
    this.y = json['y'];
  }
}
void main(){
  // 使用命名建構函數 origin 來創建實例
  var p1 = Point.origin();
  print('p1:(${p1.x},${p1.y})');
  // 創建 json 資料，很多情況下 Map 類型等於 json
  Map data = {'x':1,'y':2};
  // 使用命名建構函數 fromJson 來創建實例
  var p2 = Point.fromJson(data);
  print('p2:(${p2.x},${p2.y})');
}
```

建構函數不會被繼承，這表示父類別的命名建構函數不會被子類別繼承。如果要使用父類別中定義的命名建構函數創建子類別，則必須在子類別中實現該建構函數。

6.2.3 初始化清單

可以在建構函數主體執行之前初始化實例屬性。在建構函數的括號後面使用冒號啟動初始化清單，用逗點分隔初始化敘述，初始化清單運算式等號右邊不能使用 this 關鍵字。範例程式如下：

```
//chapter6/bin/example_06.dart
class Point{
  num x, y;
  // 名稱相同建構函數
  Point(this.x,this.y);
  // 命名建構函數 origin
  Point.origin(){
    x = 0;
    y = 0;
  }
```

```
    // 命名建構函數 fromJson 使用初始化清單為屬性設定值
    Point.fromJson(Map json):x = json['x'],y = json['y'];
}
void main(){
    Map data= {'x':1,'y':2};
    var p1 = Point.fromJson(data);
    print('p1:(${p1.x},${p1.y})');
}
```

在開發模式中，可以在初始化清單中使用應用聲明驗證輸入。

```
Point.withAssert(this.x,this.y):assert(x >= 0){
    print(' 在 Point.withAssert() 中 :($x, $y)');
}
```

設定 final 修飾的屬性時，初始化清單很方便。下面的範例在初始化清單中初始化 4 個 final 屬性。範例程式如下：

```
//chapter6/bin/example_07.dart
class Cube{
    final num l;
    final num w;
    final num h;
    final num volume;
    // 宣告方法時可以省略參數前面的資料類型
    Cube(l,w,h):l = l,w = w,h = h,volume = l*w*h;
}
main(){
    var c = Cube(2,3,6);
    print(' 長方體的體積為 :${c.volume}');
}
```

6.2.4 重新導向建構函數

有時，建構函數的唯一目的是重新導向到另一個建構函數。重新導向建構函數的主體為空，建構函數呼叫出現在冒號後面。範例程式如下：

```
//chapter6/bin/example_08.dart
class Point{
```

```
    num x,y;
    // 該類別的主建構函數
    Point(this.x, this.y);
    // 委託實現給主建構函數
    Point.alongXAxis(num x) : this(x,0);
}
main(){
    var p = Point.alongXAxis(6);
    print('p:(${p.x},${p.y})');
}
```

6.2.5 常數建構函數

如果類別生成的物件不會改變，則可以在生成這些物件時將其定義為編譯時常數。可以在建構函數前使用 const 修飾，並確保所有實例屬性都由 final 修飾。範例程式如下：

```
class ImmutablePoint{
    static final ImmutablePoint origin = const ImmutablePoint(0,0);
    final num x,y;
    const ImmutablePoint(this.x,this.y);
}
```

要使用常數建構函數創建編譯時常數，需將 const 關鍵字放在建構函數名稱之前。範例程式如下：

```
//chapter6/bin/example_09.dart
class ImmutablePoint{
    static final ImmutablePoint origin = const ImmutablePoint(0,0);
    final num x,y;
    const ImmutablePoint(this.x,this.y);
}
main(){
    // 使用建構函數創建編譯時常數
    var p = const ImmutablePoint(2,2);
    print('p:(${p.x},${p.y})');
}
```

構造兩個相同的編譯時常數會產生一個規範的實例。範例程式如下：

```
//chapter6/bin/example_10.dart
class ImmutablePoint{
   static final ImmutablePoint origin = const ImmutablePoint(0,0);
   final num x,y;
   const ImmutablePoint(this.x,this.y);
}
main(){
   var a = const ImmutablePoint(1,1);
   var b = const ImmutablePoint(1,1);
   // 它們是同一個實例
   print(identical(a,b));
}
```

6.2.6 工廠建構函數

當一個類別不需要總是創建新實例時可以使用 factory 關鍵字來修飾建構函數。舉例來說，工廠建構函數可以從快取中返回實例，或返回子類型的實例。

以下範例演示了工廠建構函數從快取中返回物件的方法。工廠建構函數接收一個參數值，如果在快取物件 _cache 中存在與該值相等的鍵，則返回與鍵連結的值，即 Logger 類別的實例。如果不存在，則使用該值作為鍵，並創建一個 Logger 類別的實例作為值，並將此鍵值對存入快取物件 _cache 中。範例程式如下：

```
//chapter6/bin/example_11.dart
class Logger{
   final String name;
   bool mute = false;
   // 在名字 _cache 前有底線代表函數庫是私有的
   // 用於儲存鍵時字串值是 Logger 物件的鍵值對集合
   static final Map<String, Logger> _cache =
   <String, Logger>{};
   // 定義工廠建構函數
   // 在工廠建構函數中無法存取 this
```

```
    factory Logger(String name){
        // 判斷 _cache 的鍵中是否存在參數值
        // 若存在則直接返回對應值
        // 若不存在則創建與參數值連結的鍵值對並返回連結的值
        return _cache.putIfAbsent(
        name, () => Logger._internal(name));
    }
    // 私有建構函數
    Logger._internal(this.name);
    // 方法
    void log(String msg){
        if (!mute) print(msg);
    }
}
main(){
    // 透過工廠建構函數創建實例
    // 像普通建構函數一樣呼叫工廠建構函數
    var logger1 = Logger('UI');
    var logger2 = Logger('UI');
    // 判斷是否是同一個實例
    print(identical(logger1,logger2));
    logger1.log('Button clicked');
}
```

6.3 方法

方法是指在類別中提供物件行為的函數，方法與函數是等值的，只是習慣將類別中定義的函數稱為方法。有 static 關鍵字修飾的方法就是類別方法，無 static 關鍵字修飾的方法就是實例方法。

6.3.1 實例方法

宣告實例方法同普通函數一樣，只是將它放在類別中。物件的實例方法可以存取執行個體變數和 this。範例程式如下：

```
//chapter6/bin/example_12.dart
import 'dart:math';
class Point{
   num x, y;
   Point(this.x, this.y);
   // 實例方法
   num distanceTo(Point other){
      var dx = x - other.x;
      var dy = y - other.y;
      return sqrt(dx * dx + dy * dy);
   }
}
```

6.3.2 類別方法

在方法前面使用 static 關鍵字修飾，該方法就是類別方法或靜態方法。類別方法不能在實例上操作，因此實例無法存取它，只能透過類別直接引用類別方法。範例程式如下：

```
//chapter6/bin/example_13.dart
import 'dart:math';
class Point{
   num x, y;
   Point(this.x, this.y);
   // 類別方法又稱靜態方法
   static num distanceBetween(Point a, Point b) {
      var dx = a.x - b.x;
      var dy = a.y - b.y;
      return sqrt(dx * dx + dy * dy);
   }
}

void main(){
   var a = Point(2, 2);
   var b = Point(4, 4);
   // 直接透過類別呼叫類別方法
   var distance = Point.distanceBetween(a, b);
   print(distance);
}
```

6.3.3 方法 getter 和 setter

getter 和 setter 是一組特殊的方法，它們提供了對物件屬性讀和寫的能力。在 Dart 中對屬性的存取，實際上都是呼叫 getter 方法，對屬性的設定值實際上都是呼叫 setter 方法。每個屬性都有一個與之連結的 getter 方法，如果是非 final 修飾的屬性還有一個連結的 setter 方法。

```
class Circle{
   final pi = 3.14;
   num r;
}
```

上例中，Circle 類別的 pi 屬性由 final 修飾，因此只有一個 setter 方法與之連結。r 屬性則有一對 setter 和 getter 方法與之連結。

屬性的 getter 和 setter 方法可以透過使用 get 和 set 關鍵字修飾來顯性宣告，方法名稱是需要定義的屬性名稱。getter 方法不需要參數清單，getter 的呼叫語法與變數的存取沒有區別。setter 方法只接收一個參數，setter 的呼叫語法與變數設定值一致。範例程式如下：

```
//chapter6/bin/example_14.dart
class Circle{
   final pi = 3.14;
   num r;
   // 定義 getter 方法
   num get area => pi*r*r;
   // 定義 setter 方法
   set area(num area)=> r = area/(2*pi);
   Circle(this.r);
}

void main(){
   // 創建 Circle 類別的實例
   var c = Circle(9);
   // 呼叫屬性 area 的 getter 方法
   print(' 面積：${c.area}');
   // 呼叫屬性 area 的 setter 方法
```

```
   c.area = 30;
   print('半徑：${c.r}');
}
```

上例中對 Circle 類別透過顯性定義 getter 和 setter 方法增加 area 屬性。從 getter 方法的返回值可以看出 area 的值是由 pi 和 r 的屬性決定的，setter 方法則是更改 r 屬性的值。應當仔細分析此例中的 area 屬性與 r 屬性和 pi 屬性設定值與設定值的區別，這是瞭解與使用 getter 和 setter 方法的關鍵。通常來說，如果一個屬性的值與其他屬性有關，則可為其顯性定義 getter 方法。如果修改一個屬性的值會影響其他屬性的值，則可為其顯性定義 setter 方法。

6.4 繼承

繼承是重用程式的一種途徑，繼承可以使子類別獲得父類別的實例成員，即實例屬性和實例方法。繼承也會使子類別繼承父類別的介面，後續會詳細說明。在 Dart 語言中採用單繼承的方式，除 Object 類別之外的所有類別都只有一個父類別，因為 Object 類別沒有父類別，因此在 Dart 中類別層次結構是以 Object 類別為根的樹。

先定義一個類別 Person：

```
class Person{
   String name;
   int age;
}
```

再定義一個類別 Employee：

```
class Employee{
   String name;
   int age;
   num salary;
}
```

可以從上面兩個類別看出來它們的屬性有很多相同之處，Employee 只比 Person 多一個 salary 屬性。這樣重複性的工作顯然是極其沒有效率的，因此採用繼承機制完成程式的重用。

在 Employee 類別的基礎上附加 extends 關鍵字和需要繼承的 Person 類別，這樣子類別 Employee 就擁有了父類別 Person 的實例屬性和實例方法。這裡我們沒有定義建構函數，因此兩個類別都只有預設建構函數。範例程式如下：

```
//chapter6/bin/example_15.dart
class Person{
   String name;
   int age;
}
// 繼承 Person 類別
class Employee extends Person{
   num salary;
}
void main(){
   var emp = Employee();
   // 判斷 emp 是否是類別 Person 子類別的實例
   print(emp is Person);
}
```

建構函數、類別屬性和類別方法不會被子類別繼承，即子類別不會從其父類別繼承建構函數，宣告沒有建構函數的子類別僅具有預設建構函數。

預設情況下，子類別會呼叫父類別的預設建構函數，父類別的建構函數在子類別建構函數本體執行之前呼叫。如果存在初始化清單，它將在父類別預設建構函數被執行之前執行，執行順序如下：

(1) 初始化清單。
(2) 父類別的預設建構函數。
(3) 子類別的預設建構函數。

6.4.1 呼叫父類別的非預設建構函數

如果在父類別中提供了建構函數，那麼必須主動呼叫父類別中所有建構函數中的建構函數。在建構函數主體之前，在冒號 (:) 之後指定父類別建構函數。

在 Employee 類別的命名建構函數的初始化清單中呼叫父類別 Person 的命名建構函數 fromJson。範例程式如下：

```
//chapter6/bin/example_16.dart
class Person{
  String name;
  int age;
  // 命名建構函數 fromJson
  Person.fromJson(Map json){
    this.name = json['name'];
    this.age = json['age'];
    print('Person.fromJson 建構函數 ');
  }
}

class Employee extends Person{
  num salary;
  //Person 提供了建構函數
  // 因此必須呼叫父類別中所有建構函數中的建構函數
  Employee.fromJson(Map json) : super.fromJson(json){
    print('Employee.fromJson 建構函數 ');
  }
}

main(){
  Map json = {'name':' 張三 ','age':23};
  // 透過命名建構函數創建實例
  var e = Employee.fromJson(json);
  e.salary = 10968;
  print(' 名字 :${e.name}，年齡 :${e.age}, 薪資 :${e.salary}');
}
```

因為父類別建構函數的參數是在呼叫建構函數之前求值的，所以參數可以是運算式。範例程式如下：

```
//chapter6/bin/example_17.dart
class Person{
  String name;
  int age;
  Person.fromJson(Map json){
    this.name = json['name'];
    this.age = json['age'];
  }
}
class Employee extends Person{
  static Map defaultData = {'name':' 張三 ','age':23};
  num salary;
  // 向父類別建構函數提供運算式
  Employee() : super.fromJson(defaultData);
}
main(){
  var e = Employee();
  e.salary = 10968;
  print(' 名字 :${e.name}，年齡 :${e.age}, 薪資 :${e.salary}');
}
```

傳遞給父類別建構函數的參數不能使用 this 關鍵字，在此之前子類別建構函數尚未被執行，子類別的實例物件也就還未被初始化，因此實例成員都無法存取，但子類別的類別成員可以被存取。

6.4.2 覆寫類別成員

子類別可以覆寫實例方法、getter 方法和 setter 方法。可以使用 @override 註釋指示有意覆寫父類別的實例成員，在子類別中不使用 @override 註釋也不影響執行結果，其主要作用是為了說明該成員在父類別中定義過。範例程式如下：

```
//chapter6/bin/example_18.dart
class Person{
```

```
String name;
   int age;
   void say(String msg){
      print('Person：$msg');
   }
}
class Employee extends Person{
   num salary;
   // 覆寫父類別 say 方法
   @override
   void say(String msg){
      print('Employee：$msg');
   }
}
main(){
   var e = Employee();
   e.say("hello");
}
```

6.4.3 覆寫運算符號

可以覆寫如表 6-1 所示的運算子。舉例來說，如果定義一個 Vector 類別，則可以定義一個 + 方法來增加兩個向量。

表 6-1 可覆寫運算符號

<	+	|	[]	>	/	^	[]=	<=
~/	&	~	>=	*	<<	==	-	%
>>								

類別覆寫 + 和 - 運算子的範例程式如下：

```
//chapter6/bin/example_19.dart
class Vector{
   final int x, y;
   Vector(this.x, this.y);
   // 覆寫 + 運算符號
   Vector operator +(Vector v) => Vector(x + v.x, y + v.y);
```

```
  // 覆寫 - 運算符號
  Vector operator -(Vector v) => Vector(x - v.x, y - v.y);
  toString(){
    return 'x:$x,y:$y';
  }
}

void main(){
  final v = Vector(2, 3);
  final w = Vector(2, 2);
  // 呼叫 Vector 類別的 + 運算符號
  print(v + w);
  // 呼叫 Vector 類別的 - 運算符號
  print(v - w);
}
```

如果覆寫 == 運算符號，則還應該覆寫物件的屬性 hashCode 的 getter 方法。覆寫 == 和 hashCode 的範例程式如下：

```
//chapter6/bin/example_20.dart
class Person{
  final String firstName, lastName;
  Person(this.firstName, this.lastName);
  // 覆寫類別的 hashCode 的 getter 方法
  @override
  int get hashCode{
    int result = 17;
    result = 37 * result + firstName.hashCode;
    result = 37 * result + lastName.hashCode;
    return result;
  }
  // 覆寫類別的 == 運算符號
  @override
  bool operator ==(dynamic other){
    if (other is! Person) return false;
    Person person = other;
    return (person.firstName == firstName &&
      person.lastName == lastName);
  }
```

```
}

void main(){
  var p1 = Person('Bob', 'Smith');
  var p2 = Person('Bob', 'Smith');
  var p3 = 'not a person';
  // 比較實例 p1 和 p2 的 hashCode 是否相等
  print(p1.hashCode == p2.hashCode);
  // 比較實例 p1 和 p2 是否相等
  print(p1 == p2);
  print(p1 != p3);
}
```

6.4.4 未定義函數

需要在程式嘗試使用不存在的方法或執行個體變數時進行檢測或做出反應，可以重新定義 noSuchMethod 方法。範例程式如下：

```
//chapter6/bin/example_21.dart
class A{
  // 覆寫 noSuchMethod 方法
  // 否則當嘗試呼叫一個不存在的成員時將拋出異常 NoSuchMethodError
  @override
  noSuchMethod(Invocation invocation){
    print(' 你嘗試使用一個不存在的成員 :${invocation.memberName}');
  }
}
void main(){
  // 這裡必須使用 dynamic
  dynamic a = A();
  // 呼叫類別中不存在的成員
  a.w;
}
```

6.5 抽象類別和介面

6.5.1 抽象類別

抽象類別是使用 abstract 關鍵字修飾的類別，抽象類別是無法實例化的類別。抽象類別通常具有抽象方法，直接使用分號 (;) 替代方法區塊即可宣告一個抽象方法，抽象方法只能存在於抽象類別中。抽象類別對於定義介面很有用且可以帶有一些方法實現。

宣告具有抽象方法的抽象類別的範例程式如下：

```
//chapter6/bin/example_22.dart
// 定義一個抽象類別
abstract class Person{
   // 定義實例屬性
   String name;
   int age;
   // 定義實例方法
   void say(String msg){
      print('Person：$msg');
   }
   // 宣告一個抽象方法
   void doSomething(String job);
}
// 繼承抽象類別
class Employee extends Person{
   // 在子類別中提供抽象方法的具體實現
   void doSomething(String job){
      print('Employee $job');
   }
}
main(){
   var e = Employee();
   e.doSomething('打掃衛生');
}
```

實例方法、getter 方法和 setter 方法都可以是抽象的，只需將它們定義為一個介面方法，並將其實現留給其他類別。

6.5.2 隱式介面

在 Dart 語言中每個類別都隱式定義了一個介面，該介面包含該類別所有實現的介面和實例成員。如果想創建一個類別 A，它支援類別 B 的所有實例屬性和實例方法，但又不繼承類別 B 的實現，那麼類別 A 應該實現類別 B 的介面，然後類別 A 再提供自己的實現。

類別透過在 implements 子句中提供一個或多個介面，然後提供介面所需的實例屬性和實例方法實現一個或多個介面。範例程式如下：

```
//chapter6/bin/example_23.dart
class Person {
   // 在介面中，但僅對此函數庫可見，因為它是私有的
   final _name;
   // 不在介面中，因為它是建構函數
   Person(this._name);
   // 在介面中
   String greet(String who) => ' 你好 $who. 我是 $_name.';
}

//Person 介面的一種實現
class Impostor implements Person {
   get _name => '';
   // 提供自己的實現
   String greet(String who) => ' 你好 $who.';
}
// 定義頂層方法
String greetBob(Person person) => person.greet('Bob');

void main() {
   print(greetBob(Person('Kathy')));
   print(greetBob(Impostor()));
}
```

上述程式中 Impostor 類別不是 Person 類別的子類別，它沒有繼承 Person 類別的任何成員。implements 子句的作用是在介面間建立連結，而非共用實現。

下面是指定一個類別實現多個介面的範例：

```
class Point implements Comparable,Location {...}
```

因為每個類別都提供了隱式介面，因此在 Dart 語言中沒有正式的介面宣告，但是在 Dart 中可以定義一個純抽象類別來描述傳統意義上的介面。

6.6 在類別增加特徵

mixin 是一種在多個類別層次結構中重用類別程式的方式，定義類時使用關鍵字 mixin 代替 class，且不為該類別提供建構函數，mixin 類別屬於抽象類別，因此不能被實例化。範例程式如下：

```
mixin Fly{
   bool canFly = true;
   void flying(){
      if(canFly)
         print('飛行中');
   }
}
```

要使用 mixin，需使用 with 關鍵字，後跟一個或多個 mixin 類別名稱。該類別會獲得 mixin 類別定義的實例屬性和實例方法，使用 mixin 的範例程式如下：

```
//chapter6/bin/example_24.dart
mixin Fly{
   bool canFly = true;
   void flying(){
      if(canFly)
```

```
         print('飛行中');
   }
}
mixin Walk{
   bool canWalk = true;
   void walking(){
      if(canWalk)
      print('行走中');
   }
}
//使用mixin
class Dove with Fly,Walk{
}
void main(){
   var d = Dove();
   //使用從mixin類別Fly獲得的方法
   d.flying();
   //使用從mixin類別Walk獲得的方法
   d.walking();
}
```

有時希望指定類型才能使用 mixin 類別，可以使用 on 子句指定父類別約束。也就是說 on 子句後面的類別是使用該 mixin 類別的父類別，並且 on 子句後面的類別也是該 mixin 類別的父類別，在 mixin 中可以使用 super 呼叫父類別中的實例成員。

定義一個類別 Bird，讓 Dove 類別繼承 Bird 類別，然後透過 on 子句在 mixin 類別 Fly 增加父類別約束。範例程式如下：

```
//chapter6/bin/example_25.dart
//用on關鍵字指定使用此mixin類別的父類別約束Bird
//Bird類同時是此mixin類別的父類別
mixin Fly on Bird{
   bool canFly = true;
   void flying(){
      if(canFly)
         print('飛行中');
   }
```

```
}
// 此 mixin 類別的使用沒有任何限制
mixin Walk{
   bool canWalk = true;
   void walking(){
      if(canWalk)
         print('行走中');
   }
}
// 定義 Bird 類別
class Bird{
}
// 讓 Dove 類別繼承 Bird 類別，使 Dove 類別可以使用 mixin 類別 Fly
class Dove extends Bird with Fly,Walk{
}
void main() {
   var d = Dove();
   d.flying();
}
```

使用 mixin 的類別是其父類別的子類別，也是 mixin 表示的類型的子類別。

mixin 的工作原理是生成一個新類別，該新類別將 mixin 的實現置於父類別之上，這表示在新類別中 mixin 類別的實現會覆寫父類別中的名稱相同實例屬性和實例方法，這裡的父類別是指 extends 關鍵字修飾的類別。如果有多個 mixin，則依次創建新類別。

當前類別沒有除 Object 之外的父類別時，範例程式如下：

```
mixin Walk{
   void walking(){
      print('行走中');
   }
}
class Dog with Walk{
}
```

在語義上等於以下程式：

```
mixin Walk{
   void walking(){
      print('行走中');
   }
}
class Dog extends Object with Walk{
}
```

它實際上是 Object 與 mixin 類別 Walk 產生一個新類別，然後 Dog 再繼承這個新類別。

如果當前類別存在除 Object 之外的父類別，範例程式如下：

```
class Dove extends Bird with Fly,Walk{
}
```

它實際上分為三步驟完成。

步驟 1：Bird 類別和 mixin 類別 Fly 產生新類別，我們把它看作類別 BirdFly。

```
class Bird with Fly{
}
```

步驟 2：產生的新類別 BirdFly 再和 mixin 類別 Walk 產生新類別，我們把它看作類別 BirdFlyWalk。

```
class BirdFly with Walk{
}
```

步驟 3：產生的新類別 BirdFlyWalk 與 Dove 完成繼承關係。

```
class Dove extends BirdFlyWalk{
}
```

繼承順序依次是 Bird 類別、Fly 類別、Walk 類別、Dove 類別。這就是它們生成新類別的順序和邏輯。如果多個類別有相同的屬性或方法，排在後面的類別會覆寫前面的類別。最終繼承關係如圖 6-1 所示。

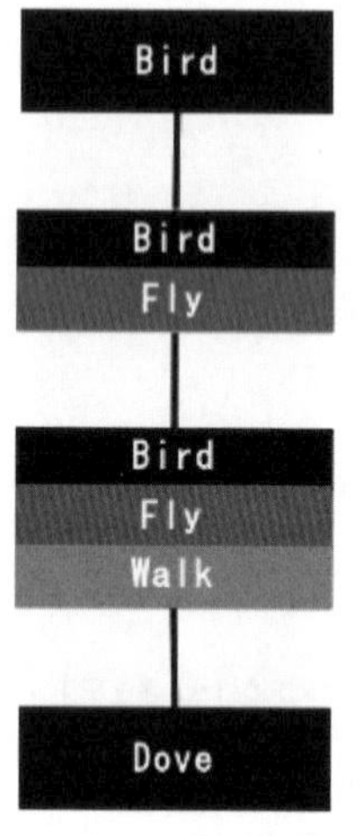

▲ 圖 6-1 繼承關係圖

6.7 列舉類別

列舉類別是一種特殊的類別，用於表示固定數量的常數值。

使用 enum 關鍵字宣告列舉類別型的結構如下：

```
enum 類別名稱 {
   值 1,
   值 2,
   值 3
}
```

每個列舉值都有一個名為 index 的 getter 方法，該方法將返回以 0 為基準索引的位置值。例如：第一個值的索引為 0，第二個值的索引為 1。範例程式如下：

```
//chapter6/bin/example_26.dart
// 定義列舉類別
```

```
enum Color{
  // 列舉值清單
  red,
  green,
  blue
}
void main(){
  // 列印列舉值在列舉類別中的索引
  print(Color.red.index);
  print(Color.green.index);
  print(Color.blue.index)
}
```

要獲取列舉中所有值的清單，需使用列舉的 values 常數。

```
var values = Color.values
```

可以在 switch 敘述中使用列舉，如果不處理所有列舉值，則會收到警告，範例程式如下：

```
//chapter6/bin/example_27.dart
// 定義列舉類別
enum Color{
  // 列舉值清單
  red,
  green,
  blue
}
void main(){
  // 列印列舉類別中的所有列舉值
  print(Color.values);
  // 為變數設定值一個列舉值
  var aColor = Color.blue;
  //switch 敘述是列舉常見的使用場景
  switch (aColor) {
    case Color.red:
      print(' 紅色 ');
      break;
    case Color.green:
      print(' 綠色 ');
```

```
        break;
      default:
      print('藍色');
  }
}
```

列舉類別有以下限制：

(1) 不能作為子類別、mixin 或實現列舉類別。

(2) 無法顯性實例化列舉。

Chapter

07

異常

Dart 程式可以拋出和捕捉異常 (或稱之例外)，異常表示發生了某些意外的錯誤。如果異常未被捕捉，則引起異常的 isolate 將被暫停，並且 isolate 及其程式將被中止。

在 Dart 語言中所有的異常都是非必檢異常，方法不宣告它們可能會引發哪些異常，並且不需要捕捉任何異常。

Dart 語言提供了 Exception 和 Error 類型，以及許多預先定義的子類型，也可以自訂異常。Error 是程式無法恢復的嚴重錯誤，表示程式出現較嚴重問題，而又無法透過程式設計處理，只能終止程式。大多數錯誤與程式編寫者執行的操作無關。Exception 是程式可以恢復的異常，它可以透過程式設計解決。Exception 可能是由參數錯誤、網路連接中斷或字串解析失敗等引起的。本章所關注的是對 Exception 及其子類別的異常處理。

7.1 拋出異常

透過 throw 敘述來主動拋出異常。範例程式如下：

```
//chapter7/bin/example_01.dart
void main(){
  throw FormatException('格式轉換異常');
}
```

異常不僅是 Exception 和 Error 物件，還可以將任何不可為空白物件作為異常拋出，例如：字串。範例程式如下：

```
//chapter7/bin/example_02.dart
void main(){
  throw '這是一個字串';
}
```

因為 throw 敘述是一個運算式，因此可以在「=>」敘述或其他使用運算式的地方出現。範例程式如下：

```
//chapter7/bin/example_03.dart
void main(){
  void distanceTo() => throw UnimplementedError();
  distanceTo();
}
```

7.2 捕捉異常

捕捉異常會阻止異常傳播，除非重新拋出異常，捕捉一個異常以便能夠對其做進一步處理。

使用 try 敘述來捕捉異常，使用 catch 敘述處理異常，catch 敘述有兩個參數，第一個是必選的異常物件，第二個是可選的堆疊物件。

try-catch 敘述宣告格式如下：

```
try{
   // 可能引起異常的敘述
}catch(e,s){
   // 處理異常
}
```

範例程式如下：

```
//chapter7/bin/example_04.dart
void main(){
   try {
      // 拋出異常
      throw Exception;
   }catch (e,s) {
      // 捕捉異常
      print(' 異常詳情 :\n $e');
      print(' 堆疊追蹤 :\n $s');
   }
}
```

可以透過 on 加異常名來捕捉指定類型的異常。範例程式如下：

```
//chapter7/bin/example_05.dart
void main(){
   try {
      throw FormatException();
   }on FormatException{
      print(' 格式轉換異常 ');
   }
}
```

try 子句中的程式區塊可能引發多個異常，此時可以使用 on 子句捕捉指定異常。如果需要異常物件，則可以同時使用 on 和 catch 子句。與拋出的異常物件類型相匹配的第一個 on 或 catch 子句會處理異常，如果 catch 子句未指定異常類型，則該子句可以處理引發的任何類型的異常。範例程式如下：

```
//chapter7/bin/example_06.dart
void main(){
    try{
        // 拋出異常
        throw IntegerDivisionByZeroException();
    }on FormatException{
        // 處理 FormatException 類型的異常
        print(' 格式轉換異常 ');
    }on Exception catch(e){
        // 處理 Exception 類型的異常
        print(' 異常：$e');
    }catch(e){
        // 未指定類型，可以處理所有異常
        print(' 其他異常：$e');
    }
}
```

對異常進行部分處理後，如果需要異常繼續傳播，則可以使用 rethrow 關鍵字。範例程式如下：

```
//chapter7/bin/example_07.dart
void action(){
    try {
        throw FormatException();
    }catch(e){
        // 部分處理異常
        print('action() 部分處理 ${e.runtimeType}');
        // 重新拋出異常
        rethrow;
    }
}
void main(){
    try {
        action();
    } catch (e) {
        print('main() 完成處理 ${e.runtimeType}.');
    }
}
```

7.3 最終操作

finally 敘述在 try 和 catch 敘述之後，無論是否觸發異常，該敘述都會被執行。如果提供該敘述，則其語法格式如下：

```
try{
   // 可能引起異常的敘述
}catch(e,s){
   // 處理異常
}finally{
   // 無論是否引發異常都需要執行的程式區塊
}
```

範例程式如下：

```
//chapter7/bin/example_08.dart
void main(){
   try{
      // 此處不會有異常拋出
      var i = 1;
   }catch(e){
      // 有異常時才會執行
      print('Error: $e');
   }finally{
      // 始終會得到執行
      print('finally 敘述區塊 ');
   }
}
```

如果沒有 catch 子句，則在 finally 子句執行後異常繼續傳播。範例程式如下：

```
//chapter7/bin/example_09.dart
void main(){
   try {
      // 此處拋出異常
      throw Exception();
   }finally{
```

```
        // 始終會得到執行
        print('finally 敘述區塊 ');
    }
}
```

7.4 自訂異常

可以透過實現 Exception 介面來定義自訂異常。範例程式如下：

```
//chapter7/bin/example_10.dart
// 實現 Exception 介面來自訂異常
class MyException implements Exception{
    // 接收訊息的變數
    final String msg;
    // 常數建構函數
    const MyException([this.msg]);
    // 覆寫 toString 方法
    @override
    String toString() => msg ?? 'MyException';
}
void main(){
    // 拋出自訂異常
    throw MyException(' 自訂異常 ');
}
```

泛型

查看集合類型 List 的 API 文件時會發現其類型實際上是 List<E>。符號 <> 將 List 標記為可泛型化的類別，即類型可參數化。通常使用一個字母來表示類型參數，例如 E、T、S、K 和 V 等。

泛型常用於要求類型一致的情況，它還可以減少程式重複。例如現在宣告一個只包含 String 類型的陣列，可以宣告為 List<String>，讀作字串類型的 List。這樣宣告可以避免因在該陣列中放入非 String 類型的值而出現異常，同時編譯器也可以及時發現並定位問題。導致異常的範例程式如下：

```
var names = List<String>();
names.addAll(['Seth', 'Kathy', 'Lars']);
// 錯誤
names.add(42);
```

創建命令列應用程式，將專案命名為 chapter8。本章所有檔案都在該專案的 bin 資料夾下創建與執行。

8.1 使用泛型

List、Set 和 Map 字面量可以被參數化。參數化字面量與已經介紹的字面量一樣，只不過在左括號之前增加了 <type>(用於 List 和 Set) 或 <keyType,valueType>(用於 Map)。使用類型字面量的範例程式如下：

```
//chapter8/bin/example_01.dart
// 創建 List<String> 的實例
var names = <String>['Seth','Kathy','Lars'];
// 創建 Set<String> 的實例
var uniqueNames = <String>{'Seth','Kathy','Lars'};
// 創建 Map<String,String> 的實例
var website = <String, String>{
'qq.com': ' 騰訊 QQ',
'aliyun.com': ' 阿里雲 ',
'toutiao.com': ' 頭條新聞 '
};
```

可以參數化集合的建構函數以便使用泛型。在使用建構函數時指定一種或多種類型，將類型放在類別名稱之後的中括號 (<type> 或 <keyType,valueType>) 中。範例程式如下：

```
//chapter8/bin/example_02.dart
// 創建 List<String> 的實例
var values = List<String>();
// 創建 Set<String> 的實例
var setValues = Set<String>();
// 創建 Map<String,String> 的實例
var mapValues = Map<String,String>();
```

Dart 泛型類型已經過規範化，這表示它們會在執行時期攜帶其類型資訊。因此可以測試集合的類型，範例程式如下：

```
//chapter8/bin/example_03.dart
print(names is List<String>);
print(uniqueNames is Set<String>);
print(website is Map<String,String>);
```

8.2 自訂泛型

使用泛型的另一個原因是減少程式重複。泛型可以在多種類型之間共用單一介面和實現，同時可以利用靜態分析。

泛型類別的定義結構如下：

```
//chapter8/bin/example_04.dart
// 定義泛型類別，泛型參數 T
class ClassName<T>{
  // 使用泛型參數定義執行個體變數
  T t;
  // 定義實例方法，方法的返回值和參數都可以使用泛型參數 T
  T method(List<T> ts){
    // 泛型參數 T 定義區域變數
    T ts1;
    ts1 = ts[0];
    return ts1;
  }
}
```

類別名稱後面加上泛型參數 T，這樣在類別的執行個體變數和實例方法中也可以使用泛型參數 T。

8.2.1 泛型類別

創建一個用於快取物件的介面，範例程式如下：

```
//chapter8/bin/example_05.dart
// 快取 Object 的介面
abstract class ObjectCache{
  Object getByKey(String key);
  void setByKey(String key, Object value);
}
```

需要字串版本的相同介面，因此創建了另一個介面，範例程式如下：

```
//chapter8/bin/example_06.dart
// 快取字串的介面
abstract class StringCache{
   String getByKey(String key);
   void setByKey(String key, String value);
}
```

當需要更多其他類型的類似介面時，會發現非常煩瑣。泛型就可以省去創建這些相似介面的麻煩，此時只需創建一個帶有類型參數的介面。範例程式如下：

```
//chapter8/bin/example_07.dart
// 快取任意物件的泛型介面
abstract class Cache<T>{
   T getByKey(String key);
   void setByKey(String key,T value);
}
```

在此程式中，T 是替代類型。可以將其視為預留位置，作為開發人員以後定義的類型。

提供介面的具體實現並創建實例。範例程式如下：

```
//chapter8/bin/example_08.dart
// 實現 Cache 介面
class CachePool<U> implements Cache<U>{
   final Map pool = Map();
   // 實現 getByKey 方法，返回類型為泛型參數 U
   @override
   U getByKey(String key){
      return pool[key];
   }
   // 實現 setByKey 方法，返回類型為泛型參數 U
   @override
   void setByKey(String key,U value){
      pool[key]= value;
   }
}
void main(){
```

```
    // 創建實例，泛型參數為 int
    var intMap = CachePool<int>();
    intMap.setByKey('first',1);
    print(intMap.getByKey('first'));

    // 創建實例，泛型參數為 String
    var stringMap = CachePool<String>();
    stringMap.setByKey('hi','hello');
    print(stringMap.getByKey('hi'));
}
```

在上例中 CachePool 類別的實例等於 Map<String,dynamic> 的實例，這裡主要是為了說明泛型介面的定義與實現。

8.2.2 泛型方法

最初 Dart 的泛型支持僅限於類別。一種稱為泛型方法的較新語法，允許在方法上使用類型參數。範例程式如下：

```
//chapter8/bin/example_09.dart
// 定義泛型方法，可為方法提供泛型參數
T first<T>(List<T> ts) {
    // 用泛型參數定義區域變數
    T tmp;
    tmp = ts[0];
    return tmp;
}
void main(){
    // 使用泛型方法，泛型參數為 int
    var intList = [1,2,3];
    print(first<int>(intList));
    // 使用泛型方法，泛型參數為 String
    var strList = ['one','two','three'];
    print(first<String>(strList));
}
```

這是使用泛型方法的比較極端的例子，其主要意圖在於演示各個可以使用泛型參數的位置，在實際使用中只會在少量位置使用泛型參數。

從 first 方法可以看出可以在多個地方使用類型參數 T：方法的泛型 (first<T>)、函數的返回數值類型 (T)、參數的類型 (List<T>)、區域變數 (T tmp)。

8.2.3 限制類型

在實現泛型類型時，可以限制其參數的類型，可以使用 extends 做到這一點。範例程式如下：

```
//chapter8/bin/example_10.dart
// 定義基礎類別
class BaseClass{}
// 定義子類別
class Extender extends BaseClass{}
// 限制泛型參數的類型
class Generic<T extends BaseClass>{
   // 覆寫 toString 方法
   @override
   String toString() {
      return 'Generic<$T> 的實例 ';
   }
}
void main(){
   // 使用基礎類別作為泛型參數
   var baseGen = Generic<BaseClass>();
   print(baseGen);

   // 使用子類別作為泛型參數
   var exteGen = Generic<Extender>();
   print(exteGen);
}
```

在為 Generic 類別指定泛型參數時，可以將 BaseClass 或其任何子類別用作通用參數。指定任何非 BaseClass 類型都會導致錯誤。

函數庫

Dart 程式是由被稱為函數庫的模組化單元組成的，一個函數庫由多個頂層宣告組成，這些宣告可能包含函數、變數及類別。

9.1 宣告與使用

使用 library 關鍵字顯性宣告函數庫。範例程式如下：

```
//chapter9/bin/example_01.dart
// 宣告函數庫
library stack;

// 宣告頂層變數，帶底線代表此變數是函數庫私有的，不對外公開
final _contents = [];

// 宣告頂級函數
// 判斷堆疊是否為空
get isEmpty => _contents.isEmpty;
// 獲取堆疊最上面的元素
get top => isEmpty ? throw '堆疊為空，不能獲取元素' : _contents.last;
```

```
//彈出元素
get pop => isEmpty ? throw '堆疊為空，不能彈出元素' : _contents.removeLast();
//推入元素
dynamic push(ele){
   _contents.add(ele);
   return ele;
}

//宣告函數庫範圍的類別
class StackTool{
   //清空堆疊
   void clear(){
      _contents.clear();
      _contents.length;
   }
   //獲取堆疊的長度
   int length(){
      return _contents.length;
   }
}
```

由關鍵字 library 修飾的單字 stack 就是函數庫名稱。

程式中 _contents 是頂層變數，它的初值是空清單。頂層變數是延遲初始化的，在它們的 getter 第一次被呼叫時才初始化。因此變數 _contents 在某個存取它的方法被呼叫時才被設定成 []。

頂層變數屬於靜態變數，因此也被稱為函數庫變數。頂層變數的作用域覆寫了宣告它們的整個函數庫，函數庫的作用域通常由多個類別與頂層函數組成。頂層變數也可以由 final 修飾，這表示它們沒有定義 setter 並且必須在宣告時就初始化。

程式中 isEmpty、top、pop 和 push 都是頂層函數，頂層函數的作用域覆寫整個函數庫，它們可以是普通方法、setter 和 getter 方法。

在函數庫中還可以宣告頂層類別，如程式中的 StackTool 類別。在 Dart 中類別都是頂層的，因為 Dart 不支持巢狀結構類別。

以底線 (_) 開頭的變數都是函數庫私有的，頂層變數 _contents 便是函數庫私有的。它只對函數庫可見，因此只能在函數庫內部存取變數 _contents。

下面是一個簡單的程式，與 stack 函數庫不一樣，這裡沒有顯性的函數庫宣告且功能單一，但它也是一個函數庫。範例程式如下：

```
//chapter9/bin/example_02.dart
void main(){
  print("Hello World");
}
```

上述程式同時也是一個指令稿，指令稿是從 main() 函數開始執行的。表示這是一個可被直接執行的函數庫，對於快速、簡單的任務，透過指令稿可以更方便地編寫實驗性程式。

9.1.1 匯入函數庫

使用 import 指令指定一個函數庫的命名空間，唯一必須指定的參數是函數庫的 URI。對於內建函數庫使用 dart 字首，再加上函數庫名稱。

應用程式需要使用內建 math 函數庫來生成隨機數，可以在檔案中匯入並使用。範例程式如下：

```
//chapter9/bin/example_03.dart
// 匯入內建函數庫 math，dart: 字首代表內建函數庫
import 'dart:math';

void main() {
  //Random 是 math 函數庫中的類別
  var random = Random();
  //Random 類別中的方法 nextInt(int max) 有一個參數，用於指定能隨機生成的最大整數
  var randomInt = random.nextInt(100);
  print("0~100 的隨機數：$randomInt");
}
```

Dart 軟體生態採用軟體套件來管理和共用軟體，例如函數庫和工具。獲取軟體套件，需要採用 pub 套件管理器。對於第三方函數庫，匯入套件

前需要增加依賴，Dart 應用程式和函數庫的根目錄都有 pubspec.yaml 檔案，其中列出了軟體套件的依賴關係，並包括其他中繼資料，例如版本編號等。

pubspec.yaml 檔案支持的依賴如表 9-1 所示。

表 9-1 依賴檔案欄位

欄位	選填	說明
name	必選	套件名，全部小寫且以底線分割單字，應當是有效的 Dart 識別符號，建議在 pub.dev 上搜索軟體套件，避免名稱重複
version	可選	版本編號，由點分割的 3 個數字，例如 1.24.3 它可以擁有建構版本 (+1，+2) 或預發行版本 (-dev.1、-alpha.1、-beta.1、-rc.1) 尾碼。預設 0.0.0，若發佈軟體套件則是必選
environment	必選	執行環境，配合 sdk 約束或 flutter sdk 約束指定執行環境版本
description	可選	描述，說明軟體套件的作用，應為純英文，60~180 個字元。若發佈軟體套件則是必選
homepage	可選	首頁，指向軟體套件的網址，對於託管的軟體套件指向軟體套件頁面的連結
repository	可選	軟體套件原始程式碼的儲存庫連結，例如：https://GitHub.com/<user>/<repository>
issue tracker	可選	軟體套件問題追蹤連結，若存在儲存庫並指向 GitHub，則 pub.dev 站將使用預設的問題追蹤器 https://GitHub.com/<user>/<repository>/issues
documentation	可選	文件連結
dependencies	可選	正常依賴項，存在依賴項時必選
dev_dependencies	可選	開發時，所需依賴項，存在依賴項時必選
dependency_overrides	可選	在開發過程中，可能需要臨時覆寫依賴項，存在依賴項時必選
executables	可選	軟體套件可以將一個或多個指令稿公開為可執行檔
publish_to	可選	預設使用 pub.dev 網站。不指定任何內容以防止發佈套裝程式。此設定可用於指定要發佈的自訂發佈套件伺服器

最簡單的 pubspec.yaml 檔案僅列出專案名稱和平台依賴資訊。

```
name: my_app
environment:
  sdk: '>=2.7.0 <3.0.0'
```

在依賴檔案中增加第三方套件依賴。

```
name: my_app
environment:
  sdk: '>=2.7.0 <3.0.0'
dependencies:
  json_string: ^2.0.1
```

當修改過 pubspec.yaml 檔案後需要在應用程式根目錄執行 pub get 命令獲取新增軟體套件，pub 工具將在應用根目錄下創建 pubspec.lock 檔案，表明依賴資訊已經更新，可以在專案中正常使用新增套件。同時還會更新 .packages 檔案，該檔案會將專案所依賴的每個套件名稱映射到系統快取中的對應套件目錄。

在檔案中引入第三方套件，需要使用 package:scheme 指定函數庫，指定函數庫後可以使用函數庫中提供的功能。範例程式如下：

```
//chapter9/bin/example_04.dart
// 匯入 pub 套件管理器提供的第三方套件
import 'package:json_string/json_string.dart';
// 使用 json_string 函數庫中定義的 mixin 類別 Jsonable
class Person with Jsonable{
  String name;
  int age;
  Person(this.name,this.age);
  // 提供 toJson 方法的實現
  @override
  Map<String, dynamic> toJson() {
    return {'name':name,'age':age};
  }
}
void main(){
```

```
    var p = Person('jobs',66);
    //呼叫 toJson 方法
    print(p.toJson());
}
```

函數庫的組成至少包含 lib 目錄和 pubspec.yaml 檔案。函數庫程式位於 lib 目錄下，可以根據需要在 lib 下創建任何層次結構。按照歸約，實現程式放在 lib/src 目錄下，該目錄是函數庫私有的，如果要公開 lib/src 下的 API，則可以在 lib 目錄下創建主函數庫檔案 <package-name>.dart，在該檔案匯出該函數庫的所有公共 API。

對於本地函數庫，當兩個檔案都在 lib 目錄內或兩個檔案都在 lib 目錄外時，可以使用相對路徑匯入。但是當 lib 目錄中的檔案被外部的檔案引用時必須使用 package: 字首。

在本專案的 bin 目錄下的 main.dart 檔案中，可以發現已經使用 package: 字首匯入了 lib 目錄下的 chapter9.dart 檔案。現在以相對路徑的形式匯入目前的目錄下的 example_01.dart 檔案。範例程式如下：

```
//chapter9/bin/main.dart
//使用 package: 字首匯入本專案 lib 目錄下的檔案
import 'package:chapter9/chapter9.dart' as chapter9;
//使用相對路徑匯入目前的目錄下的檔案
import 'example_01.dart';
void main(List<String> arguments) {
    print('Hello world: ${chapter9.calculate()}!');
}
```

9.1.2 指定函數庫字首

如果匯入兩個識別符號衝突的函數庫，則可以使用 as 為一個或兩個函數庫指定字首。使用衝突的函數庫成員時，則需要在函數庫成員前指定字首。例如：如果 library1 和 library2 都具有 Element 類別，則需要為其中一個或兩者指定函數庫字首。範例程式如下：

```
import 'package:lib1/lib1.dart';
import 'package:lib2/lib2.dart' as lib2;

// 使用來自函數庫 lib1 的 Element 類別
Element element1 = Element();

// 使用來自函數庫 lib2 的 Element 類別
lib2.Element element2 = lib2.Element();
```

實際使用中並不一定在函數庫衝突時才使用，可以為任何函數庫指定字首，只要它們的名字不衝突就可以。範例程式如下：

```
//chapter9/bin/example_05.dart

// 指定函數庫字首，也可以叫別名
// 使用該函數庫的成員必須指定字首
import 'dart:collection' as co;

void main(){
  // 透過函數庫字首 co 使用函數庫成員 Queue 類別
  co.Queue<int> queue;
}
```

9.1.3 匯入函數庫的一部分

函數庫中通常定義了大量可用的函數庫成員，而使用時只需一小部分就能滿足要求，顯然匯入整個函數庫會影響應用程式的性能。如果只想使用函數庫的一部分，則可以有選擇地匯入該函數庫。與函數庫相關的關鍵字中可以使用 show 匯入庫的一部分成員，使用 hide 以表明匯入某些成員。範例程式如下：

```
//chapter9/bin/example_06.dart
// 只匯入 collection 函數庫中的成員 Queue 和 LinkedList
import 'dart:collection' show Queue,LinkedList;
// 匯入 math 函數庫中除 sin 和 cos 之外的成員
import 'dart:math' hide sin,cos;
```

9.1.4 匯出函數庫

當定義了很多功能且位於多個不同的檔案中時，如果要使用它們就需要匯入所有的檔案。這非常麻煩且容易出錯，因此可以透過在主函數庫檔案中使用 export 關鍵字彙出所有 API。範例程式如下：

```
//chapter9/bin/example_07.dart
library example;
// 匯出函數庫
export 'example_01.dart';
export 'example_02.dart';
```

9.2 核心函數庫

核心函數庫 dart:core 包含內建類型、集合和其他核心功能，該函數庫會自動被匯入每個 Dart 程式中。

主控台列印：頂層 print() 方法接收一個任意物件的參數，並在主控台輸出該物件的字串值，實際上是呼叫該物件的 toString() 方法。範例程式如下：

```
print(anObject);
print('I drink $tea.');
```

9.2.1 數字

dart:core 函數庫定義了 num、int 和 double 類別，它們具有一些用於處理數字的基本實用方法。

1. int 常用屬性

(1) sign：返回此整數的符號，對於 0 返回 0，對於小於 0 的數返回 -1，對於大於 0 的數返回 +1。

(2) bitLength：返回儲存此整數所需的最大位元數，位元數是指二進位位元的個數。位數不包括符號位元，有號的數需要加 1，即 x.bitLength+1。

(3) isEven：判斷此整數是不是偶數，當此整數為偶數時，返回 true。

(4) isOdd：判斷此整數是不是奇數，當此整數為奇數時，返回 true。

屬性使用範例程式如下：

```
//chapter9/bin/example_08.dart
void main(){
  // 查看符號
  print('4 的符號：${4.sign}');
  print('-3 的符號：${(-3).sign}');

  // 查看儲存所需長度
  //00000100
  print(' 儲存 4 所需的位數：${4.bitLength}');
  //00000001
  print(' 儲存 1 所需的位數：${1.bitLength}');
  //11111111
  print(' 儲存 -1 所需的位數：${(-1).bitLength+1}');
  //11111100
  print(' 儲存 -4 所需的位數：${(-4).bitLength+1}');

  // 判斷交錯
  print('4 是偶數嗎：${4.isEven}');
  print('3 是奇數嗎：${3.isOdd}');
}
```

2. int 常用方法

(1) toRadixString(int radix)：將此整數根據參數 radix 指定的進位轉化為字串的表示形式，參數 radix 的範圍在整數 2~36。在字串表示中，小寫字母用於表示 9 以上的數字，a 表示 10，依次類推，z 表示 35。

(2) int tryParse(String source, {int radix})：將字串解析為可能有號的整數文字，然後返回其值。傳入的字串不能為 null，當解析出現異常時返回 null。可透過可選參數 radix 指定採用的進位，參數 radix 的範圍在整數 2~36。

(3) parse()：將字串轉為整數。

方法使用範例程式如下：

```
//chapter9/bin/example_09.dart
void main(){
  //64 根據十六進位轉為字串
  print('64 轉為十六進位：${64.toRadixString(16)}');
  //64 根據二進位轉為字串
  print('64 轉為二進位：${64.toRadixString(2)}');

  // 解析八進位的字串 '64' 解析為十進位的整數
  print(' 解析 64 根據二進位轉換的字串：${int.tryParse('64',radix:8)}');

  // 將字串解析為整數
  print(int.parse('42'));
  print(int.parse('0x42'));
  // 將字串解析為浮點數
  print(double.parse('0.50'));
}
```

9.2.2 字串

Dart 中的字串是 UTF-16 程式單元的不變序列，可以使用正規表示法 (RegExp 物件) 在字串中搜索並替換部分字串。

String 類別將此類方法定義為 split()、contains()、startsWith()、endsWith() 等。

1. 字串尋找

(1) endsWith(String other)：判斷此字串是否以 other 結尾，如果此字串以 other 結尾，則返回 true。

(2) contains(Pattern other, [int startIndex = 0])：判斷此字串是否包含其他匹配項，如果包含則返回 true。如果提供了可選參數 startIndex，則此方法僅在該索引處或之後匹配。

(3) indexOf(Pattern pattern, [int start])：返回此字串中模式的第一個匹配項的位置，如果提供可選參數 start，則從該索引處或之後開始匹配。如果沒有匹配到，則返回 -1。

(4) lastIndexOf(Pattern pattern, [int start])：返回此字串中模式的最後一個匹配項的位置，如果沒有匹配到，則返回 -1。

(5) startsWith(Pattern pattern, [int index = 0])：判斷此字串是否以特定模式開頭，如果此字串以模式匹配開頭，則返回 true。如果提供了可選參數 index，則此方法僅在該索引處或之後匹配。

字串尋找範例程式如下：

```
//chapter9/bin/example_10.dart
void main(){
  'Dart'.endsWith('t');

  var str1 = 'Dart strings';
  str1.contains('D');
  str1.contains(new RegExp(r'[A-Z]'));
  str1.contains('X', 1);
  str1.contains(new RegExp(r'[A-Z]'), 1);

  var str2 = 'Dartisans';
  str2.indexOf('art');
  str2.indexOf(new RegExp(r'[A-Z][a-z]'));

  var str3 = 'Dartisans';
  str3.lastIndexOf('a');
  str3.lastIndexOf(new RegExp(r'a(r|n)'));

  var str4 = 'Dart';
  str4.startsWith('D');
  str4.startsWith(new RegExp(r'[A-Z][a-z]'));
}
```

2. 字串截取

(1) substring(int startIndex, [int endIndex])：返回此字串的子字串，該子

字串從 startIndex(包含) 開始到 endIndex(不包含) 結束。如果不提供 endIndex 參數，則從索引 startIndex 處開始直到結束。

(2) split(Pattern pattern)：在 pattern 匹配項處拆分字串，並返回子字串清單。

可以從字串中提取單一字元，將其作為字串或整數。確切地說，實際上獲得了單獨的 UTF-16 程式單元。例如高音譜號符號 (\ u {1D11E}) 是兩個程式單元。

字串截取範例程式如下：

```
//chapter9/bin/example_11.dart
void main(){
  // 提取子字串
  print('Never odd or even'.substring(6,9));
  var string = 'dartlang';
  print(string.substring(1));
  print(string.substring(1,4));

  // 根據提供的模式分隔字串
  var parts = 'structured web apps'.split(' ');
  print(parts.length);
  print(parts[0]);

  // 透過索引獲取單一字元
  print('Never odd or even'[0]);
  // 將 split() 與空字串參數一起使用以獲取所有字元的清單，有利於疊代
  for (var char in 'hello'.split('')) {
    print(char);
  }

  // 獲取字串中的所有 UTF-16 程式單元
  var codeUnitList =
  'Never odd or even'.codeUnits.toList();
  print(codeUnitList[0]);
}
```

3. 大小寫轉換

(1) toLowerCase()：將此字串中的所有字元轉為小寫。如果字串已經全部小寫，則此方法返回此字串。

(2) toUpperCase()：將此字串中的所有字元轉為大寫。如果字串已經全部大寫，則此方法返回此字串。

大小寫轉換範例程式如下：

```
//chapter9/bin/example_12.dart
void main(){
   // 轉化為大寫
   print('structured web apps'.toUpperCase());
   // 轉化為小寫
   print('STRUCTURED WEB APPS'.toLowerCase());
}
```

4. 裁剪和空字串

(1) trim()：返回沒有任何前導和尾隨空格的字串。

(2) trimLeft()：返回沒有任何前導空格的字串。

(3) trimRight()：返回沒有任何尾隨空格的字串。

使用 trim() 方法移除字串頭部和尾部的所有空白，使用屬性 isEmpty 判斷字串是否為空。範例程式如下：

```
//chapter9/bin/example_13.dart
void main(){
   // 裁剪字串
   print(' hello '.trim());
   // 檢查字串是否為空
   print(''.isEmpty);
   // 字串中只含有空白而不可為空字串
   print(' '.isNotEmpty);
}
```

5. 字串替換

replaceAll(Pattern from, String replace)：使用 replace 替換所有匹配的子字串 from。

字串是不可變的物件，這表示可以創建它們，但不能更改它們。仔細查看 String API 參考，會注意到，沒有任何一種方法可以實際更改 String 的狀態。舉例來說，方法 replaceAll() 返回一個新的 String 而不更改原始的 String。範例程式如下：

```
//chapter9/bin/example_14.dart
void main(){
   var greetingTemplate = 'Hello, NAME!';
   var greeting =
   greetingTemplate.replaceAll(RegExp('NAME'), 'Bob');
   //greetingTemplate 沒有改變
   print(greetingTemplate);
   //greeting 是替換過後的結果
   print(greeting);
}
```

6. 建構字串

StringBuffer 是一個有效的串聯字串的類別，允許使用 write*() 方法增量建構字串，在呼叫 toString() 方法時，StringBuffer 才會創建新的 String 物件。

(1) clear()：清除字串緩衝區。
(2) write(Object obj)：將已轉為字串的 obj 的內容增加到緩衝區。
(3) writeAll(Iterable objects, [String separator = ""])：遍歷指定物件並按順序寫入它們，可選參數 separator 用於在寫入時提供分隔符號。

writeAll() 方法的第二個參數是可選的，可用於指定分隔符號，在本例中為空格。範例程式如下：

```
//chapter9/bin/example_15.dart
void main(){
   var sb = StringBuffer();
   //向字串緩衝區寫入資料
   sb..write('Use a StringBuffer for ')
      ..writeAll(['efficient', 'string', 'creation'], ' ')
      ..write('.');
```

```
    var fullString = sb.toString();
    // 列印最終字串
    print(fullString);
    // 清空字串緩衝區
    sb.clear();
}
```

7. 正規表示法

(1) Pattern 類別有兩個實現，一個是 String，另一個是 RegExp。

(2) RegExp 類別提供與 JavaScript 正規表示法相同的功能。使用正規表示法進行有效字串搜索和模式匹配。

使用正規表示法的範例程式如下：

```
//chapter9/bin/example_16.dart
void main(){
    // 這是一個或多個數字的正規表示法
    var numbers = RegExp(r'\d+');

    var allCharacters = 'llamas live fifteen to twenty years';
    var someDigits = 'llamas live 15 to 20 years';

    //contains() 可以使用正規表示法
    print(allCharacters.contains(numbers));
    print(someDigits.contains(numbers));

    // 將每個匹配項替換為另一個字串
    var exedOut = someDigits.replaceAll(numbers, 'XX');
    print(exedOut);
}
```

也可以直接使用 RegExp 和 Match 類別提供的功能，對正規表示法的匹配項進行存取和控制。範例程式如下：

```
//chapter9/bin/example_17.dart
void main(){
    var numbers = RegExp(r'\d+');
    var someDigits = 'llamas live 15 to 20 years';
```

```
    // 檢查一個正規表示法在字串中是否有匹配項
    print(numbers.hasMatch(someDigits));
    // 迴圈所有匹配項
    for (var match in numbers.allMatches(someDigits)) {
       print(match.group(0));
    }
}
```

9.2.3 URIs

Uri 類別提供了用於對 URI 中使用的字串進行編碼和解碼的函數。這些函數處理 URI 專用的字元，例如 "&" 和 "="。Uri 類別還解析並公開 URI 的元件，例如主機、通訊埠、方案等。

1. 編碼和解碼標準 URI

(1) decodeFull(String uri)：uri 使用百分比編碼對字串進行編碼，以使其可以安全地用作完整的 URI。除大寫和小寫字母、數字和字元外的所有字元 !#$&'()*+,-./:;=?@_~ 均按百分比編碼。這是 ECMA-262 版本 5.1 中為 encodeURI 函數指定的字元集。

(2) encodeFull(String uri)：解碼 uri 中的百分比編碼。

這些方法非常適合編碼或解碼完全標準的 URI，而保留完整的特殊 URI 字元。範例程式如下：

```
//chapter9/bin/example_18.dart
void main(){
   var uri = 'https://example.org/api?foo=some message';
   // 編碼 URI
   var encoded = Uri.encodeFull(uri);
   print(encoded);

   // 解碼 URI
   var decoded = Uri.decodeFull(encoded);
   print(decoded);
}
```

注意：只有 some 和 message 之間的空格被編碼。

2. 編碼和解碼 URI 元件

(1) encodeComponent(String component)：使用百分比編碼對字串 component 進行編碼，使其可以安全地用作 URI 元件。

(2) decodeComponent(String encodedComponent)：解碼 encodeComponent 中的百分比編碼。

注意：對 URI 元件進行解碼可能會更改其含義，因為某些解碼的字元可能具有指定 URI 元件類型的分隔符號的字元。在解碼各個部分之前，需始終使用該元件的分隔符號來拆分 URI 元件。

編碼與解碼 URL 元件的範例程式如下：

```
//chapter9/bin/example_19.dart
void main(){
  var uri = 'https://example.org/api?foo=some message';
  // 編碼 URI 及元件
  var encoded = Uri.encodeComponent(uri);
  print(encoded);

  // 解碼 URI 及元件
  var decoded = Uri.decodeComponent(encoded);
  print(uri);
}
```

3. 解析 URIs

(1) parse(String uri, [int start = 0, int end])：透過解析 URI 字串創建一個 Uri 物件。如果提供了 start 和 end，則它們必須指定 uri 的有效子字串，並且只有從 start 到 end 的子字串才被解析為 URI。

(2) Uri({String scheme, String userInfo, String host, int port, String path, Iterable<String> pathSegments, String query, Map<String, dynamic> queryParameters, String fragment})：使用 Uri 元件建構 Uri。

如果有一個 Uri 物件或 URI 字串，可以使用 Uri 屬性獲取其組成部分，例如：path。要從字串創建 Uri，需使用 parse() 靜態方法。範例程式如下：

```
//chapter9/bin/example_20.dart
void main(){
  // 解析 URIs
  var uri =Uri.parse('https://example.org:8080/foo/bar#frag');
  // 存取各個元件
  print(uri.scheme);
  print(uri.host);
  print(uri.path);
  print(uri.fragment);
  print(uri.origin);
}
```

可以使用 Uri() 建構函數從各個部分建構 URI。範例程式如下：

```
//chapter9/bin/example_21.dart
void main(){
  // 建構 URI
  var uri = Uri(
    scheme: 'https',
    host: 'example.org',
    path: '/foo/bar',
    fragment: 'frag');
  print(uri);
}
```

9.2.4 時間和日期

時間和日期（DateTime）物件是一個時間點，可以採用 UTC(世界統一時間) 或本地時區來生成時間。可以使用幾種建構函數創建 DateTime 物件：

(1) DateTime(int year, [int month = 1, int day = 1, int hour = 0, int minute = 0, int second = 0, int millisecond = 0, int microsecond = 0])：根據本地時區來創建 DateTime 實例。

(2) DateTime.now()：使用本地時區中的當前日期和時間構造一個 DateTime 實例。

(3) DateTime.utc(int year, [int month = 1, int day = 1, int hour = 0, int minute = 0, int second = 0, int millisecond = 0, int microsecond = 0])：使用 UTC 構造一個 DateTime 實例。

(4) tryParse(String formattedString)：解析字串 formattedString 來構造一個 DateTime 實例。如果解析出錯，則返回 null。

日期創建與解析範例程式如下：

```
//chapter9/bin/example_22.dart
void main(){
  // 獲取當前日期和時間
  var now = DateTime.now();
  print(now);
  // 使用本地時區創建一個新的 DateTime
  var t1 = DateTime(2000);
  print(t1);
  // 指定月份和日期
  var t2 = DateTime(2000, 1, 2);
  print(t2);
  // 指定日期使用 UTC 時間
  var t3 = DateTime.utc(2000);
  print(t3);
  // 解析 ISO 8601 日期
  var t4 = DateTime.tryParse('2000-01-01T00:00:00Z');
  print(t4);
}
```

9.3 數學函數庫

數學函數庫（dart:math）提供了常用功能，例如正弦和餘弦、最大值和最小值，以及常數，例如 pi 和 e。數學函數庫中的大多數功能都作為頂層方法使用。

要在應用中使用此函數庫，需匯入 dart:math。範例程式如下：

```
import 'dart:math';
```

1. 常用函數

(1) max<T extends num>(T a, T b)：返回兩個數字中的較大者。

(2) min<T extends num>(T a, T b)：返回兩個數字中的較小者。

(3) sqrt(num x)：將 x 轉為 double 類型並返回值的正平方根。如果 x 為 -0.0，則返回 -0.0；不然如果 x 為負或 NaN，則返回 NaN。

範例程式如下：

```
//chapter9/bin/example_23.dart
import 'dart:math';

void main(){
  // 取最大或最小值的函數
  print(max(1,1000));
  print(min(1, -1000));
}
```

2. 數學常數

此函數庫中還定義了常用的數學常數，包括 e、pi、sqrt2 等。範例程式如下：

```
//2.718281828459045
print(e);
//3.141592653589793
print(pi);
//1.4142135623730951
print(sqrt2);
```

3. 隨機數

(1) Random([int seed])：創建一個亂數產生器，可選的 seed 參數用於初始化生成器的內部狀態。

(2) nextBool()：隨機生成一個布林值。

(3) nextDouble()：生成非負的隨機浮點值，該值均勻地分佈在從 0.0(含) 到 1.0(不含)。

(4) nextInt(int max)：生成一個非負的隨機整數，其範圍為 0(含) 到 max(不包括) 均勻分佈。

使用 Random 類別生成隨機數時，可以選擇提供一個種子給 Random 建構函數。範例程式如下：

```
var random = Random();
//0.0~1.0
random.nextDouble();
//0~9
random.nextInt(10);
```

甚至可以生成隨機布林值。範例程式如下：

```
var random = Random();
//true 或 false
random.nextBool();
```

9.4 轉換函數庫

dart:convert 函數庫具有編碼器和解碼器，用於在不同的資料表示形式之間進行轉換。包括常見的資料形式 JSON 和 UTF-8，並支持創建其他轉換器。轉換器是指對資料進行編碼和解碼。JSON 是一種簡單的文字格式，用於表示結構化物件和集合。UTF-8 是一種常見的可變寬度編碼，可以表示 Unicode 字元集中的每個字元。

dart:convert 函數庫可在 Web 應用程式和命令列應用程式中使用。要使用它，需匯入 dart:convert。範例程式如下：

```
import 'dart:convert';
```

9.4.1 編碼和解碼 JSON

(1) jsonEncode(Object value, {Object toEncodable(Object nonEncodable)})：將值轉為 JSON 字串，如果 value 包含無法直接編碼為 JSON 字串的物件，則可使用 toEncodable 函數將其轉為直接可編碼的物件。如果省略 toEncodable，則預設呼叫該不可呼叫物件的 toJson() 方法。

(2) jsonDecode(String source, {Object reviver(Object key, Object value)})：解析字串並返回生成的 Json 物件。對於在解碼過程中已解析的每個物件或清單屬性，都會呼叫一次可選的 reviver 函數。key 參數可以是清單屬性的整數清單索引、物件屬性的字串映射鍵，或最終結果為 null。

使用 jsonDecode() 將 JSON 編碼的字串解碼為 Dart 物件。範例程式如下：

```
//chapter9/bin/example_24.dart
import 'dart:convert';
void main(){
  // 確保在 JSON 字串中使用雙引號 (")，不能使用單引號 (')
  var jsonString = '''[{"score": 40},{"score": 80}]''';
  // 將 JSON 字串解碼為清單
  var scores1 = jsonDecode(jsonString);
  print(scores1 is List);

  var firstScore = scores1[0];
  print(firstScore is Map);
  print(firstScore['score']);
}
```

使用 jsonEncode() 將受支援的 Dart 物件編碼為 JSON 格式的字串。範例程式如下：

```
var scores2 = [
  {'score': 40},
  {'score': 80},
  {'score': 100, 'overtime': true, 'special_guest': null}
```

```
];
// 將清單 scores2 編碼為 JSON 格式字串
var jsonText = jsonEncode(scores2);
print(jsonText);
```

只有類型為 int、double、String、bool、null、List 或 Map(帶有字串鍵) 的物件可以直接編碼為 JSON，而 List 和 Map 物件是遞迴編碼的。

9.4.2 解碼和編碼 UTF-8 字元

(1) encode(String input)：將字串 input 轉為 UTF-8 編碼的位元組清單。
(2) decode(List<int> codeUnits, {bool allowMalformed})：將 UTF-8 程式單元解碼為對應的字串。如果 allowMalformed 為 true，解碼器使用 Unicode 替換字元 U + FFFD 替換無效 (或未終止) 的字元序列。不然它將引發 FormatException。

使用 utf8.decode() 方法將 UTF-8 編碼的位元組解碼為 Dart 字串。範例程式如下：

```
//chapter9/bin/example_25.dart
import 'dart:convert';
void main(){
  //ignore: omit_local_variable_types
  List<int> utf8Bytes = [
    0xc3, 0x8e, 0xc3, 0xb1, 0xc5, 0xa3, 0xc3, 0xa9,
    0x72, 0xc3, 0xb1, 0xc3, 0xa5, 0xc5, 0xa3, 0xc3,
    0xae, 0xc3, 0xb6, 0xc3, 0xb1, 0xc3, 0xa5, 0xc4,
    0xbc, 0xc3, 0xae, 0xc5, 0xbe, 0xc3, 0xa5, 0xc5,
    0xa3, 0xc3, 0xae, 0xe1, 0xbb, 0x9d, 0xc3, 0xb1
  ];
  // 對位元組資料進行解碼
  var funnyWord = utf8.decode(utf8Bytes);
  print(funnyWord);
}
```

使用 utf8.encode() 將 Dart 字串編碼為 UTF-8 編碼的位元組清單。範例程式如下：

```
List < int > encoded = utf8.encode('Îñţérñåţîöñ~åļîžåţîòñ');
print(encoded.length == utf8Bytes.length);
for (int i = 0; i < encoded.length; i++) {
  // 比較位元組是否相等
  print(encoded[i]== utf8Bytes[i]);
}
```

9.5 輸入和輸出函數庫

dart:io 函數庫提供用於處理檔案、目錄、處理程序、sockets、WebSocket、HTTP 用戶端和伺服器的 API。

一般來說 dart:io 函數庫實現並提升了非同步 API，同步方法很容易阻塞應用程式，從而難以擴充。因此，大多數操作都是透過 Future 或 Stream 物件返回結果的，Future 或 Stream 物件是現代伺服器平台常見的模式。

要使用 dart:io 函數庫，必須將其匯入。範例程式如下：

```
import 'dart:io';
```

透過 I/O 函數庫，命令列應用程式可以讀取和寫入檔案及瀏覽目錄。有兩種選擇來讀取檔案的內容：一次全部讀取和流式傳輸。一次讀取一個檔案需要足夠的記憶體來儲存檔案的所有內容。如果檔案很大，或想在讀取檔案時處理，則應使用串流式傳輸。

1. 檔案類別常用建構函數和方法

(1) File(String path)：創建 File(檔案) 物件，如果 path 是相對路徑，則在使用時相對當前工作目錄。如果 path 是絕對路徑，則與當前工作目錄無關。

(2) File.fromUri(Uri uri)：從 URI 創建一個 File 物件。

(3) openRead([int start,int end])：為此檔案的內容創建一個新的獨立 Stream。如果存在 start，則從位元組偏移量開始讀取檔案。否則從頭

開始 (索引 0)。如果存在 end，則僅讀取直到位元組索引的 end。不然直到檔案結束。為了確保釋放系統資源，必須將串流讀取完整，否則必須取消對流的訂閱。

(4) openWrite({FileMode mode: FileMode.write,Encoding encoding: utf8})：為該檔案創建一個新的獨立 IOSink。當該 IOSink 不再使用時，必須關閉，以釋放系統資源。IOSink 支援兩種 FileMode 值：FileMode.write，將初始寫入位置設定為檔案的開頭；FileMode.append，將初始寫入位置設定為檔案的尾端。

(5) readAsString({Encoding encoding: utf8})：使用指定的 Encoding 以字串形式讀取整數個檔案內容。讀取檔案內容後，返回一個以字串結尾的 Future <String>。

(6) readAsLines({Encoding encoding: utf8})：使用指定的 Encoding 以行讀取整數個檔案的內容。返回一個 Future <List <String >>。

(7) readAsBytes()：以位元組清單的形式讀取整數個檔案的內容。返回 Future <Uint8List>。

2. 以文字形式讀取檔案

讀取使用 UTF-8 編碼的文字檔時，可以使用 readAsString() 讀取整數個檔案內容。當各行很重要時，可以使用 readAsLines()。在這兩種情況下，都將返回一個 Future 物件，該物件以一個或多個字串的形式提供檔案的內容。範例程式如下：

```
//chapter9/bin/example_26.dart
import 'dart:io';
Future main() async{
   var config = File('config.txt');
   var contents;
   // 將整個檔案的內容放在單一字串中
   contents = await config.readAsString();
   print(' 檔案內容的字元長度：${contents.length}');
   // 將檔案內容以行作為分割，拆分成多個字串
   contents = await config.readAsLines();
```

```
    print('檔案內容的行數 ${contents.length}');
}
```

3. 以二進位形式讀取檔案

以下程式將整個檔案作為位元組讀取到位元組清單中，對 readAsBytes() 的呼叫返回一個 Future 物件，當可用時會提供結果。範例程式如下：

```
//chapter9/bin/example_27.dart
import 'dart:io';

Future main() async{
    var config = File('config.txt');
    try {
        var contents = await config.readAsBytes();
        print('檔案內容的位元組長度為 ${contents.length}');
    } catch (e) {
        print(e);
    }
}
```

為了捕捉錯誤，以免導致未捕捉的異常，可以在 Future 上註冊 catchError 處理常式，或在非同步函數中使用 try-catch。範例程式如下：

```
Future main() async{
    var config = File('config.txt');
    try {
        var contents = await config.readAsString();
        print(contents);
    } catch (e) {
        print(e);
    }
}
```

4. 以串流讀取檔案內容

可以使用 Stream 讀取檔案，每次讀取一點。可以使用 Stream API 或 await for。範例程式如下：

```
//chapter9/bin/example_28.dart
import 'dart:io';
import 'dart:convert';

Future main() async{
   var config = File('config.txt');
   //ignore: omit_local_variable_types
   Stream<List<int>> inputStream = config.openRead();

   var lines =
   utf8.decoder.bind(inputStream).transform(LineSplitter());
   try {
      await for (var line in lines) {
         print(' 從串流中獲取的字元長度：${line.length}');
      }
      print(' 檔案關閉 ');
   } catch (e) {
      print(e);
   }
}
```

5. 寫入檔案內容

可以使用 IOSink 將資料寫入檔案。使用 File 的 openWrite() 方法獲取可以寫入的 IOSink 物件。IOSink 類別有以下常用函數：

(1) IOSink(StreamConsumer<List<int>> target,{Encoding encoding: utf8})：建構函數，創建一個 IOSink 物件，StreamConsumer 是接收多個完整流的「接收器」的抽象介面。使用者可以使用 addStream 接收多個連續的串流，並且當不需要增加更多資料時，close 方法將告訴使用者完成其工作並關閉。

(2) add(List<int> data)：將位元組資料增加到目標使用者。

(3) close()：關閉目標使用者。

(4) flush()：返回一個 Future，一旦基礎 StreamConsumer 接收了所有緩衝的資料，該 Future 將完成。

(5) write(Object obj)：將 obj 轉為 String，並將結果的編碼增加到目標使用者。

(6) writeAll(Iterable objects, [String separator = ""])：遍歷指定 objects 並按順序寫入它們。可選的 separator 是指遍歷的物件間寫入時的分隔符號。

預設模式 FileMode.write 完全覆寫檔案中的現有資料。範例程式如下：

```
//chapter9/bin/example_29.dart
import 'dart:io';
   void main()async{
   var logFile = File('log.txt');
   var sink = logFile.openWrite();
   sink.write(' 檔案存取時間 ${DateTime.now()}\n');
   await sink.flush();
   await sink.close();
}
```

要增加到檔案的尾端，需使用可選的 mode 參數指定 FileMode.append。範例程式如下：

```
var sink = logFile.openWrite(mode: FileMode.append);
```

6. 列出目錄中的檔案

尋找目錄的所有檔案和子目錄是非同步作業，list() 方法返回一個 Stream，當遇到檔案或目錄時該 Stream 發出一個物件。Directory 類別有以下常用函數：

(1) Directory(String path)：建構函數，創建一個 Directory(目錄) 物件。如果 path 是相對路徑，則在使用時對應當前工作目錄。如果 path 是絕對路徑，則與當前工作目錄無關。

(2) list({bool recursive: false, bool followLinks: true})：列出此目錄的子目錄和檔案。可選參數 recursive 表示遞迴到子目錄。返回用於目錄、檔案和連結的 FileSystemEntity 物件串流。

目錄遍歷範例程式如下：

```
//chapter9/bin/example_30.dart
import 'dart:io';
Future main() async {
   // 當前專案的根目錄
   var dir = Directory('');
   try {
      var dirList = dir.list();
      await for (FileSystemEntity f in dirList) {
         if (f is File) {
            print(' 發現檔案 ${f.path}');
         } else if (f is Directory) {
            print(' 發現目錄 ${f.path}');
         }
      }
   } catch (e) {
      print(e.toString());
   }
}
```

Chapter 10

非同步

Dart 函數庫有許多返回 Future 或 Stream 物件的函數。這些函數是非同步的：它們在設定可功耗時的操作 (例如 I/O) 之後返回，而無須等待該操作完成。

async 和 await 關鍵字支援非同步程式設計，可以編寫出看起來類似於同步程式的非同步程式。

10.1 Future

Future 表示非同步作業的結果，它有兩種狀態：

(1) 未完成狀態：當呼叫非同步函數時，它返回未完成的 Future，並且持續到非同步函數操作完成。
(2) 完成狀態：如果非同步函數操作成功，則返回一個值；如果非同步函數操作失敗，則返回一個錯誤。

10.1.1 創建 Future

可以透過建構函數創建 Future，建構函數的參數是一個函數，該函數的返回數值類型為 Future<T> 或 T，其中 T 代表的是任何類型。下例中 getInt 函數的返回值是 int 類型。範例程式如下：

```
//chapter10/bin/example_01.dart
import 'dart:math';
import 'dart:async';
// 同步函數，返回 int 類型的值
int getInt(){
   print(' 執行 getInt 方法 ');
   // 創建隨機物件實例
   Random rng = Random();
   // 返回 0~100 的隨機數
   return rng.nextInt(100);
}
void main(){
   // 透過 Future 類別的建構函數創建 Future
   // 此處呼叫的 getInt 函數返回值是 int 類型
   Future<int> future = Future(getInt);
   // 驗證是否是 Future<int> 的實例
   print(Future(getInt));
}
```

也可以使用 async 標記函數，使函數成為非同步函數，非同步函數會自動將返回值包裝成一個 Future。此時 getInt 函數的返回值是 Future<int> 類型。範例程式如下：

```
//chapter10/bin/example_02.dart
import 'dart:math';
import 'dart:async';
// 在函數增加 async 標記以表明是非同步函數
// 返回值使用 Future 包裝並提供泛型參數 int
Future<int> getInt() async{
   print(' 執行 getInt 方法 ');
   Random rng = Random();
   return rng.nextInt(100);
}
```

```
void main(){
    //getInt 方法的返回值是 Future<int> 類型
    Future<int> future = getInt();
    // 驗證是否是 Future<int> 類型
    print(getInt());
}
```

10.1.2 使用 Future

可以使用 Future 類別提供的 then、catchError 和 whenComplete 方法對 Future 物件進行進一步處理，當非同步作業成功時，執行 then 方法，then 方法接收一個參數為非同步作業返回值的回呼函數。當非同步作業失敗時，執行 catchError 方法，catchError 方法接收一個參數為錯誤物件的回呼函數。當非同步作業完成時，無論執行失敗還是成功都會執行 whenComplete 方法，whenComplete 方法接收一個無參的自訂回呼函數。範例程式如下：

```
//chapter10/bin/example_03.dart
import 'dart:math';
import 'dart:async';
Future<int> getInt() async{
    print(' 執行 getInt 方法 ');
    Random rng = Random();
    return rng.nextInt(100);
}
void main(){
    // 透過 Future 類別的建構函數創建 Future
    Future<int> future = getInt();
    future.then((Object onValue){
        // 非同步呼叫成功
        print(' 非同步作業成功，值為 $onValue');
    }).catchError((Object onError){
        // 非同步呼叫失敗，用於捕捉異常
        print(' 非同步作業失敗 : $onError');
    }).whenComplete((){
        // 非同步呼叫完成時呼叫，與是否成功無關
```

```
      print('非同步作業完成');
   });
}
```

也可以使用 await 關鍵字等待非同步作業完成，使用 await 關鍵字的函數必須使用 async 標記，並且使用 try 敘述捕捉異常。範例程式如下：

```
//chapter10/bin/example_04.dart
import 'dart:math';
import 'dart:async';
Future<int> getInt() async{
   print('執行 getInt 方法');
   Random rng = Random();
   return rng.nextInt(100);
}
void main() async{
   // 透過 Future 類別的建構函數創建 Future
   Future<int> future = getInt();
   try{
      // 使用 await 等待非同步呼叫完成，使其等於同步程式
      var onValue=await future;
      print('非同步作業成功，值為 $onValue');
   }on Exception catch(onError){
      print('非同步作業失敗：$onError');
   }finally{
      print('非同步作業完成');
   }
}
```

儘管非同步功能可能會執行耗時的操作，但它不會等待這些操作。相反，非同步函數會照常執行直到遇見第一個 await 運算式。然後，它返回 Future 物件，僅在 await 運算式完成後才恢復執行。

在 await 運算式中，運算式的值通常是 Future，如果不是，則該值將自動包裝在 Future 中。此 Future 物件表示承諾返回一個物件。await 運算式的值就是返回的物件。await 運算式使執行暫停，直到該物件可用為止。

如果非同步函數不需要返回值，則需將返回類型修改為 Future<void>。

10.2 Stream

Stream 是一系列非同步事件的來源。Stream 提供了一種接收事件序列的方式，每個事件不是資料事件（又被稱為 Stream 的元素），就是錯誤事件（發生故障時的通知），當 Stream 發出所有事件後，單一 done 事件將通知監聽器已完成。

Stream 和 Future 的很多特性類似，但也有一些區別：

(1) Future 在非同步作業完成時提供單一結果、錯誤或值。Stream 可以提供多個結果。
(2) Future 使用 then、catchError、whenComplete 方法獲取或處理結果，Stream 則只需透過 listen(監聽) 即可處理所有值。
(3) Future 發送和接收相同的值，而 Stream 可以使用輔助方法在值到達前進行處理。

10.2.1 創建 Stream

首先創建一個 StreamController 物件，然後使用 StreamController 物件的 stream 屬性返回一個 Stream 物件。範例程式如下：

```
StreamController<int> controller = StreamController<int>(
   onListen: startTimer,
   onPause: stopTimer,
   onResume: startTimer,
   onCancel: stopTimer);

Stream stream = controller.stream;
```

StreamController 建構函數支援泛型，這裡使用 int 類型。建構函數提供了多個可選參數：

(1) onListen：監聽 stream 時呼叫的回呼函數。

(2) onPause：stream 暫停時呼叫的回呼函數。

(3) onResume：stream 恢復時呼叫的回呼函數。

(4) onCancel：取消 stream 時呼叫的回呼函數。

(5) sync：布林值，預設值為 false，同步 stream 標記。

10.2.2 使用 Stream

使用 Stream 類別的 listen 方法監聽 stream，該方法提供以下參數：

(1) onData：必選參數，回呼函數，該函數的參數是 Stream 事件發出的值。

(2) onError：可選參數，回呼函數，來自 Stream 的錯誤。該回呼函數類型必須是 void onError(error) 或 void onError(error,StackTrace stackTrace)，該函數的兩個參數一個是錯誤物件，另一個是可選的堆疊追蹤資訊。如果省略此函數且 stream 發生錯誤，則會將錯誤訊息向外傳遞。

(3) onDone：可選參數，回呼函數，當此 stream 關閉並發送完成事件時，將呼叫此回呼函數。

(4) cancelOnError：可選參數，布林值，預設值為 false。如果值為 true，則在 stream 傳遞第一個錯誤事件時自動取消訂閱。

listen 方法返回用於訂閱 Stream 中事件的物件 StreamSubscription，該物件保留上述處理事件的回呼函數，還可以發出取消訂閱事件和臨時暫停 stream 中的事件。完整範例程式如下：

```
//chapter10/bin/example_05.dart
import 'dart:async';
// 用於返回 stream 物件，stream 又稱為串流
Stream<int> createStream(Duration interval,int maxCount){
  // 定義流量控制器
  StreamController<int> controller;
  // 定義計時器
  Timer timer;
```

```
    //計數變數
    int counter = 0;

    void tick(_){
       counter++;
       //將counter的值作為事件發送給stream
       controller.add(counter);
       //判斷計數變數是否達到最大值
       if (counter == maxCount) {
          //終止計時器
          timer.cancel();
          //關閉stream並通知監聽器
          controller.close();
       }
    }
    //啟動計時器
    void startTimer(){
       print('createStream開始執行');
       //interval是呼叫tick函數的時間間隔
       timer = Timer.periodic(interval,tick);
    }
    //終止計時器
    void stopTimer(){
       if (timer != null){
          timer.cancel();
          timer = null;
       }
       print('createStream結束執行');
    }
    //創建流量控制器
    controller = StreamController<int>(
       onListen: startTimer,
       onPause: stopTimer,
       onResume: startTimer,
       onCancel: stopTimer);
    //返回流
    return controller.stream;
}
void main() async{
```

```
    // 接收返回的 stream 物件
    // 時間間隔為 1s，最大值為 10s
    Stream<int> stream= createStream(const Duration(seconds:1),10);
    // 監聽 stream
    stream.listen((int value) {
      print(' 來自 createStream 的值：$value');
    });
}
```

函數 createStream 返回 Stream 物件，接收兩個參數，Duration 類型的參數表示持續時間，參數 maxCount 表示發起事件的最大次數。當對 stream 執行 listen 方法時，回呼函數 startTimer 開始執行。

Timer 是計時器類別，Timer.periodic 命名建構函數創建一個重複的計時器，該建構函數有兩個參數，第一個參數表示持續時間，第二個參數是回呼函數，該回呼函數接收一個 Timer 類型的參數。計時器從指定的持續時間倒計時到 0。當計時器達到 0 時，計時器將呼叫指定的回呼函數。

因為在回呼函數 tick 內部可以直接使用物件 timer，所以在傳入參數處使用 "_" 做預留位置。StreamController 物件使用 add 方法向 stream 發送事件，使用 close 方法關閉 stream。

10.3 生成器函數

傳統函數只會返回單一值，生成器函數生成值的序列。生成器函數可以採用同步返回帶有值的 Iterable 物件，在非同步中返回 Stream 物件。

關鍵字 yield 返回單一值到序列，但是不會停止生成器函數。

生成器函數隨選生成值，當開始疊代 iterator 或開始監聽 stream 時才生成值。

10.3.1 同步生成器

要實現同步生成器功能，需將函數主體標記為 sync*，並使用 yield 敘述傳遞值到序列。範例程式如下：

```
//chapter10/bin/example_06.dart
Iterable<int> getNumbers(int number) sync*{
   print('生成器開始執行');
   int k = 0;
   while (k < number) yield k++;
   print('生成器執行結束');
}
void main() {
   print('創建 iterator');
   Iterable<int> numbers = getNumbers(10);
   print('開始疊代');
   for (int val in numbers) {
      print('$val');
   }
}
```

10.3.2 非同步生成器

要實現非同步生成器函數，需將函數主體標記為 async*，並使用 yield 敘述傳遞值到 stream。範例程式如下：

```
//chapter10/bin/example_07.dart
Stream<int> getStream(int number) async*{
   print('非同步生成器開始執行');
   int k = 0;
   while (k < number) yield k++;
   print('非同步生成器執行結束');
}
void main() {
   //創建 stream
   Stream<int> stream = getStream(10);
   //監聽 stream
   stream.listen((int value) {
```

```
      print(' 非同步生成器生成的值： $value');
  });
}
```

10.3.3 遞迴生成器

遞迴是指在函數內部呼叫函數本身，如果生成器是遞迴的，則可以使用 yield* 來提高其性能。範例程式如下：

```
//chapter10/bin/example_08.dart
Iterable<int> getNumbersRecursive(int number) sync* {
  print(' 生成 $number 開始 ');
  if (number > 0) {
    yield* getNumbersRecursive(number - 1);
  }
  yield number;
  print(' 生成 $number 結束 ');
}
void main() {
  print(' 創建 iterator');
  Iterable<int> numbers = getNumbersRecursive(3);
  print(' 開始疊代 ');
  for (int val in numbers) {
    print('$val');
  }
  print('main 函數結束 ');
}
```

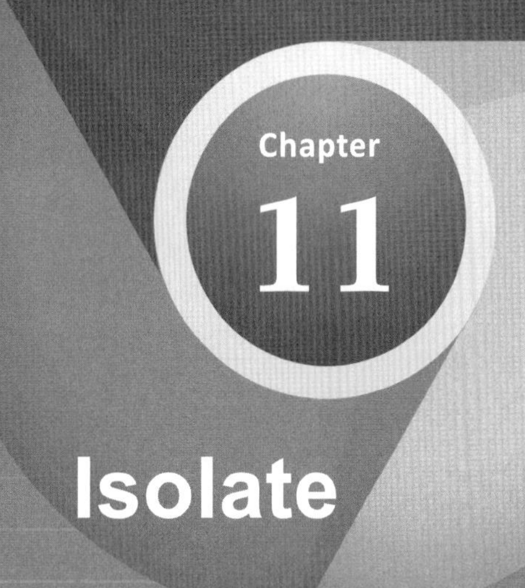

Isolate

大多數電腦，甚至在行動平台上，都具有多核心 CPU。為了利用所有這些核心，開發人員通常使用併發執行的共用記憶體執行緒。但是，共用狀態併發易於出錯，並且可能導致複雜的程式。

11.1 什麼是 Isolate

所有 Dart 程式始終在 Isolate 中執行，一個 Isolate 只有一個執行緒，一個專用的記憶體區域和它自己的事件循環，完全獨立於其他 Isolate。如圖 11-1 所示的是單一 Isolate 的示意圖，長方形區塊代表區塊，由循環箭頭圖示代表事件循環管理。

▲ 圖 11-1 Isolate

Dart 應用具有 main() 方法，該方法具有標準的 Isolate，並且從 main() 執行的程式都在該 Isolate 上執行，其內部的事件由事件循環管理執行。絕大多數 Dart 應用程式都在此單一標準 Isolate 上執行，如果需要在後台執行耗時任務，就需要透過 Isolate.spawn() 創建新的 Isolate。不同的 Isolate 彼此完全獨立，並且如果需要在它們之間存取資料或交換資訊，則唯一的方法是透過訊息，如圖 11-2 所示。

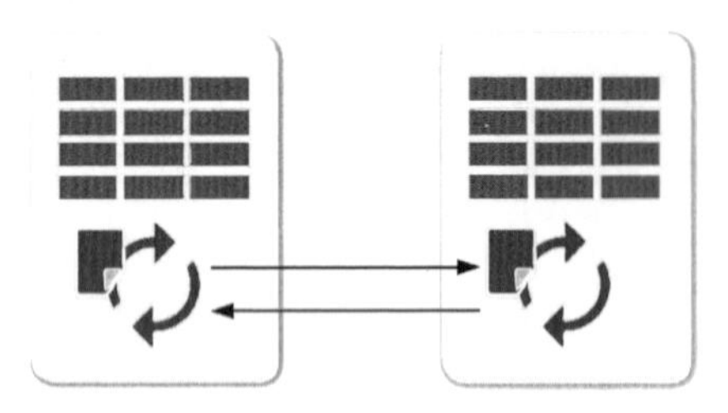

▲ 圖 11-2 Isolate 間傳遞訊息

11.2 事件循環

每個 Isolate 都用一個自己的事件循環，Dart 中的事件循環包含兩個佇列：事件佇列和微任務佇列。

事件佇列是 Dart 處理此 Isolate 外部事件的唯一途徑，例如：I/O 操作、Stream、使用者的點擊、網路請求的回呼、來自其他 Isolate 的訊息或由使用者互動觸發的任何操作。這些被稱為事件，並且會被增加到事件佇列，事件循環將根據佇列連續處理事件。

微任務是在使用頂層函數 scheduleMicrotask() 時傳入的函數所包含的任務，該函數通常執行耗時工作，常用於非同步任務，並且該微任務將被增加到微任務佇列。scheduleMicrotask() 會非同步執行一個函數，透過此函數註冊的回呼始終按循序執行，並保證在其他非同步事件 (如 Timer 事件或 DOM 事件) 之前執行。微任務佇列不處理外部事件。

啟動應用程式時，將創建並啟動一個新執行緒。創建主執行緒後，Dart 會將微任務和事件佇列初始化為空，同時執行 main() 方法，執行過程中將微任務放入微任務佇列，事件放入事件佇列，一直到執行完 main() 方法中的最後一行程式，然後啟動事件循環。值得注意的是，微任務佇列的執行優先順序高於事件佇列，只有當微任務佇列中的任務全部執行完成，並且微任務佇列為空時，事件循環才開始處理事件佇列中的事件，執行流程如圖 11-3 所示。

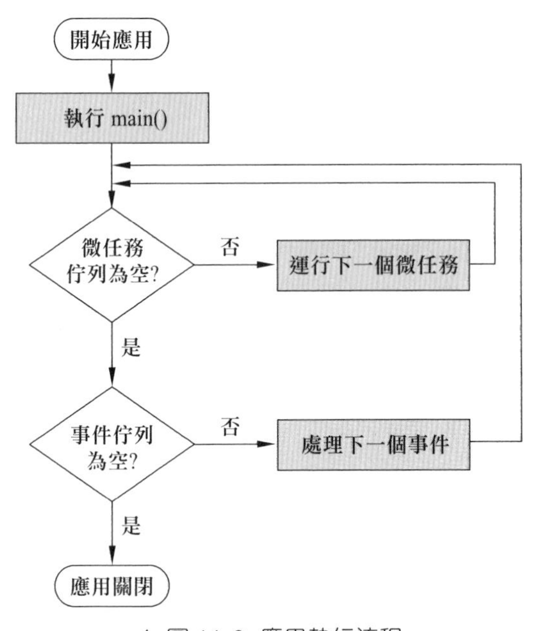

▲ 圖 11-3 應用執行流程

Dart 採用單執行緒，事件循環透過併發存取執行緒來允許平行任務執行。執行緒負責執行程式，一旦程式進入該執行緒，它的執行順序就是線性的，並且直到所有程式執行完畢後才會停止。事件循環控制執行緒將按循序執行內容，它不會更改或執行 Dart 程式，僅用來控制其執行。事件循環會將事件佇列和任務佇列中的項目放入執行緒的呼叫堆疊中，由於外部事件可以不斷擴充事件佇列，進而使得 Dart 中的非同步和響應成為可能。

為了更加直觀地瞭解事件循環控製程式執行的順序，範例程式如下：

```
//chapter11/bin/example_01.dart
import 'dart:async';
main(){
  print('main 方法 #1');
  scheduleMicrotask(() => print(' 微任務 #1'));

  Future.delayed(Duration(seconds:1),
    () => print('future 發起的事件 #1 (delayed)'));
  Future(() => print('future 發起的事件 #2'));
  Future(() => print('future 發起的事件 #3'));

  scheduleMicrotask(() => print(' 微任務 #2 '));
  print('main 方法 #2');
}
```

這裡使用了 Future() 和 Future.delayed() 在事件佇列中安排任務，可以查看執行結果：

```
main 方法 #1
main 方法 #2
微任務 #1
微任務 #2
future 發起的事件 #2
future 發起的事件 #3
future 發起的事件 #1 (delayed)
```

當在執行事件佇列中的某一個事件時，如果事件中的某個任務必須被執行完成後才能執行下一個事件，但是該事件中某個任務又比較消耗時間，則可以使用 scheduleMicrotask() 將其增加到微任務佇列。範例程式如下：

```
//chapter11/bin/example_02.dart
import 'dart:async';
main() {
  print('main #1');
  scheduleMicrotask(() => print(' 微任務 #1'));
```

```
  Future.delayed(Duration(seconds:1),
        () => print('事件 #1 (延遲)'));

  Future(() => print('事件 #2 '))
        .then((_) => print('事件 #2a'))
        .then((_) {
     print('事件 #2b');
     scheduleMicrotask(() => print('微任務 #0 來自事件 #2b'));
  }).then((_) => print('事件 #2c'));

  scheduleMicrotask(() => print('微任務 #2'));

  Future(() => print('事件 #3'))
     .then((_) => Future(
        () => print('事件 #3a 新的 Future')))
     .then((_) => print('事件 #3b'));

  Future(() => print('事件 #4'));
  scheduleMicrotask(() => print('微任務 #3'));
  print('main #2');
}
```

執行結果如下：

```
main #1
main #2
微任務 #1
微任務 #2
微任務 #3
事件 #2
事件 #2a
事件 #2b
事件 #2c
微任務 #0 來自事件 #2b
事件 #3
事件 #4
事件 #3a 新的 Future
事件 #3b
事件 #1 (延遲)
```

11.3 創建 Isolate

與傳統多執行緒語言不同，Dart 採用 Isolate 實現多執行緒。可以使用 Isolate 執行非同步、後台、耗時的任務，可以在單一應用程式中創建和執行多個 Isolate。

所有 Dart 程式都執行在 Isolate 中，當呼叫 main 函數時，其實該函數就位於一個 Isolate 中，被稱作 main Isolate。可以從以下幾方面來深入瞭解 Isolate：

(1) 用於實現併發程式設計。
(2) 擁有自己獨立的區塊。
(3) 每個 Isolate 都有獨立的事件循環，該循環始終執行並捕捉事件 (例如：從網路獲取資料) 且加以處理。
(4) 與執行緒相似但不共用記憶體。
(5) 只能透過訊息與其他 Isolate 通訊。

使用 Isolate 類別的 spawn 方法創建新的 Isolate，它包含兩個參數，第一個參數是函數名稱，第二個參數可以是任何類型，它是當前 Isolate 傳遞給新 Isolate 的訊息，該參數會被第一個參數提供的函數捕捉。範例程式如下：

```
Isolate.spawn(newIsolate, ' 執行耗時任務 ');
```

這裡傳遞名為 newIsolate 的函數，並且傳遞了一個字串類型的訊息。

可以使用 Isolate 的 kill 方法關閉創建的 Isolate。範例程式如下：

```
isolate.kill(priority: Isolate.immediate);
```

完整範例程式如下：

```
//chapter11/bin/example_03.dart
import 'dart:io';
import 'dart:isolate';
```

```
void newIsolate(String arg) {
  print(' 新的 Isolate 被創建並帶有訊息 : $arg');
  for(int progress=0; progress<500 ;progress++){
    print(' 新的 Isolate 的任務進度 : $progress');
  }
  print(' 新的 Isolate 任務完成 ');
}

void main() async{
  print(' 創建新的 Isolate');
  Isolate isolate=await Isolate.spawn(newIsolate, ' 執行耗時任務 ');
  print(' 新的 Isolate 被創建，main Isolate 開始執行自己的任務 ');
  for(int progress=0;progress<500;progress++){
    print('main Isolate 任務進度 : $progress');
  }
  //stdin 允許在命令列中同步或非同步讀取輸入串流
  //stdin.first 表示在收到輸入串流中的第一個元素後，停止監聽此串流，繼續執行後
  續程式
  await stdin.first;
  // 關閉新的 Isolate
  isolate.kill(priority: Isolate.immediate);
  isolate=null;
  print(' 新的 Isolate 被關閉 ');
}
```

11.4 獲取訊息

Isolate 之間通訊需要使用 ReceivePort 類別，該類別有兩個重要方法：

(1) receivePort.sendPort：返回發送到此接收通訊埠的 SendPort 物件，SendPort 物件用於將訊息發送到 ReceivePort 物件。

(2) receivePort.listen((dynamic receivedData) {})：用於接收 sendPort 發送的訊息。

範例程式如下：

```
//chapter11/bin/example_04.dart
import 'dart:isolate';

void newIsolate(SendPort sendPort) {
  print('新的 Isolate 被創建並帶有訊息：$sendPort');
  for(int progress=0; progress<500 ;progress++){
    print('新的 Isolate 的任務進度：$progress');
  }
  sendPort.send("任務完成");
  print('新的 Isolate 任務完成');
}

void main() async{
  ReceivePort receivePort= ReceivePort();
  print('創建新的 Isolate');
  Isolate isolate=await Isolate.spawn(newIsolate,receivePort.sendPort);
  print('新的 Isolate 被創建，main Isolate 開始執行自己的任務');
  for(int progress=0;progress<500;progress++){
    print('main Isolate 任務進度：$progress');
  }

  receivePort.listen((dynamic receivedData) {
    print('資料來自新的 Isolate：$receivedData');

    //有條件地關閉新的 Isolate
    if(receivedData is String && receivedData.toString().contains('任務完成
')){
      //關閉新的 Isolate
      isolate.kill(priority: Isolate.immediate);
      isolate=null;
    }
  });
}
```

11.5 相互通訊

在 Isolates 間傳遞訊息有兩種方式：第一種，透過使用 ReceivePort 類別，使用方式同獲取訊息一致；第二種，使用 Dart 團隊開發的 stream_channel 套件。

11.5.1 使用 ReceivePort

與獲取訊息一樣，只是將 ReceivePort 物件在兩個 Isolates 間相互傳遞。範例程式如下：

```
//chapter11/bin/example_05.dart
import 'dart:isolate';

void newIsolate(SendPort sendPortOfMainIsolate) {
  ReceivePort receivePort=ReceivePort();
  print('新的 Isolate 被創建並帶有訊息：$sendPortOfMainIsolate');
  //監聽 main Isolate 發送的訊息
  receivePort.listen((dynamic receivedData) {
    print('新的 Isolate 任務命令：$receivedData');
    sendPortOfMainIsolate.send(receivedData);
  });
  //將新 Isolate 的 sendPort 發送到 main Isolate，使其可以發送訊息到此新 Isolate
  sendPortOfMainIsolate.send(receivePort.sendPort);
}

void main() async{
  ReceivePort receivePort= ReceivePort();
  print('創建新的 Isolate');
  Isolate isolate=await Isolate.spawn(newIsolate,receivePort.sendPort);
  print('新的 Isolate 被創建，main Isolate 開始執行自己的任務');
  for(int progress=0;progress<100;progress++){
    print('main Isolate 任務進度：$progress');
  }
```

```
  // 監聽新的 Isolate 發送的訊息
  receivePort.listen((dynamic receivedData) {
    print(' 資料來自新的 Isolate : $receivedData');
    if (receivedData is SendPort) {
      SendPort sendPortOfNewIsolate = receivedData;
      for(int commond=0;commond<50;commond++){
        print('main Isolate 命令 : $commond');
        sendPortOfNewIsolate.send(commond);
      }
    }else{
      print(' 資料來自新的 Isolate, 命令 : $receivedData');
    }
  });
}
```

11.5.2 使用 stream_channel

在 pubspec.yaml 檔案中增加 stream_channel 依賴項。

```
dependencies:
  stream_channel: ^2.0.0
```

將 receivePort 物件設定為 IsolateChannel.connectReceive()，並將同一個 receivePort 物件的 sendPort 設定為 IsolateChannel.connectSend()。

使用 channel.sink.add('message') 發送訊息，使用 channel.stream.listen((dynamic receiveData){}); 接收訊息。範例程式如下：

```
//chapter11/bin/example_06.dart
import 'dart:isolate';
import 'package:stream_channel/isolate_channel.dart';

void elIsolate(SendPort sPort) {
  print(" 新的 Isolate 被創建 ");
  IsolateChannel channel = IsolateChannel.connectSend(sPort);
  // 監聽根 Isolate 發送過來的訊息
```

```
  channel.stream.listen((data) {
    print('新的 Isolate 收到訊息：$data');
  });
  // 向根 Isolate 發送訊息
  channel.sink.add("hi");
}
void main() async {
  ReceivePort rPort = ReceivePort();
  IsolateChannel channel = IsolateChannel.connectReceive(rPort);
// 監聽新的 Isolate 發送過來的訊息
  channel.stream.listen((data) {
    print('根 Isolate 收到訊息：$data');
    // 向新的 Isolate 發送訊息
    channel.sink.add('How are you');
  });

  await Isolate.spawn(elIsolate, rPort.sendPort);
}
```

Chapter 12

拓展閱讀

12.1 可呼叫類別

為了允許像函數一樣呼叫 Dart 類別的實例，需實現 call() 方法。

在 WannabeFunction 類別定義了一個 call() 方法，該函數接收 3 個字串並將它們串聯起來，每個字串之間用空格分隔，並附加一個驚嘆號。範例程式如下：

```
//chapter12/bin/example_01.dart
// 宣告類別
class WannabeFunction{
   // 實現 call 方法
   String call(String a, String b, String c) => '$a $b $c!';
}
main(){
   // 創建類別的實例
   var wf = WannabeFunction();
   // 像方法一樣呼叫類別
   var out = wf('Hi', 'there,', 'gang');
   print(out);
}
```

12.2 擴充方法

當使用其他人的 API 或被廣泛使用的函數庫時，更改 API 通常是不切實際或不可能的。但是，如果仍想增加一些功能，那麼擴充方法將是一種途徑。擴充方法允許對已有 API 進行覆寫或拓展。擴充的成員可以是方法、getters、setters、運算子，擴充還可以具有靜態欄位和靜態幫助方法。

擴充方法的宣告語法如下：

```
extension <extension name> on <type>{
  (<member definition>)*
}
```

例如：可以對 String 類別進行擴充，覆寫其原 API 中的 parseInt() 方法，並拓展一個返回重複一次值的方法 doubleWrite()。

擴充方法宣告的範例程式如下：

```
// 對內建類型 String 進行擴充
extension parsingString on String {
  int parseInt() {
    // 在解析的值的基礎上加 1
    return int.parse(this)+1;
  }
  String doubleIt(){
    // 複寫原值
    return this+this;
  }
}
```

將上述程式保持在 string_apis.dart 檔案中，假設現在使用它的程式與其在同一目錄下，匯入並使用。範例程式如下：

```
//chapter12/bin/example_02.dart
// 匯入擴充函數庫
import 'string_apis.dart';
main(){
```

```
    // 使用擴充方法 parseInt()，該方法覆寫原有 API
    print('${'12'.parseInt()}');
    // 使用擴充方法 doubleIt()，該方法覆寫原有 API
    print('${'go'.doubleIt()}');
}
```

在使用擴充方法時，不能對類型為 dynamic 的變數呼叫控制方法，從而導致異常的範例程式如下：

```
import 'string_apis.dart';
main(){
    dynamic d = '2';
    // 執行時期異常：NoSuchMethodError
    print(d.parseInt());
}
```

擴充方法可以和類型推斷一起使用。範例程式如下：

```
//chapter12/bin/example_03.dart
import 'string_apis.dart';
main(){
    var v = '12';
    // 類型推斷
    print(v.parseInt());
}
```

使用時也可以使用副檔名顯性呼叫。範例程式如下：

```
//chapter12/bin/example_04.dart
import 'string_apis.dart';
main(){
    // 透過副檔名顯性呼叫
    print(parsingString('Go').doubleIt());
}
```

擴充方法還可以對支持泛型的類別進行擴充。

12.3 類型定義

在 Dart 中函數是物件，就像字串和數字是物件一樣。typedef 用於宣告函數類型，它為函數類型提供一個名稱，可以在宣告欄位和返回類型時使用該名稱。當將 typedef 定義的函數類型分配給變數時，會保留類型資訊。

不使用 typedef 宣告的函數類型，使用範例程式如下：

```
//chapter12/bin/example_05.dart
class SortedCollection{
   Function compare;

   SortedCollection(int f(Object a, Object b)) {
      compare = f;
   }
}
int sort(Object a, Object b) => 0;
void main(){
   SortedCollection coll = SortedCollection(sort);
   // 我們知道 compare 是一個函數
   // 但是它的函數類型是宣告
   print(coll.compare is Function);
}
```

分配 f 給 compare 時，類型資訊會遺失。參數 f 的類型是 (Object，Object) → int，而 compare 的類型是 Function。如果為函數提供類型定義，並保留類型資訊，則開發人員和工具都可以使用該資訊。範例程式如下：

```
//chapter12/bin/example_06.dart
// 定義類型名稱
typedef Compare = int Function(Object a, Object b);
class SortedCollection {
   Compare compare;
   SortedCollection(this.compare);
```

```
}
int sort(Object a, Object b) => 0;

void main() {
  SortedCollection coll = SortedCollection(sort);
  print(coll.compare is Function);
  // 判斷是不是 Compare 類型
  print(coll.compare is Compare);
}
```

因為 typedef 只是定義類型名稱，所以它們提供了一種檢查任何函數類型的方法。範例程式如下：

```
//chapter12/bin/example_07.dart
// 宣告函數別名且可提供泛型參數
typedef Compare<T> = int Function(T a, T b);

int sort(int a, int b) => a - b;

void main() {
  // 判斷 sort 是否是 Compare<int> 類型的函數
  print(sort is Compare<int>);
}
```

12.4 中繼資料

使用中繼資料提供有關程式的附加資訊。中繼資料註釋以字元 @ 開頭，跟著字元 @ 的是對編譯時常數的引用 (例如 deprecated) 或對常數建構函數的呼叫。

所有 Dart 程式都有兩個註釋：@deprecated 和 @override。有關使用 @override 的範例，參見擴充一個類別。

註釋 @deprecated 表示棄用，使用 @deprecated 註釋的範例程式如下：

```
//chapter12/bin/example_08.dart
class Television {
  // 棄用：使用 [turnOn] 代替
  @deprecated
  void activate() {
    turnOn();
  }
  // 打開電視電源
  void turnOn() {
    // 實現細節
  }
}
```

可以定義自己的中繼資料註釋。定義帶有兩個參數的 @todo 註釋的範例程式如下：

```
//chapter12/bin/todo.dart
library todo;
// 定義註釋
class Todo{
  final String who;
  final String what;
  const Todo(this.who, this.what);
}
```

使用 @todo 註釋的範例程式如下：

```
//chapter12/bin/example_09.dart
// 匯入註釋函數庫
import 'todo.dart';
// 使用註釋
@Todo('jobs', ' 做家務 ')
void doSomething() {
  print(' 做一些事 ');
}
```

中繼資料可以出現在函數庫、類別、類別類型名稱、類型參數、建構函數、工廠、函數、欄位、參數或變數宣告之前，也可以出現在 import 或 export 指令之前。可以在執行時期使用反射來檢索中繼資料。

12.5 註釋

Dart 支持單行註釋、多行註釋、文件註釋。

單行註釋以 "//" 開頭，Dart 編譯器會忽略 "//" 和行尾之間的所有內容。範例程式如下：

```
//chapter12/bin/example_10.dart
void main(){
  // 列印當前時間
  print(DateTime.now());
}
```

多行註釋以 "/*" 開頭，以 "*/" 結尾。Dart 編譯器將忽略 "/*" 和 "*/" 之間的所有內容，多行註釋可以巢狀結構。範例程式如下：

```
//chapter12/bin/example_11.dart
void main(){
  /*
  * 列印當前時間
  * 使用 UTC 時間
  * */
  print(DateTime.now().toUtc());
}
```

文件註釋是以 "///" 或 "/**" 開頭的多行或單行註釋。在連續的行上使用 "///" 與多行文件註釋具有相同的效果。

在文件註釋中，Dart 編譯器將忽略所有文字，除非將其括在中括號中。使用中括號，可以引用類別、方法、欄位、頂級變數、函數和參數。中括號中的名稱在文件化的程式元素的詞法範圍內解析。

參考其他類別和參數的文件註釋範例程式如下：

```
//chapter12/bin/example_12.dart
/// 連接 MySQL 資料庫設定類別
class ConnectionSettings{
```

```
  //...
}
/// 連接 MySQL 資料庫的類別
class MySQLConnection{
  /// 關閉資料庫連接
  Future close(){
    //...
  }
  /// 連接指定通訊埠上的 MySQL 伺服器
  ///[ConnectionSettings] 類別包含用戶名、密碼、資料庫等資訊
  static Future<MySQLConnection> connect(ConnectionSettings c) async{
    //...
  }
  /// 為 [values] 中每組 sql 參數執行多次 [sql] 敘述
  Future<List<Object>> queryMultiString(String sql,Iterable<List<Object>>
values){
    //...
  }
}
```

在生成的文件中，[ConnectionSettings] 成為指向 ConnectionSettings 類別的 API 文件的連結。

要解析 Dart 程式並生成 HTML 檔案，可以使用 SDK 的文件生成工具。

Chapter 13

服務端開發

13.1 HTTP 請求與回應

13.1.1 服務端

HttpServer 是使用 HTTP 傳輸內容的服務端物件。使用 HttpServer 類別的靜態方法 bind 創建 HttpServer 物件，使用時需要向該方法傳遞主機和通訊埠兩個必選參數，該 HttpServer 物件就綁定到了對應的主機和通訊埠。範例程式如下：

```
import 'dart:io';
main()async{
  // 創建 HttpServer 實例，並綁定主機位址和通訊埠
  var server = await HttpServer.bind('localhost',80);
}
```

方法 bind 的第一個參數是主機名稱，可以是特定的主機名稱或字串。也可以使用 InternetAddress 類別提供的預先定義值指定主機：

(1) loopbackIPv6：使用位址 localhost，IP 版本為 6。

(2) loopbackIPv4：使用位址 localhost，IP 版本為 4。
(3) anyIPv6：將自動分配 IP 位址，IP 版本為 6。
(4) anyIPv4：將自動分配 IP 位址，IP 版本為 4。

方法 bind 的第二個參數是 int 類型的通訊埠名稱，該通訊埠是該服務在電腦上的唯一標識。1024 以下是為系統標準服務保留的通訊埠，HTTP 的服務通訊埠為 80，也可以指定為 1024 及以上的通訊埠編號。

HttpServer 是提供 HttpRequest 物件的 Stream，因此可以監聽。監聽函數會接收一個回呼函數，回呼函數的參數是 HttpRequest 物件。可以透過 HttpRequest 物件獲取請求相關的資訊。範例程式如下：

```
// 監聽請求串流
server.listen((HttpRequest request){
   // 列印請求位址
   print('請求位址 ${request.requestedUri}');
   // 寫入回應資訊
   request.response.write('存取成功！');
   // 關閉當前請求的回應物件
   request.response.close();
});
```

HttpRequest 物件包含一些有用的屬性：

(1) method：請求方法。可能的值：GET、POST、PUT、DELETE 等。
(2) uri：請求位址。一個 Uri 物件，包含請求相關的主機、通訊埠、查詢字串等資訊。
(3) headers：請求的標頭。HttpHeaders 物件包含內容類別型、內容長度、日期等資訊。
(4) response：返回與 HttpRequest 連結的 HttpResponse 物件，伺服器利用該物件回應請求。

使用 method 屬性獲取請求方法名稱，並針對對應方法處理相關交易。範例程式如下：

```
if(request.method == 'GET'){
    // 處理 GET 請求
}else if(request.method == 'POST'){
    // 處理 POST 請求
}
```

屬性 uri 的最大作用是從請求位址中獲取查詢參數及少量資料，常用於處理 GET 請求。範例程式如下：

```
// 獲取所有查詢參數及值組成的 Map 物件
// 常在處理 GET 請求中使用
var queryMap = request.uri.queryParameters;
// 獲取指定參數名稱對應的值
var name = queryMap['name'];
```

使用 headers 屬性獲取 HttpHeaders 物件，常用該物件獲取請求內容類別型。範例程式如下：

```
// 請求的內容類別型
var contentType = request.headers.contentType;
```

HttpResponse 物件包含針對回應的一些屬性和方法：

(1) encoding：回應資料採用的編碼，常用的編碼方式是 JSON 和 UTF-8。
(2) headers：響應的標頭，它返回一個 HttpHeaders 物件。
(3) statusCode：回應狀態碼，用於指示請求成功或失敗。
(4) write(Object obj)：將 obj 轉化為字串，並寫入回應資料中。
(5) writeln([Object obj = ""])：將 obj 轉化為字串，並在響應資料新的一行中寫入該字串。
(6) writeAll(Iterable objects, [String separator = ""])：遍歷可疊代物件 objects 並按順序寫入回應資料中，可以為 objects 中的元素間指定分隔符號 separator。
(7) addStream(Stream<List<int>> stream)：將串流增加到回應資料中，返回 Future 物件。

回應狀態碼是由 HttpStatus 物件提供的，它預先定義了一部分狀態碼。常用的有以下常數：

(1) HttpStatus.ok：表示請求成功。

(2) HttpStatus.notFound：表示不存在該頁面。

(3) HttpStatus.methodNotAllowed：表示不回應該方法。

設定回應碼和資料的範例程式如下：

```
// 回應 GET 請求
// 設定回應狀態碼
request.response.statusCode= HttpStatus.ok;
// 向回應物件 HttpResponse 主體寫入資料
request.response.write('GET 請求成功！');
```

除了處理請求位址上的查詢參數，有時還需要處理請求主體中的資料。HttpRequest 的主體資料是 List<int> 類型的 Stream，可以使用 Utf8Decoder 的 bind 方法將其轉為 String 類型的 Stream。POST 請求對於可以發送的資料量沒有限制，並且可以分區塊傳輸，Stream 的 join 方法可以將這些區塊拼湊在一起。範例程式如下：

```
// 處理 POST 請求
// 對請求解碼
var content = await utf8.decoder.bind(request).join();
// 將字串解析為 json 資料
var data = jsonDecode(content) as Map;
// 疊代 json 資料
data.forEach((k,v){
  print('鍵：$k, 值：$v');
});
```

在請求使用 UTF-8 編碼格式時，才可以使用 Utf8Decoder 解碼請求主體資料。

如果當前是在本地執行服務端和用戶端程式，則它們的存取與回應不受影響。在實際部署應用時，可能需要接收不同位址或不同通訊埠的請求，此時就需要設定跨域參數。範例程式如下：

```
//跨域請求設定
void addCorsHeaders(HttpResponse response){
  //允許的來源，即主機位址
  response.headers.add('Access-Control-Allow-Origin','*');
  //允許請求使用的方法
  response.headers.add('Access-Control-Allow-Methods',
    'GET,POST,OPTIONS');
  //允許請求使用的標頭
  response.headers.add('Access-Control-Allow-Headers',
    'Origin,X-Requested-With,Content-Type,Accept');
}
```

服務端程式需要與瀏覽器或用戶端配合使用，此處只提供了接收和回應請求的服務，發出請求和處理回應將使用用戶端來完成。服務端完整程式如下：

```
//chapter13/bin/http_server.dart
import 'dart:io';
import 'dart:convert';

main()async{
  //創建HttpServer實例並綁定主機位址和通訊埠
  var server = await HttpServer.bind('localhost',80);
  //監聽請求串流
  server.listen((HttpRequest request)async{
    //快取請求的內容類別型contentType
    var contentType = request.headers.contentType;
    //根據請求方法名稱進入不同的處理程式區塊
    if(request.method == 'GET'){
      await handleGet(request);
    }else if(request.method == 'POST'
        && contentType?.mimeType == 'application/json'){
      await handlePost(request);
    }else{
      request.response.statusCode = HttpStatus.methodNotAllowed;
      request.response.write('${request.method}請求不被服務端支援！');
    }
```

```
        // 關閉當前請求的回應物件
        await request.response.close();
    });
}
// 處理 GET 請求的實現程式
void handleGet(HttpRequest request)async{
    // 應用跨域請求設定
    addCorsHeaders(request.response);
    try{
        // 處理 GET 請求
        // 獲取所有查詢參數及值組成的 Map 物件
        // 常在處理 GET 請求中使用
        var queryMap = request.uri.queryParameters;
        // 獲取指定參數名稱對應的值
        var name = queryMap['name'];
        // 列印參數 name 的值
        if(name != null) print('參數 name 的值：$name');
        // 回應 GET 請求
        request.response.statusCode= HttpStatus.ok;
        request.response.write('GET 請求成功！');
    }catch(e){
        request.response.statusCode= HttpStatus.internalServerError;
        request.response.write('GET 請求異常：$e');
    }
}
// 處理 POST 請求的實現程式
void handlePost(HttpRequest request)async{
    // 應用跨域請求設定
    addCorsHeaders(request.response);
    try{
        // 處理 POST 請求
        // 對請求解碼
        var content = await utf8.decoder.bind(request).join();
        // 將字串解析為 json 資料
        var data = jsonDecode(content) as Map;
        // 疊代 json 資料
        data.forEach((k,v){
            print('鍵：$k, 值：$v');
        });
```

```
      // 回應 POST 請求
      // 使用串聯運算符號設定物件的多個值
      request.response
        ..statusCode= HttpStatus.ok
        ..write('POST 請求成功！')
        ..write(' 這是新的一行 ');
    }catch(e){
      request.response.statusCode= HttpStatus.internalServerError;
      request.response.write('POST 請求異常：$e');
    }
}

// 跨域請求設定
void addCorsHeaders(HttpResponse response){
    // 允許的來源，即主機位址
    response.headers.add('Access-Control-Allow-Origin','*');
    // 允許的方法
    response.headers.add('Access-Control-Allow-Methods',
      'GET,POST,OPTIONS');
    // 允許的標頭
    response.headers.add('Access-Control-Allow-Headers',
      'Origin,X-Requested-With,Content-Type,Accept');
}
```

13.1.2 用戶端

HttpClient 是採用 HTTP 向伺服器發出請求並接收回應資料的用戶端。它可以透過一系列方法將 HttpClientRequest 物件發送到伺服器，並從伺服器接收 HttpClientResponse 物件。

創建用戶端實例：

```
// 創建 HttpClient 實例
var client = HttpClient();
```

當需要關閉用戶端實例時可以呼叫 close 方法，它有一個可選參數 force，該參數表示是否強制關閉用戶端。其預設值為 false，表示在完成所有連接後關閉用戶端。當值為 true 時，將關閉所有連接並立即釋放資源。這些被關閉的連接將收到一個錯誤事件，以指示用戶端已關閉。呼叫範例程式如下：

```
// 完成所有與伺服器的連接後關閉用戶端
client.close();
// 立即關閉用戶端並釋放資源
client.close(force:true);
```

HttpClient 定義了打開 HTTP 連接的常用方法，它們都返回一個 HttpClientRequest 並包裝在 Future 物件中。HttpClientRequest 物件可透過 headers 屬性返回的 HttpHeaders 設定請求標頭資訊，也可以透過繼承的 write、writeln、writeAll 方法向請求主體寫入資料。

常用方法如下：

(1) get(String host, int port, String path)：使用 GET 方法打開 HTTP 連接。
(2) post(String host, int port, String path)：使用 POST 方法打開 HTTP 連接。
(3) put(String host, int port, String path)：使用 PUT 方法打開 HTTP 連接。
(4) delete(String host, int port, String path)：使用 DELETE 方法打開 HTTP 連接。

透過 get 方法發出 GET 請求，需要向該方法指定伺服器主機、通訊埠編號和路徑，路徑中可能包含查詢字串。該方法返回 HttpClientRequest 物件。範例程式如下：

```
//GET 請求
var reqGet = await client.get(_host,80,'?name=jobs');
```

呼叫 HttpClientRequest 物件的 close 方法會返回一個 HttpClientResponse 物件，因為它是包裝在 Future 中的，所以可以呼叫 then 方法並提供回呼函數。HttpClientResponse 的主體來自伺服器的資料流程，因此可以使用解碼器解析回應主體資訊。範例程式如下：

```
// 呼叫 close 方法以在完成連接後獲取 HttpClientResponse 物件
await reqGet.close().then((response){
   // 使用 utf8.decoder 轉換資料，將 List<int> 轉為 String
   // 因為它們是串流，所以可以透過監聽獲取轉換後的資料
   response.transform(utf8.decoder).listen((contents) {
      // 處理資料
      print(contents);
   });
});
```

使用 post 方法發出 POST 請求，其參數與 get 方法一致。指定標頭的 contentType 為 json 格式，並透過 write 方法寫入 json 資料。範例程式如下：

```
//POST 請求
var reqPost = await client.post(_host,80,'')
   // 指定內容類別型為 json
   ..headers.contentType = ContentType.json
   // 寫入資料
   ..write(jsonEncode(jsonData));
```

使用 delete 發起一個 DELETE 請求，因為該請求不被伺服器支持，所以請求會失敗。可以透過 HttpClientResponse 的 statusCode 得到請求狀態碼以判斷是否成功。範例程式如下：

```
// 發出 DELETE 請求
var reqDelete = await client.delete(_host,80,'');
await reqDelete.close().then((response){
   // 判斷請求是否成功
   if(response.statusCode != HttpStatus.ok){
      print('DELETE 請求失敗 ');
   }
});
```

用戶端完整程式如下：

```
//chapter13/bin/http_client.dart
import 'dart:io';
import 'dart:convert';
```

```
// 主機名稱等於 localhost
String _host = InternetAddress.loopbackIPv6.host;

//json 資料
Map jsonData = {
   'city': ' 上海 ',
   'area': '16800 平方公里 ',
   'population': '1600 萬 '
};

Future main() async{
   // 創建 HttpClient 實例
   var client = HttpClient();

   // 發出 GET 請求
   var reqGet = await client.get(_host,80,'?name=jobs');
   // 呼叫 close 方法以在完成連接後獲得 HttpClientResponse 物件
   await reqGet.close().then((response){
      // 使用 utf8.decoder 轉換資料，將 List<int> 轉為 String
      // 因為它們是串流，所以可以透過監聽獲取轉換後的資料
      response.transform(utf8.decoder).listen((contents) {
         // 處理資料
         print(contents);
      });
   });

   // 發出 POST 請求
   var reqPost = await client.post(_host,80,'')
      // 指定內容類別型為 json
      ..headers.contentType = ContentType.json
      // 寫入資料
      ..write(jsonEncode(jsonData));
   await reqPost.close().then((response){
      response.transform(utf8.decoder).listen((contents) {
         print(contents);
      });
   });

   // 發出 DELETE 請求
```

```
    var reqDelete = await client.delete(_host,80,'');
    await reqDelete.close().then((response){
       // 判斷請求是否成功
       if(response.statusCode != HttpStatus.ok){
          print('DELETE 請求失敗 ');
          }
          response.transform(utf8.decoder).listen((contents) {
             print(contents);
       });
    });
}
```

到此已經編寫了服務端和用戶端的程式，首先執行服務端程式以啟動服務，然後執行用戶端程式，並觀察主控台列印資訊。

13.2 shelf 框架

shelf 框架使得創建和編寫 Web 服務的各個部分變得容易。

(1) 公開一小組簡單類型。

(2) 將服務邏輯映射為一個簡單函數：擁有單一參數 request 和返回值 response。

(3) 輕鬆混合並匹配同步和非同步處理。

(4) 靈活地返回具有相同模型的簡單字串或位元組流。

簡單範例程式如下：

```
//chapter13/bin/shelf_basic_server.dart
import 'package:shelf/shelf.dart';
import 'package:shelf/shelf_io.dart' as io;
void main(){
   // 創建處理常式
   Response echoRequest(Request request) {
      // 返回請求 URL 和 method 作為回應
```

```
    return Response.ok(' 請求 URL:${request.URL}, 請求方法：${request.method}');
  }
  // 創建介面卡實例
  var server = io.serve(echoRequest,'localhost',1024);
  // 成功得到實例後列印服務主機位址與通訊埠
  server.then((server){
    print(' 服務位址 http://${server.address.host}:${server.port}');
  });
}
```

在上述程式中涉及許多重要概念：處理常式、中介軟體、介面卡。

13.2.1 處理常式

處理常式 handler 是可以處理 Request 物件並返回 Response 物件的函數。範例程式如下：

```
// 創建處理常式
Response echoRequest(Request request) {
  // 返回請求位址作為回應
  return Response.ok(' 請求位址 :${request.URL}');
}
```

該處理常式將請求的 URL 字串作為回應並返回到請求的發出者。

Request 物件常用屬性和方法：

(1) encoding：訊息文字的編碼。
(2) headers：HTTP 標頭。
(3) method：HTTP 請求方法，例如：GET、POST 等。
(4) requestedUri：請求的原始 Uri。
(5) handlerPath：當前處理常式的 URL 路徑，與 URL 組成 requestedUri。
(6) URL：當前處理常式到請求資源的 URL 路徑及查詢參數，相對於 handlerPath。

(7) change({Map<String, Object> headers, Map<String, Object> context, String path, dynamic body})：透過複製當前 Request 並應用指定的更改來創建新的 Request。
(8) read ()：返回表示正文的 Stream，類型為 Stream<List<int>>。
(9) readAsString([Encoding encoding])：將正文作為字串的 Future。

Response 物件常用屬性和方法：

(1) statusCode：回應的 HTTP 狀態碼。
(2) headers：HTTP 標頭。
(3) encoding：訊息文字的編碼。
(4) read()：返回表示訊息文字的 Stream，類型為 Stream<List<int>>。
(5) readAsString([Encoding encoding])：將正文作為字串的 Future。
(6) Response(int statusCode, {dynamic body, Map<String, Object>headers, Encoding encoding, Map<String, Object> context})：使用指定的 status Code 創建一個 HTTP 響應。
(7) Response.ok(dynamic body, {Map<String, Object> headers, Encoding encoding, Map<String, Object> context})：創建一個狀態碼為 200 的成功回應。
(8) Response.notFound(dynamic body, {Map<String, Object> headers, Encoding encoding, Map<String, Object> context})：創建一個狀態碼為 404 的未找到響應。
(9) Response.internalServerError({dynamic body, Map<String, Object> headers, Encoding encoding, Map<String, Object> context})：創建一個狀態碼為 500 的內部伺服器錯誤響應。
(10) change({Map<String, Object> headers, Map<String, Object> context, dynamic body})：透過複製當前 Response 並應用指定的更改來創建新的 Response。

13.2.2 介面卡

介面卡可以創建 Request 物件，將它們傳遞給處理常式，並處理由處理常式生成的 Response 物件。在大多數情況下，介面卡轉發來自底層 HTTP 伺服器的請求，並且轉發及回應到底層 HTTP 伺服器。shelf_io.serve 就是這種介面卡。介面卡也可能在瀏覽器中使用 window.location 和 window.history 合成 HTTP 請求，或可以直接從 HTTP 用戶端將請求傳遞到 Shelf 處理常式。

serve(Handler handler, dynamic address, int port,{SecurityContext securityContext, int backlog, bool shared: false}) → Future<HttpServer>：啟動一個 HttpServer，它監聽指定的位址和通訊埠，並將請求發送到處理常式。

在使用 serve 方法時處理常式可以是單獨的處理常式，也可以是透過 Pipeline 類別的實例返回的處理常式，在這裡使用的是單獨的處理常式。

範例程式如下：

```
io.serve(echoRequest,'localhost',1024);
```

執行程式，當服務啟動成功後會在主控台列印服務位址和通訊埠。

```
服務位址 http://localhost:1024
```

編寫用戶端程式，程式使用 GET 請求存取指定主機位址和通訊埠的服務。範例程式如下：

```
//chapter13/bin/shelf_basic_client.dart
import 'dart:io';
import 'dart:convert';

Future main() async{
  // 創建 HttpClient 實例
  var client = HttpClient();
  // 發出 GET 請求
```

```
    var reqGet = await client.get('localhost',1024, '?msg=hello');
    // 呼叫 close 方法以在完成連接後獲得 HttpClientResponse 物件
    await reqGet.close().then((response){
        // 使用 utf8.decoder 轉換資料，將 List<int> 轉為 String
        // 因為它們是串流，所以可以透過監聽獲取轉換後的資料
        response.transform(utf8.decoder).listen((contents){
            // 處理資料
                print(contents);
        });
    });
}
```

執行用戶端程式，可以觀察到服務端返回的回應。

```
請求 URL:?msg=hello, 請求方法：GET
```

13.2.3 中介軟體

處理常式也可以進行部分處理，然後將請求轉發給另一個處理常式，這種處理常式被稱為中介軟體（middleware），因為它位於服務堆疊的中間。可以把中介軟體視作傳入一個 handler，然後返回一個新 handler 的函數。示意如下：

```
Handler Middleware(Handler innerHandler);
```

一個 shelf 應用通常由多層中介軟體組成，每個中介軟體中又有一個或多個處理常式。shelf.Pipeline 類別使這類應用程式易於建構。

Pipeline 是一個幫助程式，可以將一組中介軟體和一個處理常式很容易地組合在一起。

(1) addMiddleware(Middleware middleware)：將中介軟體增加到中介軟體組中，並返回一個新的 Pipeline 物件。中介軟體組中的最後一個中介軟體將是處理請求的最後一個中介軟體，且是處理回應的第一個中介軟體。

(2) addHandler(Handler handler)：如果 Pipeline 中的所有中介軟體均已透過請求，將傳入的 handler 作為請求的最終處理常式，並返回一個的新的處理常式。一個 shelf 應用程式只能有一個處理常式，因此此方法只能呼叫一次。

使用範例程式如下：

```
var handler = Pipeline()
    // 增加中介軟體
    .addMiddleware(logRequests())
    // 增加處理常式
    .addHandler(echoRequest);
// 創建介面卡實例
var server = io.serve(handler,'localhost',1024);
```

logRequests() 方法是 shelf 框架內建的方法，它的作用是列印請求的時間、內部處理常式所用時間、請求方法、回應狀態碼和請求 Uri。它將返回一個 Middleware 物件，因此可以作為中介軟體在應用程式中使用。使用 addHandler 方法增加處理常式 echoRequest，該方法會返回一個新的處理常式，可以直接將返回值用作參數傳遞到介面卡。

執行此程式並執行用戶端程式將可以在主控台看到 logRequests 方法列印的日誌資訊。

```
服務位址 http://localhost:1024
2020-07-26T15:23:43.901771  0:00:00.014704 GET    [200]/?msg=hello
```

可以使用 createMiddleware 方法創建中介軟體，該方法包含 3 個可選參數：

(1) requestHandler：請求處理常式，接收 Request 物件，返回 Response 物件的函數。該處理常式用於對 Request 物件進行進一步處理，如果需要繼續傳遞 Request 物件，則返回 null。不然正常返回 Response 物件。
(2) responseHandler：回應處理常式，接收 Response 物件，返回 Response 物件的函數。它的作用是對 Response 物件進行進一步處理。

(3) errorHandler：錯誤處理常式，接收錯誤參數 error 和堆疊追蹤參數 StackTrace，返回 Response 物件的函數。在程式執行出錯時對錯誤進行處理。

通常來說只會向中介軟體傳遞一個參數，因為部分中介軟體只對請求進行處理，部分只對回應進行處理。這裡為了演示創建和使用的中介軟體提供了所有參數，在實際使用中需要傳遞所有參數的中介軟體極少。範例程式如下：

```
// 創建中介軟體
var middleware = createMiddleware(requestHandler: reqHandler,
   responseHandler: resHandler,errorHandler: errHandler);
// 請求處理常式

// 如果希望直接回應請求則透過回應資訊
Response reqHandler(Request request){
   print(' 請求處理常式請求 Uri:${request.requestedUri}');
   // 返回 null 表示使 request 繼續傳遞到下一個中介軟體或處理常式
   // 否則將直接回應請求
   return null;
}
// 回應處理常式
Response resHandler(Response response){
   print(' 回應處理常式回應碼 :${response.statusCode}');
   return response;
}
// 錯誤處理常式
Response errHandler(error,StackTrace st){
   if (error is HijackException) throw error;
   print(' 錯誤處理常式：$error');
   throw error;
}
```

完整範例程式如下：

```
//chapter13/bin/shelf_pipeline_server.dart
import 'package:shelf/shelf.dart';
import 'package:shelf/shelf_io.dart' as io;
```

```
void main(){
  var handler = Pipeline()
    // 增加中介軟體
    .addMiddleware(logRequests())
    // 增加自訂中介軟體
    .addMiddleware(middleware)
    // 增加處理常式
    .addHandler(echoRequest);

  // 創建介面卡實例
  var server = io.serve(handler,'localhost',1024);
  // 成功得到實例後列印服務主機位址與通訊埠
  server.then((server){
    print(' 服務位址 http://${server.address.host}:${server.port}');
  });
}
// 創建處理常式
Response echoRequest(Request request){
  // 返回請求 URL 和 method 作為回應
  return Response.ok(' 請求 URL:${request.URL}, 請求方法：${request.method}');
}
// 創建中介軟體
var middleware = createMiddleware(requestHandler: reqHandler,
  responseHandler: resHandler,errorHandler: errHandler);
// 請求處理常式

// 如果希望直接回應請求則透過回應資訊
Response reqHandler(Request request){
  print(' 請求處理常式，請求 Uri:${request.requestedUri}');
  // 返回 null 表示使 request 繼續傳遞到下一個中介軟體或處理常式
  // 否則將直接回應請求
  return null;
}
// 回應處理常式
Response resHandler(Response response){
  print(' 回應處理常式，回應碼 :${response.statusCode}');
  return response;
}
// 錯誤處理常式
```

```
Response errHandler(error,StackTrace st){
   if (error is HijackException) throw error;
   print('錯誤處理常式：$error');
   throw error;
}
```

執行此服務，並執行用戶端應用。查看服務在主控台的輸出資訊：

```
服務位址 http://localhost:1024
請求處理常式請求 Uri:http://localhost:1024/?msg=hello
回應處理常式回應碼 :200
2020-07-26T15:23:43.901771  0:00:00.014704 GET    [200]/?msg=hello
```

13.3 路由套件

一些中介軟體還可以採用多個處理常式，並為每個請求呼叫一個或多個處理常式。例如：路由器中介軟體可能會根據請求的 Uri 和 HTTP 方法選擇要呼叫的處理常式。shelf_router 套件就提供了路由的功能。

shelf_router 套件可以透過組合請求處理常式，簡化在 Dart 中建構 Web 應用程式的過程。使用 shelf_router 套件前需先增加依賴並執行 pub get 命令：

```
dependencies:
   shelf: ^0.7.7
   shelf_router: ^0.7.2
```

13.3.1 定義路由

該套件有一個 Router 類別，以 HTTP 方法和路由模式將請求匹配到為基礎對應的處理常式。它包含以下方法和屬性：

(1) Router()：無參建構函數，創建一個 Router 物件。

(2) handler → Handler：獲取一個處理常式，該處理常式會將傳入的請求路由傳到已註冊的處理常式。

(3) get(String route,Function handler)：使用 handler 處理到路徑的 GET 請求。

(4) post(String route,Function handler)：使用 handler 處理到路徑的 POST 請求。

(5) put(String route,Function handler)：使用 handler 處理到路徑的 PUT 請求。

(6) delete(String route, Function handler)：使用 handler 處理到路徑的 DELETE 請求。

(7) all(String route, Function handler)：使用 handler 處理到路徑的所有請求。

(8) mount(String prefix, Router router)：在字首 prefix 下附加一個路由器，透過這樣的形式可以將多個路由器物件組合在一起形成完整的 API。

該套件還包含一個頂層函數：

params(Request request, String name)：透過路徑捕捉 URL 中參數的值。

這裡先創建一個 Router 物件，然後使用 get 方法，範例程式如下：

```
//chapter13/bin/shelf_router_basic_server.dart
import 'package:shelf/shelf.dart';
import 'package:shelf/shelf_io.dart' as io;
import 'package:shelf_router/shelf_router.dart';

main(){
  // 創建 Router 物件
  var app = Router();
  // 使用 Router 類別的 get 方法創建可以捕捉路徑 /hello 上的 GET 請求
  // 這裡的處理常式是一個匿名函數
  app.get('/hello',(Request request){
    return Response.ok('Hello World');
```

```
    });
    // 透過 shelf_io.serve 方法啟動一個 HttpServer
    var server = io.serve(app.handler, 'localhost', 1024);
    // 成功得到實例後列印服務主機位址與通訊埠
    server.then((server){
      print('服務位址 http://${server.address.host}:${server.port}');
    });
}
```

啟動服務，並創建用戶端請求到路徑 /hello 下的 GET 方法。範例程式如下：

```
//chapter13/bin/shelf_router_basic_client.dart
import 'dart:io';
import 'dart:convert';

Future main() async{
    // 創建 HttpClient 實例
    var client = HttpClient();
    // 發出 GET 請求
    var reqGet = await client.get('localhost',1024,'/hello');
    // 呼叫 close 方法以在完成連接後獲得 HttpClientResponse 物件
    await reqGet.close().then((response){
      // 使用 utf8.decoder 轉換資料，將 List<int> 轉為 String
      // 因為它們是串流，所以可以透過監聽獲取轉換後的資料
      response.transform(utf8.decoder).listen((contents){
        // 處理資料
        print(contents);
      });
    });
}
```

執行用戶端程式結果如下：

```
Hello World
```

此時如果請求到除 /hello 以外的路徑將導致錯誤。

13.3.2 路由參數

在路徑中可以嵌入 URL 參數，其方法是使用中括號將參數名稱包裹起來。參數不僅作為路由匹配的一部分，還可以在處理常式中使用。範例程式如下：

```
// 可以將參數名稱 paramName 放在 <> 中以在路徑中指定參數
app.get('/users/<userName>/whoami',(Request request) async{
  // 可以使用 params(request,param) 讀取匹配參數的值
  var userName = params(request, 'userName');
  return Response.ok(' 你是 ${userName}');
});
```

也可以將 URL 參數傳入處理常式的參數清單中。範例程式如下：

```
// 可以將 URL 參數傳入處理常式
app.get('/users/<userName>/sayhello', (Request request, String userName)
async{
  return Response.ok(' 你好 ${userName}');
});
```

可以同時傳遞多個參數，此時處理常式必須接收所有參數或一個也不接收，且必須按照 URL 參數順序接收。範例程式如下：

```
// 可以在 URL 中指定多個參數
// 處理常式必須按 URL 中出現的順序接收所有參數或一個也不接收
app.get('/users/<userName>/<userId>',(Request request, String userName,String
userId) async{
  return Response.ok(' 你好 ${userName},${userId}');
});
```

可以透過 <paramName|REGEXP> 的形式指定滿足正規表示法 REGEXP 的參數 paramName。如果未指定正規表示法，則預設使用正規表示法 '[^/]+'，它表示匹配除 / 之外的所有字串。範例程式如下：

```
// 可以使用 <paramName|REGEXP> 指定自訂正規表示法，其中 REGEXP 是正規表示法 ( 省略
^ 和 $)
// 如果未指定正規表示法，將使用 '[^ /]'
app.get('/users/<userName>/messages/<msgId|[0-9]+>',(Request request)
```

```
async{
   var userName = params(request, 'userName');
   var msgId = int.parse(params(request, 'msgId'));
   return Response.ok('userName:${userName},msgId:${msgId}');
});
```

正規表示法 "[0-9]+" 表示匹配一串數字。

13.3.3 組合路由

定義一個 Router 物件，然後使用 mount 方法將其安裝到現有 Router 物件中。範例程式如下：

```
// 創建新路由器
var art = Router();
art.get('/<articleId>', (Request request) async {
   var articleId = params(request, 'articleId');
   return Response.ok('article ID:${articleId}');
});
// 將 art 路由器安裝到現有路由器 app 上
// 字首可以使用根路徑 /
app.mount('/article/',art);
```

在實際使用中需要將每個路由器物件封裝在類別中。範例程式如下：

```
//chapter13/bin/shelf_router_server.dart
class App{
   // 返回處理常式
   Handler get handler{
      // 創建 Router 物件
      final app = Router();

      // 可以將參數名稱 paramName 放在 <> 中以在路徑中指定參數
      app.get('/users/<userName>/whoami',(Request request) async{
         // 可以使用 params(request,param) 讀取匹配參數的值
         var userName = params(request, 'userName');
         return Response.ok(' 你是 ${userName}');
      });
```

```
        // 可以將 URL 參數傳入處理常式
        app.get('/users/<userName>/sayhello', (Request request, String userName)
async{
            return Response.ok(' 你好 ${userName}');
        });

        // 可以在 URL 中指定多個參數
        // 處理常式必須按 URL 中出現的順序接收所有參數或一個也不接收
        app.get('/users/<userName>/<userId>',(Request request, String
userName,String userId) async{
            return Response.ok(' 你好 ${userName},${userId}');
        });

        // 可以使用 <paramName|REGEXP> 指定自訂正規表示法，其中 REGEXP 是正規表示法
( 省略 ^ 和 $)
        // 如果未指定正規表示法，將使用 '[^ /]'
        app.get('/users/<userName>/messages/<msgId|[0-9]+>',(Request request)
async{
            var userName = params(request, 'userName');
            var msgId = int.parse(params(request, 'msgId'));
            return Response.ok('userName:${userName},msgId:${msgId}');
        });

        // 將 art 路由器安裝到現有路由器上
        // 字首可以使用根路徑 /
        app.mount('/article/', Art().router);
        // 捕捉所有請求並過濾掉已定義的路由
        app.all('/<ignored|.*>', (Request request) {
            return Response.notFound(' 頁面未發現 ');
        });

        // 返回處理常式，它會將請求路由匹配到已註冊的處理常式上
        return app.handler;
    }
}
class Art{
    // 返回路由器
    Router get router{
        // 創建新路由
```

```
    var art = Router();
    // 捕捉 /article/String 下所有請求
    art.get('/<articleId>', (Request request) async{
      var articleId = params(request, 'articleId');
      return Response.ok(' 文字 ID:${articleId}');
    });

    // 捕捉 /article/ 下所有請求並過濾掉已定義的路由
    art.all('/<ignored|.*>',(Request request){
      return Response.notFound(' 頁面未發現 ');
    });
    return art;
  }
}
```

然後透過 Pipeline 組合中介軟體和 Router 物件返回的 handler。範例程式如下：

```
//chapter13/bin/shelf_router_server.dart
import 'package:shelf/shelf.dart';
import 'package:shelf/shelf_io.dart' as io;
import 'package:shelf_router/shelf_router.dart';

main(){
  // 創建 Router 物件
  var app = App();
  // 透過 shelf_io.serve 方法啟動一個 HttpServer
  var server = io.serve(app.handler, 'localhost', 1024);
  // 成功得到實例後列印服務主機位址與通訊埠
  server.then((server){
    print(' 服務位址 http://${server.address.host}:${server.port}');
  });
}
```

啟動服務後就可以存取路由器中定義的路由了。

13.3.4 路由註釋

還可以透過註釋使用路由，首先增加開發依賴項並執行 pub get 命令：

```
dev_dependencies:
  shelf_router_generator: ^0.7.0+1
```

每個方法都有對應的註釋，這些註釋在 shelf 包中定義，使用時需要依賴開發依賴套件 shelf_router_generator。以下列出常用註釋：

(1) Route.get(String route)：將路徑 route 的 GET 請求與註釋方法匹配。
(2) Route.post(String route)：將路徑 route 的 POST 請求與註釋方法匹配。
(3) Route.put(String route)：將路徑 route 的 PUT 請求與註釋方法匹配。
(4) Route.delete(String route)：將路徑 route 的 DELETE 請求與註釋方法匹配。
(5) Route.mount(String prefix)：將字首 prefix 的請求路由與註釋方法匹配。
(6) Route.all(String route)：將路徑 route 的使用請求與註釋方法匹配。

使用註釋時將一組相關的處理常式封裝在類別中，處理常式需要顯性宣告，函數名稱以底線開頭。在處理常式上使用合適的註釋，並設定路由參數。

使用 part 指令指示路由器生成程式生成的檔案名稱並作為當前函數庫的一部分：

```
part 'shelf_router_generator.g.dart';
```

路由器生成程式會按類別為單位生成返回 Router 物件的函數，可以透過生成的函數獲取當前類別的 Router 物件和處理常式。生成函數的名字遵循一定規律，以 _$ 開頭，中間是封裝處理常式的類別名稱，然後以 Router 結尾。

完整範例程式如下：

```
//chapter13/bin/shelf_router_generator.dart
import 'dart:async' show Future;
import 'package:shelf/shelf.dart';
import 'package:shelf/shelf_io.dart' as io;
import 'package:shelf_router/shelf_router.dart';

// 由 pub run build_runner build 生成的檔案
part 'shelf_router_generator.g.dart';

main() async{
   final service = App();
   // 透過 Pipeline 物件組合中介軟體和單一處理常式
   var handler = Pipeline()
      .addMiddleware(logRequests())
      .addHandler(service.handler);
   // 透過 shelf_io.serve 方法啟動一個 HttpServer
   var server = await io.serve(handler, 'localhost', 1024);
   print('Serving at http://${server.address.host}:${server.port}');
}

class App{
   @Route.get('/users/<userName>/whoami')
   Future<Response> _users(Request request) async{
      var userName = params(request, 'userName');
      return Response.ok('You are ${userName}');
   }

   @Route.get('/users/<userName>/<userId>')
   Future<Response> _usersId(Request request, String userName,String userId)
async {
      return Response.ok('Hello ${userName},${userId}');
   }

   @Route.get('/users/<userName>/messages/<msgId|[0-9]+>')
   Future<Response> _msgId(Request request) async {
      var userName = params(request, 'userName');
      var msgId = int.parse(params(request, 'msgId'));
```

```
      return Response.ok('userName:${userName},msgId:${msgId}');
   }

   @Route.mount('/article/')
   Router get _art => Art().router;

   @Route.all('/<ignored|.*>')
   Future<Response> _notFound(Request request) async{
      return Response.notFound('Page not found');
   }
   // 生成的 _$AppRouter 函數可用於獲取此物件的 handler
   Handler get handler => _$AppRouter(this).handler;
}

class Art{
   @Route.get('/<articleId>')
   Future<Response> _articleId(Request request) async {
      var articleId = params(request, 'articleId');
      return Response.ok('article ID:${articleId}');
   }
   @Route.all('/<ignored|.*>')
   Future<Response> _notFound(Request request)async => Response.notFound
('null');
   // 生成的 _$ArtRouter 函數用於公開此物件的 Router
   Router get router => _$ArtRouter(this);
}
```

因為此時函數還沒有生成，所以需要根據函數名稱生成規律拼湊函數名稱，並透過函數獲得當前類別的 Router 物件或處理常式。

然後在編輯器的命令列中輸入以下命令：

```
pub run build_runner build
```

生成的檔案內容如下：

```
//chapter13/bin/shelf_router_generator.g.dart
//GENERATED CODE - DO NOT MODIFY BY HAND

part of 'shelf_router_generator.dart';
```

```
//*******************************************************************
//ShelfRouterGenerator
//*******************************************************************
// 與路由定義類 App 連結的函數
Router _$AppRouter(App service) {
   final router = Router();
   router.add('GET', r'/users/<userName>/whoami', service._users);
   router.add('GET', r'/users/<userName>/<userId>', service._usersId);
   router.add(
      'GET', r'/users/<userName>/messages/<msgId|[0-9]+>', service._msgId);
   router.mount(r'/article/', service._art);
   router.all(r'/<ignored|.*>', service._notFound);
   return router;
}

// 與路由定義類 Art 連結的函數
Router _$ArtRouter(Art service) {
   final router = Router();
   router.add('GET', r'/<articleId>', service._articleId);
   router.all(r'/<ignored|.*>', service._notFound);
   return router;
}
```

可以從程式中看出這與手動編寫的程式非常相似。

Chapter 14

Angular 基礎

AngularDart 是前端框架 Angular 的 Dart 版本，本文簡稱 Angular。它採用 HTML 和 Dart 建構 Web 端應用程式，並且以套件的形式提供，套件名叫 angular。如果需要成系統的 UI，可以使用材質化元件庫，套件名叫 angular_components。

應用程式啟動後，Angular 框架會接管所有工作，控制顯示視圖和互動。

14.1 初始專案

初始專案是為了快速了解專案重要檔案和一些概念，本節無須掌握，重在概覽專案結構。

創建新專案時，在可選的範本清單中選擇 AngularDartWebApp 創建 Web 用戶端應用程式範例，該範本採用 angular 套件和材質化元件庫 angular_components, 如圖 14-1 所示。

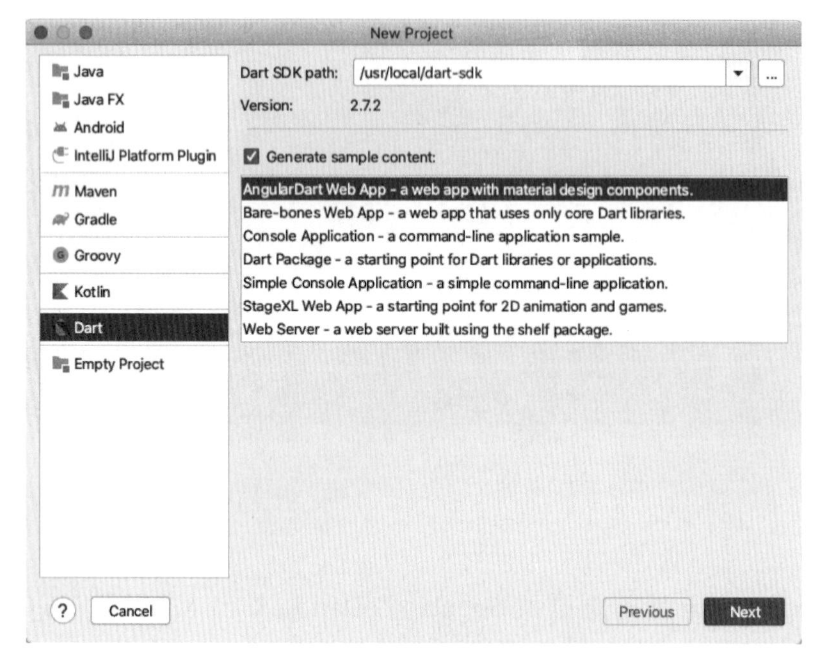

▲ 圖 14-1 應用程式範本

應用程式範本是由 stagehand 套件提供的，主要包括以下範本：

(1) Console Application：命令列應用程式。

(2) Dart Package：Dart 函數庫或應用程式。

(3) Web Server：採用 shelf 框架的 Web 服務端應用程式。

(4) AngularDart Web App：採用 Angular 框架的 Web 用戶端應用程式。

如果在創建專案時可選的範本沒有完全顯示，可以在命令列中執行以下命令：

```
pub global activate stagehand
```

該命令不僅可以獲取還可以更新 stagehand 套件。當命令執行成功後，重新啟動編輯器就可以得到完整的範本清單。

將專案命名為 angulardart，如圖 14-2 所示。該專案主要用於介紹 Angular 應用程式的基本組成要素。

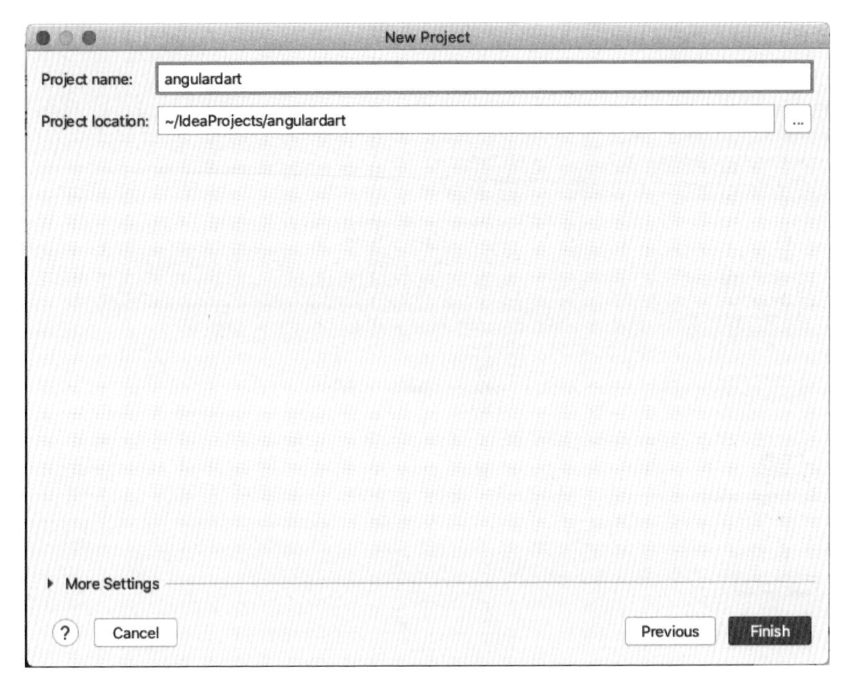

▲ 圖 14-2 專案命名

創建好的專案目錄結構如圖 14-3 所示。

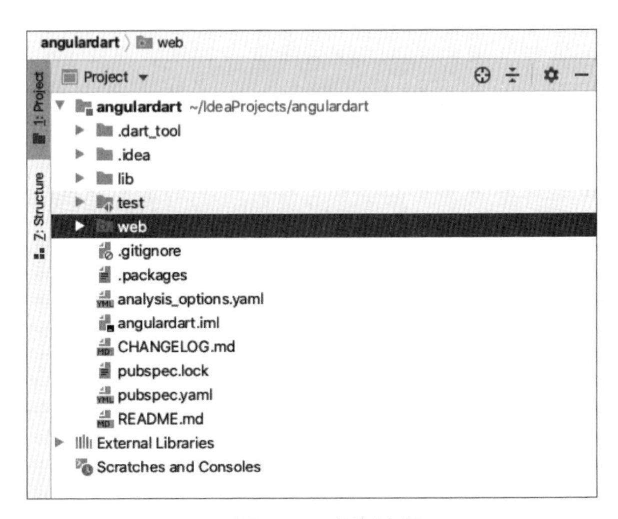

▲ 圖 14-3 專案結構

選中 web 目錄下的 index.html 檔案並點擊，檔案內容如圖 14-4 所示。

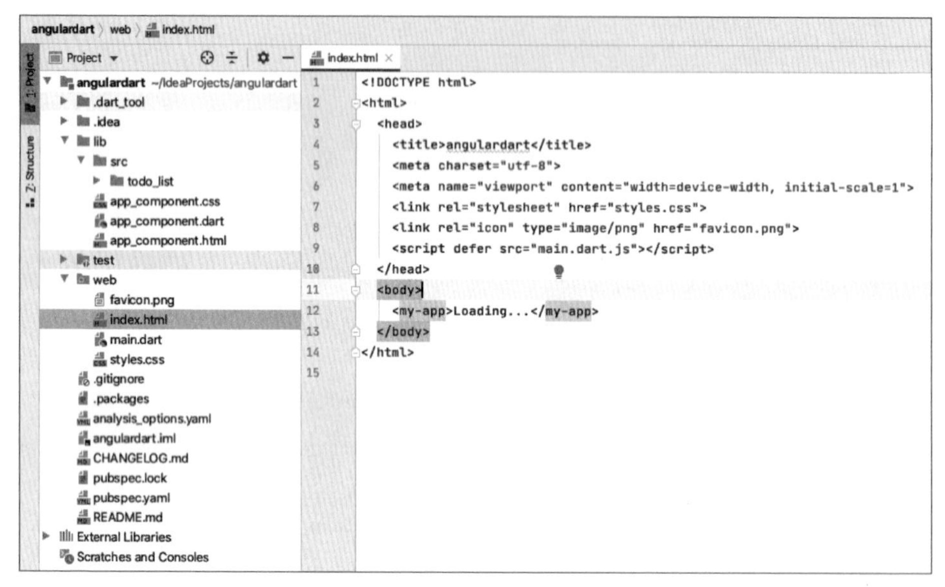

▲ 圖 14-4 首頁

在右邊的視圖中點擊右鍵，選擇 Run 'web/index.html' 選項，專案將自動編譯並執行，如圖 14-5 所示。

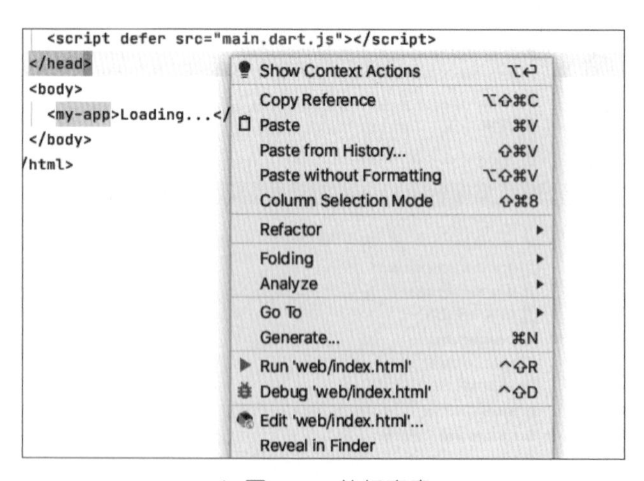

▲ 圖 14-5 執行專案

Web 專案編譯和執行需要使用工具 webdev，它採用了兩種編譯器：dartdevc 和 dart2js。在開發時 webdev 會選擇 dartdevc，因為它支持增量編譯。第一次編譯需要較多時間，之後在修改程式時編譯時間將大大縮短。在部署應用程式時 webdev 會選擇 dart2js，它做了大量針對性最佳化，並且可以保證在所有現代瀏覽器中正常執行。

初始專案執行結果如圖 14-6 所示。

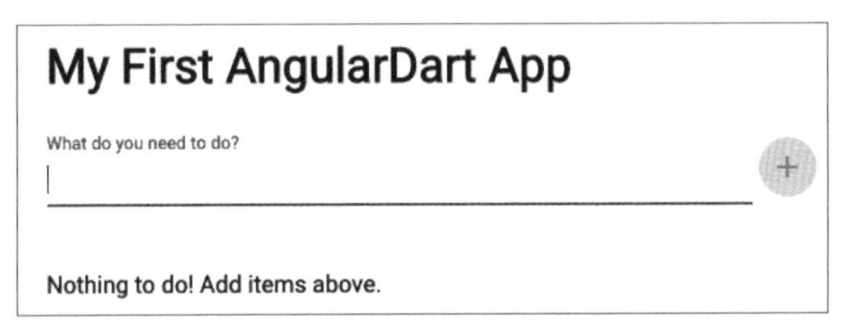

▲ 圖 14-6 執行結果

14.1.1 專案詳情

專案的所有檔案均採用底線命名法。Angular 框架中的重要概念就是元件，元件通常是由 Dart 語言編寫的元件類別檔案、HTML 語言編寫的範本檔案及 CSS 編寫的樣式檔案組成。根據約定這些檔案的名稱必須以 component 作為尾碼，例如：根元件連結的檔案 app_component.dart、app_component.html 和 app_component.css。

專案中絕大多數開發工作將在 lib 目錄下完成，web 目錄則存放應用首頁、全域樣式檔案和包含應用執行起點函數的檔案，在 pubspec.yaml 檔案中存放的是與專案設定相關的資訊。專案的重要目錄展開內容如圖 14-7 所示。

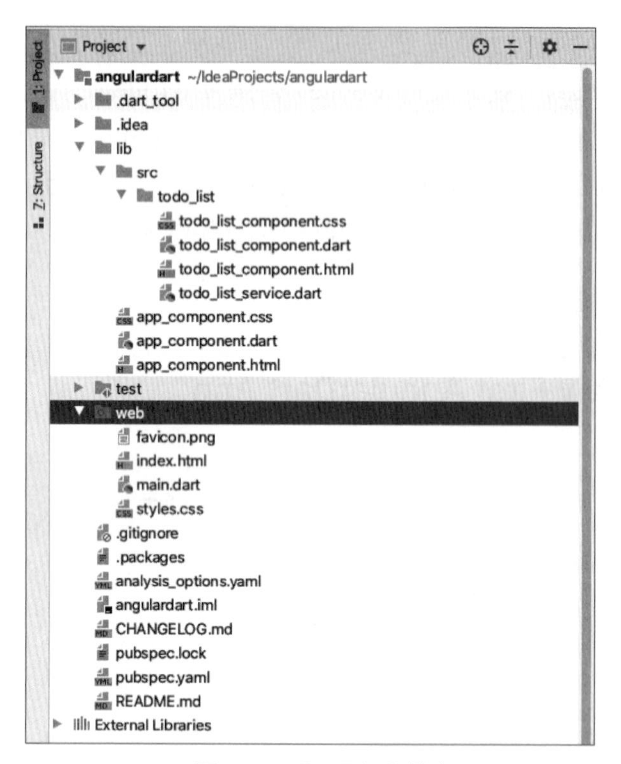

▲ 圖 14-7　重要目錄和檔案

pubspec.yaml 檔案用於指示本專案的專案名稱、平台依賴、函數庫依賴、開發依賴等資訊，可用欄位見章節函數庫。依賴資訊具體內容如下：

```
//chapter14/angulardart/pubspec.yaml
# 專案名稱
name: angulardart
# 描述資訊
description: A web app that uses AngularDart Components
#version: 1.0.0
#homepage: https://www.example.com
# 平台依賴資訊
environment:
sdk: '>=2.7.0 <3.0.0'
# 函數庫依賴資訊
dependencies:
  angular: ^5.3.0
```

```
   angular_components: ^0.13.0
# 開發依賴資訊
dev_dependencies:
   angular_test: ^2.3.0
   build_runner: ^1.6.0
   build_test: ^0.10.8
   build_web_compilers: ^2.3.0
   pedantic: ^1.8.0
   test: ^1.6.0
```

web 目錄下的 index.html 檔案是專案的首頁，用於匯入適用專案全域的樣式檔案、JavaScript 指令檔等內容。程式如下：

```
//chapter14/angulardart/web/index.html
<!DOCTYPE html>
<html>
   <head>
      <title>angulardart</title>
      <meta charset="utf-8">
      <meta name="viewport" content="width=device-width, initial-scale=1">
      <link rel="stylesheet" href="styles.css">
      <link rel="icon" type="image/png" href="favicon.png">
      <script defer src="main.dart.js"></script>
   </head>
   <body>
      <my-app>Loading...</my-app>
   </body>
</html>
```

Angular 會將 Dart 程式編譯成 JavaScript 以確保能在瀏覽器中執行，編譯後的檔案名稱為 main.dart.js。指令稿引用程式如下：

```
<script defer src="main.dart.js"></script>
```

在頁面主體部分插入專案根元件在註釋中提供的選擇器 my-app。

```
<body>
   <my-app>Loading...</my-app>
</body>
```

style.css 是專案全域樣式檔案，可以匯入外部樣式表或定義元素的全域樣式。

```
//chapter14/angulardart/web/style.css
/* 匯入外部樣式表 */
@import URL(https://fonts.googleapis.com/css?family=Roboto);
@import URL(https://fonts.googleapis.com/css?family=Material+Icons);
/* 頁面主體樣式 */
body {
   max-width: 600px;
   margin: 0 auto;
   padding: 5vw;
}
/* 頁面字型 */
* {
   font-family: Roboto, Helvetica, Arial, sans-serif;
}
```

main.dart 檔案中的 main 函數是應用執行的起點，在函數本體內部執行 angular 套件定義的頂層函數 runApp。

runApp <T>(ComponentFactory<T> componentFactory,{InjectorFactory createInjector: _identityInjector})：以 componentFactory 為根啟動一個新的 AngularDart 應用程式。

ComponentFactory 實例是使用 @Component 註釋的類別的實現，例如：example.dart 檔案程式如下：

```
@Component(
   selector: 'example',
   template: '...',
)
class Example{}
```

然後編譯器在 example.template.dart 中生成 ExampleNgFactory，可以透過匯入此生成的檔案來存取。

InjectorFactory 實例是可選的，它是一個函數，用於向根目錄提供服務，例如：路由服務。

main.dart 檔案程式如下：

```
//chapter14/angulardart/web/main.dart
// 匯入 angular 函數庫
import 'package:angular/angular.dart';
// 匯入專案根元件，該檔案是由 AngularDart 編譯器生成的
import 'package:angulardart/app_component.template.dart' as ng;
//main() 是所有 dart 應用程式的執行起點函數
void main(){
   //runApp() 是 Angular 應用程式的啟動函數
   // 將 AppComponent 作為根元件來啟動應用程式
   // 呼叫時採用類別名稱加尾碼 NgFactory
   runApp(ng.AppComponentNgFactory);
}
```

以 *.template.dart 結尾的檔案都是由名為 angular_compiler 的編譯器生成的檔案，該編譯器生成的檔案存放在專案根目錄 .dart_tool/build/generated 目錄下。

lib 目錄是存放專案資源的資料夾，例如：元件、服務、範本、指令、樣式檔案等。在該目錄下首先可以發現根元件的 3 個相關檔案：

app_component.css 是根元件引用的樣式檔案，此時檔案中沒有任何實際作用的內容。

```
//chapter14/angulardart/lib/app_component.css
:host{
}
```

app_component.html 是根元件引用的範本檔案，包含標題標籤和子元件標籤。

```
//chapter14/angulardart/lib/app_component.html
<h1>My First AngularDart App</h1>
<todo-list></todo-list>
```

app_component.dart 檔案中在 AppComponent 類別上使用元件註釋 @Component 以定義根元件。程式如下：

```
//chapter14/angulardart/lib/app_component.dart
import 'package:angular/angular.dart';
import 'src/todo_list/todo_list_component.dart';

@Component(
   selector: 'my-app',
   styleURLs: ['app_component.css'],
   templateURL: 'app_component.html',
   directives: [TodoListComponent],
)
class AppComponent{
}
```

在 lib 目錄下還有一個 src 目錄，所有的實現檔案都應放在該目錄下。src 目錄下還有一個子元件目錄 todo_list，該目錄包含子元件 TodoListComponent 相關的檔案。將與元件相關的檔案放在單獨的資料夾中是非常好的做法。

由於上述內容包含一些未學習內容，因此首先需要刪除子元件目錄 todo_list 及其包含的檔案。

刪除 app_component.html 檔案中的子元件標籤，修改結果如下：

```
<h1>My First AngularDart App</h1>
```

刪除 app_component.dart 檔案中與子元件相關的匯入資訊和註釋資訊，修改後程式如下：

```
//chapter14/angulardart/lib/app_component.dart
import 'package:angular/angular.dart';
// 元件註釋
@Component(
   selector: 'my-app',
)
// 元件類別
class AppComponent{
}
```

14.1.2 元件註釋

AppComponent 是一個普通的類別，因為有了註釋 @Component 而成為 angular 元件，按照約定元件類別的名字以 Component 作為尾碼。註釋 @Component 常用參數如表 14-1 所示。

表 14-1 元件註釋參數

參數	說明
selector	在範本中指定一個 CSS 選擇器，支持 element、[attribute]、.class、:not() 和自訂 HTML 元素
template	元件視圖的內聯範本，HTML 字串
templateURL	元件視圖的外部範本 URL
styles	用於樣式化元件視圖的內聯 CSS 樣式
styleURLs	用於樣式化元件視圖的外部樣式表 URL 清單
directives	元件範本中使用的指令清單
providers	此元件及其子級的依賴項注入提供程式的清單
pipes	元件範本中使用的管道清單
exports	匯出在範本中可能使用的識別符號清單。

14.1.3 元件範本

元件視圖採用 HTML 語言，幾乎支援所有合法的 HTML 元素，為了安全起見 <script> 元素不被支援。透過註釋向元件提供元件視圖時可以使用 template 或 templateURL 參數，template 參數接收的是字串，而 templateURL 引用的是檔案名稱。

使用單引號包裹單行 HTML 元素：

```
template: '<h1>AngularDart 應用程式 </h1>',
```

使用三重單引號包裹多行 HTML 元素：

```
template: '''
   <h1>AngularDart 應用程式 </h1>
   <p> 根元件範本 </p>''',
```

14.1.4 元件樣式

元件樣式採用標準 CSS 規則，Angular 會將元件樣式和視圖綁定。樣式僅適用於當前元件的範本，而應用程式其他位置的元素均不受影響。可以使用 styles 或 styleURLs 參數為範本提供樣式，styles 接收字串清單，而 styleURLs 引用的是檔案名稱。

styles 參數採用包含 CSS 樣式的字串清單：

```
styles: ['h1 {font-weight:normal;}','p {text-indent:2em;}']
```

重新執行應用程式查看頁面資訊。完整程式如下：

```
//chapter14/angulardart/lib/app_component.dart
import 'package:angular/angular.dart';
// 元件註釋
@Component(
  selector: 'my-app',
  template: '''
    <h1>AngularDart 應用程式 </h1>
    <p> 根元件範本 </p>
  ''',
  styles: ['h1 {font-weight:normal;}','p {text-indent:2em;}']
)
// 元件類別
class AppComponent {
}
```

14.1.5 樣式和範本檔案

在實際開發中所需的範本結構複雜，CSS 樣式也更加多樣。為了結構清晰，通常會將範本和 CSS 樣式分別放在不同的檔案中。

將 CSS 樣式放在 app_component.css 檔案中。

```
//chapter14/angulardart/lib/app_component.css
h1 { font-weight:normal;}
p { text-indent:2em;}
```

將範本放在 app_component.html 檔案中。

```
//chapter14/angulardart/lib/app_component.html
<h1>Angular Web</h1>
<p> 範本 </p>
```

在註釋中使用 styleURLs 參數引用樣式表檔案，該參數可以接收多個樣式表檔案。使用 templateURL 參數引用範本檔案，該參數只能接收單一範本檔案。最終程式如下：

```
//chapter14/angulardart/lib/app_component.dart
import 'package:angular/angular.dart';
@Component(
   selector: 'my-app',
   styleURLs: ['app_component.css'],
   templateURL: 'app_component.html'
)
class AppComponent{
}
```

14.2 資料綁定

Angular 支援資料綁定，它採用一種用於協調使用者視圖和應用程式資料值的機制。只要增加綁定標記到範本，Angular 就會明確如何連接範本和元件。

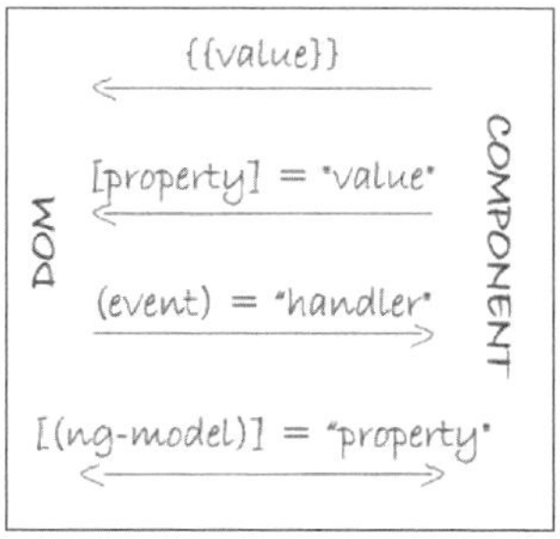

▲ 圖 14-8 資料流程向

資料綁定有 4 種形式，每種形式都有資料流程動的方向：元件到 DOM、DOM 到元件、雙向流動，如圖 14-8 所示。綁定語法和類型如表 14-2 所示。

表 14-2 資料綁定

資料方向	語法	類型
單向，從資料來源到視圖目標	{{expression}} [target]="expression" bind-target="expression"	插值 Property 綁定 Attribute 綁定 Class 綁定 Style 綁定
單向，從視圖目標到資料來源	(target)="statement" on-target="statement"	事件綁定
雙向	[(target)]="expression"	雙向資料綁定

資料綁定適用於 DOM 屬性 (property) 和事件，而非屬性 (attribute)。本節僅介紹前 3 種資料綁定的使用，雙向資料綁定將在後續合適的位置繼續介紹。

創建新的 Angular 應用程式並命名為 data_bind，lib 目錄下只留與根元件 AppComponent 相關檔案即可，打開專案 web 目錄下的 index.html 檔案，選擇合適的瀏覽器後打開即可執行本專案。

14.2.1 範本運算式和敘述

在開始資料綁定前需要瞭解表中出現的兩個重要單字：expression 和 statement。在 Angular 框架中它們分別被叫作範本運算式和範本敘述。

範本運算式會產生一個值，Angular 執行運算式並將其值分配給綁定目標的屬性，目標可能是 HTML 元素、元件或指令。

舉個例子，在下述程式中的插值大括號 ({{}}) 包含的內容 1+1 就是範本運算式：

```
{{1+1}}
```

在屬性綁定中，範本運算式會出現在等號右邊的引號中：

```
[property]="expression"
```

範本運算式與 Dart 語言中的運算式幾乎相同，但也有部分內容是不被支援或有副作用的運算式：

(1) 設定運算子 (=、+=、-=、…)。
(2) new 或 const 關鍵字。
(3) 在運算式後使用分號 (;)。
(4) 自動增加和自減符 (++ 和 --)。
(5) 位元運算符號 | 和 &。
(6) 字串插值 ($variableName 或 ${expression})。

範本敘述常見於事件綁定中，範本敘述的作用是響應事件綁定目標引發的事件，通常用於修改元件屬性或呼叫元件方法。下列程式將按鈕的點擊事件與敘述 delete() 綁定，該敘述是元件中定義的方法：

```
<button (click)="delete()"> 刪除 </button>
```

與範本運算式一樣，範本敘述與 Dart 敘述類似。它支援設定運算子 (=) 和多行敘述，這表示可以使用分號 (;)。但也有一部分是被禁止的：

(1) new 或 const 關鍵字。
(2) 自動增加和自減符 (++ 和 --)。
(3) 分配運算子 (+= 和 -= 等)。
(4) 位元運算符號 | 和 &。
(5) 管道符 (|) 和安全導覽運算子 (?.)。

14.2.2 插值

插值是將元件屬性的值傳入範本的一種方式，它適用於對資料進行簡要展示和處理的場景。語法如下：

```
{{expression}}
```

在範本中使用插值前需要在元件中定義屬性。這裡在元件 AppComponent 中定義標題（title）和名字（name）兩個屬性。範例程式如下：

```
//chapter14/data_bind/lib/app_component.dart
import 'package:angular/angular.dart';

@Component(
   selector: 'my-app',
   styleURLs: ['app_component.css'],
   templateURL: 'app_component.html',
)
class AppComponent {
   // 用於插值
   String title = 'Angular Web';
   String name = 'Jobs';
}
```

Angular 使用雙大括號 ({{}}) 作為插值符，插值使用範本運算式求值。將 title 屬性透過插值放在 h1 元素之間，name 屬性透過插值為輸入框元素的 value 屬性設定值。範例程式如下：

```
<h1>{{title}}</h1>
<input type="text" value="{{name}}">
```

插值符中的運算式也可以包含字串，使用加號連接。在範本檔案 app_component.html 中寫入以下程式：

```
//chapter14/data_bind/lib/app_component.html
<h1>{{' 標題 '+title}}</h1>
<input type="text" value="{{name}}">
```

修改程式後會自動編譯，刷新瀏覽器即可，執行效果如圖 14-9 所示。

標題Angular Web

Jobs

▲ 圖 14-9 插值

14.2.3 屬性 (property) 綁定

編寫屬性綁定以設定視圖元素的屬性，綁定將屬性設定為範本運算式的值。屬性綁定使用中括號 ([]) 作為標記，語法如下：

```
<element [property]="expression"></element>
```

最常見的屬性綁定是將元素屬性設定為元件屬性值，在元件中定義 imgPath 和 isAble 屬性。範例程式如下：

```
//chapter14/data_bind/lib/app_component.dart
import 'package:angular/angular.dart';

@Component(
   selector: 'my-app',
   styleURLs: ['app_component.css'],
   templateURL: 'app_component.html',
)
class AppComponent {
   // 用於屬性綁定
   String imgPath = '/favicon.png';
   bool isAble = true;
}
```

在範本檔案中將元件的 imgPath 屬性與 img 元素的 src 屬性綁定，isAble 屬性與按鈕元素的 disabled 屬性綁定。範例程式如下：

```
//chapter14/data_bind/lib/app_component.html
<img [src]="imgPath">
<button [disabled]="isAble"> 按鈕處於禁用狀態 </button>
```

也可以使用 bind- 字首替代中括號：

```
<button bind-disabled="!isAble"> 按鈕處於可用狀態 </button>
```

執行效果如圖 14-10 所示。

▲ 圖 14-10 屬性綁定

插值和屬性綁定在很多情況下可以相互替代，將元素屬性設定為非字串值時，必須使用屬性綁定。

14.2.4 屬性 (attribute) 綁定

屬性 (attribute) 綁定與屬性 (property) 綁定語法類似，區別在於它使用 attr 字首，後跟一個點 (.) 和該屬性 (attribute) 的名字。然後，就可以使用具體值或範本運算式設定屬性 (attribute) 值：[attr.attr-name]。

在範本檔案中建立表格，並設定 colspan 屬性。程式如下：

```
//chapter14/data_bind/lib/app_component.html
<table border="1" style="border-collapse:collapse;">
   <tr><td [attr.colspan]="2"> 表格單元跨越兩列 </td><td>1,3</td><td>1,4</td>
   </tr><tr><td>2,1</td><td>2,2</td><td>2,3</td><td>2,4</td></tr>
</table>
```

執行結果如圖 14-11 所示。

表格單元跨越兩列		1,3	1,4
2,1	2,2	2,3	2,4

▲ 圖 14-11 屬性 (attribute) 綁定

屬性 (attribute) 綁定主要使用場景是設定 aria 屬性，對於無障礙閱讀非常適用。例如：

```
<input type="text" aria-label=" 用戶名 "/>
```

14.2.5 類別綁定

此處的類別指元素屬性 class，可以使用類別綁定在元素的 class 屬性增加或移除 CSS 類別名稱。類別綁定與屬性綁定類似，區別在於類別綁定使用 class 字首，後跟點 (.) 和 class 名：[class.class-name]。

以下範例透過幾種方式在元素增加和刪除 uppercase 類別。

直接將 CSS 名增加到屬性 class 的字串中：

```
<p class="bold italic uppercase">this is a test</p>
```

使用屬性綁定 [class]，該綁定會使用元件屬性 textStyle 替換元素 class 屬性已指定的所有類別名稱，元件屬性 textStyle 是字串形式，它可以包含多個類別名稱。

```
<p class="bold italic uppercase" [class]="textStyle">this is a test</p>
```

也可以透過類別綁定 [class.class-name] 的形式綁定單個類別名稱，運算式的值為 true 時增加該類別名稱，值為 false 時移除該類別名稱。

```
<!-- 使用元件屬性增加或移除 class 名 uppercase -->
<p [class.uppercase]="isUppercase">this is a test</p>
<p [class.uppercase]="!isUppercase">this is a test</p>
```

在元件 AppComponent 中增加 textStyle 和 isUppercase 屬性。程式如下：

```
//chapter14/data_bind/lib/app_component.dart
import 'package:angular/angular.dart';

@Component(
  selector: 'my-app',
  styleURLs: ['app_component.css'],
  templateURL: 'app_component.html',
)
class AppComponent {
  // 用於 CSS 類別綁定
  String textStyle = "bold italic";
  bool isUppercase = true;
}
```

在樣式檔案 app_component.css 中增加以下樣式。程式如下：

```
//chapter14/data_bind/lib/app_component.css
.bold{
  /* 粗體字型 */
  font-weight:bold;
}
```

```
.italic{
   /* 使字型傾斜 */
   font-style:italic;
}
.uppercase{
   /* 對單字轉大寫 */
   text-transform: uppercase;
}
```

在範本檔案 app_component.html 中增加以下內容。程式如下：

```
//chapter14/data_bind/lib/app_component.html
<p class="bold italic uppercase">this is a test</p>
<p class="bold italic uppercase" [class]="textStyle">this is a test</p>
<p [class.uppercase]="isUppercase">this is a test</p>
<p [class.uppercase]="!isUppercase">this is a test</p>
```

執行結果如圖 14-12 所示。

THIS IS A TEST

this is a test

THIS IS A TEST

this is a test

▲ 圖 14-12 類別綁定

14.2.6 樣式綁定

可以使用樣式 (style) 綁定設定內聯樣式，樣式綁定語法類似屬性綁定。以 style 為字首，後跟一個點 (.) 和 CSS 樣式屬性的名字：[style.style-property]。

樣式綁定在設定時通常會和條件運算子 (expr1? expr2:expr3) 配合使用。範例程式如下：

```
<p [style.color]="isSpecial ? 'red': 'green'">this is a test</p>
<p [style.background-color]="canSave ? 'cyan': 'grey'" >this is a test</p>
```

某些樣式具有單位，以下範例有條件地以 em 和 % 為單位設定字型大小。

```
<p [style.font-size.em]="isSpecial ? 3 : 1" >this is a test</p>
<p [style.font-size.%]="!isSpecial ? 150 : 50" >this is a test</p>
```

在元件 AppComponent 中增加 isSpecial 和 canSave 屬性。範例程式如下：

```
//chapter14/data_bind/lib/app_component.dart
import 'package:angular/angular.dart';

@Component(
   selector: 'my-app',
   styleURLs: ['app_component.css'],
   templateURL: 'app_component.html',
)
class AppComponent {
   // 用於樣式綁定
   bool isSpecial = false;
   bool canSave = true;
}
```

在範本檔案 app_component.html 中增加以下內容：

```
//chapter14/data_bind/lib/app_component.html
<p [style.color]="isSpecial ? 'red': 'green'">this is a test</p>
<p [style.background-color]="canSave ? 'cyan': 'grey'" >this is a test</p>
<p [style.font-size.em]="isSpecial ? 3 : 1" >this is a test</p>
<p [style.font-size.%]="!isSpecial ? 150 : 50" >this is a test</p>
```

執行結果如圖 14-13 所示。

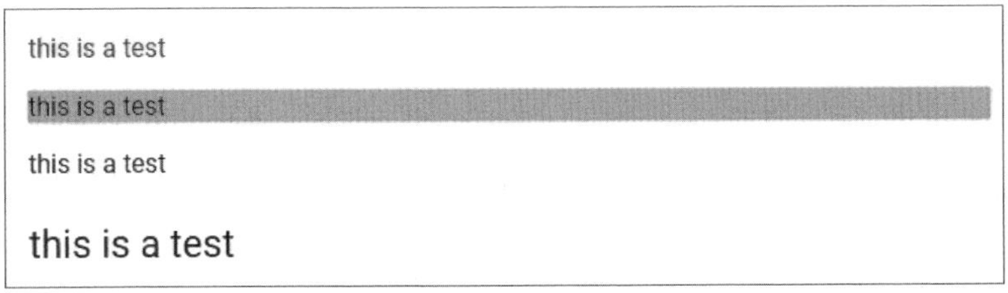

this is a test

this is a test

this is a test

this is a test

▲ 圖 14-13 樣式綁定

樣式屬性名稱可以使用半字組元線 (font-size) 或小駝峰 (fontSize) 編寫，它們都被 Angular 支援。但是只有半字組元線的形式能被 dart:html 函數庫中的某些方法存取，因此首選半字組元線作為樣式屬性名稱。

14.2.7 事件綁定

為了回應使用者的操作，時常需要監聽某些事件，例如：按鍵、滑鼠移動、點擊和觸控等。為了讓這些操作更加便利，可以使用 Angular 提供的事件綁定，在事件綁定中資料從元素傳遞到元件。

事件綁定語法組成：

```
<element (target)="statement"></element>
```

括號裡的名字 target 表示目標事件，以下範例中目標是按鈕的點擊事件。等號右邊的雙引號裡包裹的是範本敘述，以下範例中敘述是元件的 onSave() 方法。

```
<button (click)="onSave()">Save</button>
```

也可以以 on- 字首代替括號，這是規範形式：

```
<button on-click="onSave()">Save</button>
```

在事件綁定中，Angular 為目標事件設定事件處理常式。當事件觸發時，處理常式執行範本敘述。範本敘述通常包含一個接收器，接收器響應該事件並執行操作。例如將 HTML 控制項的值儲存到元件屬性。

綁定透過事件物件 $event 傳遞關於事件的所有資訊，包括資料值。事件物件的結構由目標事件決定，如果目標事件是原生 DOM 元素事件，則 $even 是 DOM 事件物件，具有諸如 target 和 target.value 之類的屬性。

在元件 AppComponent 中增加屬性 value 用於儲存值，增加 onSetValue() 方法用於接收一個值，然後將值指定給元件屬性 value。範例程式如下：

```
//chapter14/data_bind/lib/app_component.dart
import 'package:angular/angular.dart';

@Component(
   selector: 'my-app',
   styleURLs: ['app_component.css'],
   templateURL: 'app_component.html',
)
class AppComponent {
   // 用於事件綁定
   String value;
   void onSetValue(var val){
      value = val;
   }
}
```

在範本檔案 app_component.html 中增加的程式如下：

```
//chapter14/data_bind/lib/app_component.html
<input (input)="onSetValue($event.target.value)">
<p>value：{{name}}</p>
```

此程式將 value 屬性透過插值放在 p 元素中，用於顯示 value 屬性的值。然後將 input 事件綁定到輸入框，當使用者進行更改時，將觸發 input 事件，並且綁定將執行範本敘述 onSetValue($event.target.value)，其中 $event.target.value 用於獲取更改後的值。

其執行結果是每次輸入事件都會使 value 屬性值被更新並在視圖中顯示，例如在輸入框中輸入 hello，執行效果如圖 14-14 所示。

hello

value: hello

▲ 圖 14-14 事件綁定

14.3 內建指令

指令是帶有 @Directive 註釋的類別，而元件是帶有範本的指令。@Component 註釋實際上是在 @Directive 註釋的基礎上擴充了範本功能。從技術上來說元件就是指令，由於元件是 Angular 應用程式獨特的組成部分，並且是 Angular 的核心，所以將元件單獨拿出來做以區分。

Angular 範本是動態的，當 Angular 繪製它們時，將根據指令列出的指令轉換 DOM。Angular 中指令分為屬性指令和結構指令，當然也可以創建自訂指令。

14.3.1 屬性指令

屬性指令監聽並修改其他 HTML 元素、屬性 (attribute)、屬性 (property) 和元件的行為。它們通常應用於元素，就像它們就是 HTML 屬性一樣。

許多 Angular 套件都有自己的屬性指令，例如 Router 和 Forms 套件。這裡介紹常用的屬性指令：

(1) NgClass：增加或移除一組 CSS 類別。
(2) NgStyle：增加或移除一組 HTML 樣式。
(3) NgModel：雙向資料綁定到一個 HTML 表單元素，在小節表單中會詳細說明。

使用指令需要在元件註釋的參數 directives 中列出需要的指令，更新註釋 @Component。範例程式如下：

```
@Component(
   selector: 'my-app',
   styleURLs: ['app_component.css'],
   templateURL: 'app_component.html',
   directives: [NgClass,NgStyle],
)
```

1. NgClass

透過動態增加或刪除 CSS 類別可以控制元素的外觀，而綁定到 NgClass 可以同時增加或刪除多個 CSS 類別。

在使用時，需要將 NgClass 綁定到 Map 類型的控制項上。該 Map 物件的每個鍵都是一個 CSS 類別名稱，如果應該增加該 CSS 類別，則對應的值應設定為 true。如果應該移除該 CSS 類別，則對應的值應設定為 false。為了動態控制是否增加或刪除 CSS 類別，值通常是一個結果為布林值的運算式。

在元件 AppComponent 中增加 4 個布林屬性 isUppercase、isSpecial、isBold 和 isItalic，增加 Map 類型的控制項 currentClasses，創建一個方法 setCurrentClasses()，該方法設定元件屬性 currentClasses。從 Map 字面量中可以看出，鍵是 CSS 類別名稱，值中組成運算式的屬性是布林值。範例程式如下：

```
//chapter14/built_in_directives/lib/app_component.dart
import 'package:angular/angular.dart';

@Component(
   selector: 'my-app',
   styleURLs: ['app_component.css'],
   templateURL: 'app_component.html',
   directives: [NgClass,NgStyle]
)
class AppComponent {
   // 用於屬性指令 ngClass
   bool isUppercase = true;
   bool isSpecial = false;
   bool isBold = true;
   bool isItalic = false;
   Map<String, bool> currentClasses = <String, bool>{};
   void setCurrentClasses(){
      currentClasses = <String, bool>{
         'bold': isBold,
         'italic': !isItalic,
```

```
            'uppercase': isUppercase
        };
    }
}
```

在樣式檔案中增加 CSS 類別。程式如下：

```
//chapter14/built_in_directives/lib/app_component.css
.bold{
    /* 粗體字型 */
    font-weight:bold;
}
.italic{
    /* 使字型傾斜 */
    font-style:italic;
}
.uppercase{
    /* 對單字轉大寫 */
    text-transform: uppercase;
}
```

將 NgClass 指令綁定到 currentClasses 屬性會對應地設定元素的 CSS 類別。增加點擊事件綁定，範本敘述是 setCurrentClasses()。點擊按鈕後方法被執行，currentClasses 完成初始化。程式如下：

```
//chapter14/built_in_directives/lib/app_component.html
<div [ngClass]="currentClasses">this is a test</div>
<button (click)="setCurrentClasses()"> 初始化元件屬性 currentClasses</button>
```

可以在初始化元件或屬性改變時呼叫 setCurrentClasses 方法，這裡是透過按鈕來初始化。點擊按鈕執行結果如圖 14-15 所示。

THIS IS A TEST

初始化元件屬性currentClasses

▲ 圖 14-15 指令 NgClass

2. NgStyle

可以根據元件的狀態動態設定元素的內聯樣式，使用 NgStyle 指令可以同時設定多個內聯樣式。

與 NgClass 類似，將 NgStyle 綁定到 Map 類型的控制項，該 Map 物件的每個鍵都是一個 CSS 樣式名稱，其對應的值是該樣式的值。

在元件 AppComponent 中增加一個 setCurrentStyles() 方法，該方法設定元件屬性 currentStyles。在 Map 字面量的值部分使用了條件運算式，條件運算式根據布林值返回一個字串，該字串是符合對應樣式的值。範例程式如下：

```
//chapter14/built_in_directives/lib/app_component.dart
import 'package:angular/angular.dart';

@Component(
   selector: 'my-app',
   styleURLs: ['app_component.css'],
   templateURL: 'app_component.html',
   directives: [NgClass,NgStyle]
)
class AppComponent {
   // 用於屬性指令 ngStyle
   Map<String, String> currentStyles = <String, String>{};
   void setCurrentStyles() {
      currentStyles = <String, String>{
         'font-style': isItalic ? 'italic' : 'normal',
         'font-weight': isBold ? 'bold' : 'normal',
         'font-size': isSpecial ? '24px' : '36px'
      };
   }
}
```

將 NgStyle 屬性綁定到 currentStyles 可以對應地設定元素的樣式。增加點擊事件綁定，範本敘述是 setCurrentStyles()。點擊按鈕後方法被執行，currentStyles 完成初始化。程式如下：

```
//chapter14/built_in_directives/lib/app_component.html
<div [ngStyle]="currentStyles">this is a test</div>
<button (click)="setCurrentStyles()"> 初始化元件屬性 currentStyles</button>
```

可以在初始化元件或屬性改變時呼叫 setCurrentStyles 方法，這裡是透過按鈕來初始化。點擊按鈕執行結果如圖 14-16 所示。

this is a test

初始化元件屬性currentStyles

▲ 圖 14-16 指令 NgStyle

14.3.2 結構指令

結構指令用於 HTML 動態佈局，透過增加、刪除和控制它們所依附的宿主元素來建構或重構 DOM 結構。

常見的結構指令如下：

(1) NgIf：有條件地從 DOM 中增加或刪除元素。
(2) NgFor：為清單中的每個專案重複一個範本。
(3) NgSwitch：僅顯示多個可能的元素之一。

在元件註釋的參數 directives 中列出需要的結構指令，更新註釋 @Component。程式如下：

```
@Component(
   selector: 'my-app',
   styleURLs: ['app_component.css'],
   templateURL: 'app_component.html',
   directives: [NgClass,NgStyle,NgIf,NgFor,NgSwitch,NgSwitchWhen,NgSwitchDefa
ult],
)
```

1. NgIf

指令 NgIf 與布林運算式綁定，根據運算式的值從 DOM 中增加或刪除元素，使用 NgIf 指令的元素被稱為宿主元素。

在元件 AppComponent 中增加屬性 isNull 和切換其布林值的 toggle 方法。範例程式如下：

```
//chapter14/built_in_directives/lib/app_component.dart
import 'package:angular/angular.dart';

@Component(
   selector: 'my-app',
   styleURLs: ['app_component.css'],
   templateURL: 'app_component.html',
   directives：[NgClass,NgStyle,NgIf,NgFor,NgSwitch,NgSwitchWhen,
NgSwitchDefault]
)
class AppComponent {
   // 用於結構指令 *ngIf
   bool isNull = true;
   // 切換 isNull 的布林值
   toggle(){
      isNull = isNull ? false : true;
   }
}
```

在範本檔案 app_component.html 中增加的程式如下：

```
//chapter14/built_in_directives/lib/app_component.html
<button (click)="toggle()"> 切換 isNull 的真假值 </button>
<div *ngIf="isNull">isNull:{{isNull}}</div>
<div *ngIf="!isNull">!isNull:{{isNull}}</div>
```

當 isNull 運算式返回 true 時，NgIf 將 div 元素增加到 DOM。如果運算式為 false，則 NgIf 從 DOM 中刪除 div 元素。

需要說明的是 NgIf 指令工作模式與顯示或隱藏不一樣，顯示或隱藏通常會使用類別或樣式綁定來控制元素的可見性。在範本檔案中增加的程式如下：

```
//chapter14/built_in_directives/lib/app_component.html
<!-- isNull 初值為 true -->
<div [class.hidden]="!isNull"> 使用類別綁定顯示 </div>
<div [class.hidden]="isNull"> 使用類別綁定隱藏 </div>
<div [style.display]="isNull? 'block' : 'none'"> 使用樣式顯示 </div>
<div [style.display]="isNull? 'none' : 'block'"> 使用樣式隱藏 </div>
```

樣式檔案中增加以下 CSS 類別，程式如下：

```
//chapter14/built_in_directives/lib/app_component.css
.hidden{
   /* 隱藏元素 */
   display:none;
}
```

使用類別或樣式綁定隱藏元素時，該元素及其所有後代仍保留在 DOM 中。這些元素的所有元件都保留在記憶體中，Angular 可能會繼續檢查更改。應用可能會佔用大量運算資源，從而降低性能。

使用 NgIf 指令時，如果運算式為 false，則 Angular 將從 DOM 中刪除該元素及其後代。它銷毀了它們的元件，有可能釋放大量資源，從而帶來了回應速度更快的使用者體驗。

顯示或隱藏適用於子元素少的元素，而 NgIf 適用於大型元件樹。

NgIf 指令常用於防止 null 值，如果運算式嘗試存取 null 值，Angular 將拋出錯誤。因此可以透過判斷屬性是否為 null 來避免錯誤。

```
<div *ngIf="name != null">Hello, {{name}}</div>
```

2. NgFor

NgFor 是一種重複指令，一種以相同元素結構顯示當前清單中每個項目的方法。通常定義一個 HTML 元素區塊，Angular 使用該 HTML 元素區塊作為範本來呈現清單中的每個項目。在元素區塊中也可以包含子元件宣告的選擇器，在後續會介紹。

首先在 app_component.dart 檔案中增加 JieQi 類別，在元件中定義 jieqiList 清單，並使用字面量為其初始化值。範例程式如下：

```
//chapter14/built_in_directives/lib/app_component.dart
import 'package:angular/angular.dart';

@Component(
```

```
    selector: 'my-app',
    styleURLs: ['app_component.css'],
    templateURL: 'app_component.html',
    directives: [NgClass,NgStyle,NgIf,NgFor,NgSwitch,NgSwitchWhen,NgSwitchDefault]
)
class AppComponent {
    // 用於結構指令 *ngFor
    // 使用字面量初始化泛型為 JieQi 的清單
    List<JieQi> jieqiList = [
        JieQi(1,' 雨水 '),
        JieQi(2,' 穀雨 '),
        JieQi(3,' 白露 '),
        JieQi(4,' 寒露 '),
        JieQi(5,' 霜降 '),
        JieQi(6,' 小雪 '),
        JieQi(7,' 大雪 ')];
}
//JieQi 類別
class JieQi{
    int id;
    String label;
    JieQi(this.id,this.label);
}
```

將 NgFor 應用於 <div> 的範例程式如下：

```
//chapter14/built_in_directives/lib/app_component.html
<div *ngFor="let jieqi of jieqiList">
    {{jieqi.label}}
</div>
```

分配給 *ngFor 的字串不是範本運算式，這是一種 Angular 可以解析的微語法。字串 let jieqi of jieqiList 的意思是：將 jieqiList 清單中的單一項目儲存在本地循環變數 jieqi 中，並可以在每次疊代的 HTML 範本中使用循環變數 jieqi。

1) 範本輸入變數

jieqi 之前的 let 關鍵字創建了一個名為 jieqi 的範本輸入變數。NgFor 指令會疊代元件的 jieqiList 屬性返回的 jieqiList 清單，並在每次疊代期間將 jieqi 設定為清單中的當前項目。

要存取 jieqi 的屬性，需在 NgFor 宿主元素 (或其子元素) 中引用 jieqi 輸入變數。這裡在插值中引用 jieqi。

```
<div *ngFor="let jieqi of jieqiList">{{jieqi.label}}</div>
```

2) 索引

NgFor 指令上下文的 index 屬性在每次疊代中返回該項目的從 0 開始的索引。可以在範本輸入變數中捕捉索引並在範本中使用它。

範例在名為 i 的變數中捕捉索引，並顯示該索引和 jieqi 名。範例程式如下：

```
//chapter14/built_in_directives/lib/app_component.html
<div *ngFor="let jieqi of jieqiList;let i=index;trackBy:trackById">
  ({{i}})-{{jieqi.label}}
</div>
```

3) 追蹤 TrackBy

NgFor 指令的性能可能會很差，尤其是對於大型清單而言。對一項進行很小的更改，刪除一項或增加一項，都會觸發一系列 DOM 操作。

舉例來說，重新查詢伺服器可能會重置清單 jieqiList 中的所有 JieQi 物件，且其中大多數是重複的，因為每個 JieQi 物件的 id 是固定的。此時 Angular 只能檢測到新物件引用的新清單，因此只能拆除舊的 DOM 元素並插入所有新的 DOM 元素。

Angular 可以透過使用 TrackBy 避免這種混亂，使用時為其設定一個函數，該函數必須符合追蹤函數的類型定義。範例程式如下：

```
Object TrackByFn (int index,dynamic item)
```

它是一個為索引處的項目返回唯一鍵的函數。預設情況下項目本身被用作鍵實例化新的範本，如果資料發生變化，Angular 將視為不同的物件來重新繪製資料。如果可以確定疊代物件中項目的唯一性，則可以提供遵循 TrackByFn 類型定義的函數最佳化性能。

在該元件增加一個方法，該方法返回 NgFor 應該追蹤的值 id。範例程式如下：

```
//chapter14/built_in_directives/lib/app_component.dart
Object trackById(int index, dynamic o) {
   return o is JieQi ? o.id : o;
}
```

在微語法運算式中，將 TrackBy 設定為此方法。範例程式如下：

```
//chapter14/built_in_directives/lib/app_component.html
<div *ngFor="let jieqi of jieqiList;let i=index">
   {{jieqi.id}},{{jieqi.label}}
</div>
```

3. NgSwitch

NgSwitch 就像 Dart 中的 switch 敘述。它可以根據 switch 條件顯示幾個可能元素中的元素，Angular 僅將所選元素放入 DOM。

NgSwitch 實際上需要 3 個指令協作：NgSwitch、NgSwitchCase 和 NgSwitchDefault。

(1) NgSwitch 是控制器，將其綁定到返回 switch 值的運算式。NgSwitch 是屬性指令，而非結構指令。它的作用是更改其伴隨指令的行為，而不會直接操作 DOM。

(2) NgSwitchCase 和 NgSwitchDefault 是結構指令，它們在 DOM 中增加或刪除了元素。當 NgSwitchCase 的綁定值等於 switch 值時，會將其元素增加到 DOM。如果所有 NgSwitchCase 中都沒有匹配值，則 NgSwitchDefault 將其元素增加到 DOM 中，NgSwitchDefault 可以省略。

在元件中增加屬性 color，該屬性的作用是提供 switch 值。範例程式如下：

```
//chapter14/built_in_directives/lib/app_component.dart
String color = 'Green';
```

在範本中增加程式如下：

```
//chapter14/built_in_directives/lib/app_component.html
<div [ngSwitch]="color">
   <span *ngSwitchCase="'Red'">紅色 </span>
   <span *ngSwitchCase="'Green'">綠色 </span>
   <span *ngSwitchCase="'Black'">黑色 </span>
   <span *ngSwitchDefault>藍色 </span>
</div>
```

在這裡 color 值是一個字串，但 switch 值可以是任何類型。

14.4 範本引用變數

範本引用變數通常是對範本中一個 DOM 元素的引用，它還可以引用 Angular 元件或指令等。

在元件中增加屬性 number，並增加方法 callPhone()，方法 callPhone() 的作用是將傳入的值指定給 number。範例程式如下：

```
//chapter14/template_reference_variable/lib/app_component.dart
import 'package:angular/angular.dart';
@Component(
   selector: 'my-app',
   styleURLs: ['app_component.css'],
   templateURL: 'app_component.html',
)
class AppComponent {
   String number;
   callPhone(var tel) {
```

```
        number = tel;
    }
}
```

在元素上使用 # 加變數名稱宣告一個引用變數，而後在範本中的任何位置都可以使用範本引用變數。此範例中在元素 <input> 中宣告範本引用變數 phone，在元素 <button> 點擊事件的範本敘述 callPhone(phone.value) 中引用。範例程式如下：

```
//chapter14/template_reference_variable/lib/app_component.html
<!-- 宣告引用變數 phone -->
<input #phone placeholder=" 輸入電話號碼 " type="tel">
<!-- 使用引用變數 phone -->
<button (click)="callPhone(phone.value)"> 撥打 </button>
<p>{{number}}</p>
```

14.4.1 設定值

大多數情況下，Angular 會將範本引用變數的值設定為宣告該變數的元素。在上一個範例中，範本引用變數 phone 指的是輸入框。

點擊按鈕處理常式會將輸入值傳遞給元件的 callPhone 方法，但是有一些指令可以更改該行為並將其值設定為其他值。例如：NgForm 指令可以將其自身設定值給範本引用變數。

指令 NgForm 是由套件 angular_forms 提供的，這裡跟著步驟做就可以了，套件 angular_forms 的更多內容將在後續做詳細說明。首先增加依賴項並執行 pub get 命令：

```
dependencies:
   angular: ^5.3.0
   angular_components: ^0.13.0
   angular_forms: ^2.1.2
```

在元件的指令清單中增加包含 angular_forms 套件所有指令的常數值 formDirectives。屬性 name 用於與輸入控制項進行雙向資料綁定，屬性

submitMessage 用於儲存表單的值，方法 onSubmit 接收一個 NgForm 參數，並設定屬性 submitMessage 的值。範例程式如下：

```
//chapter14/template_reference_variable/lib/app_component.dart
import 'package:angular/angular.dart';
import 'package:angular_forms/angular_forms.dart';

@Component(
   selector: 'my-app',
   styleURLs: ['app_component.css'],
   templateURL: 'app_component.html',
   directives: [formDirectives],
)
class AppComponent {
   String name;
   String submitMessage = '';
   void onSubmit(NgForm form) {
      submitMessage = '提交表單的值是 ${form.value}';
   }
}
```

在範本中建立表單，在元素 form 上定義範本引用變數 newForm，並將 NgForm 指令的匯出物件 ngForm 設定值給 newForm。範例程式如下：

```
//chapter14/template_reference_variable/lib/app_component.html
<form (ngSubmit)="onSubmit(newForm)" #newForm="ngForm">
   <input ngControl="name"
      required
      [(ngModel)]="name">
   <button type="submit"
      [disabled]="!newForm.form.valid">提交</button>
</form>
<div [hidden]="!newForm.form.valid">
      {{submitMessage}}
</div>
```

在此範例中，範本引用變數 newForm 出現 3 次，並被大量 HTML 分隔。newForm 的值是對 Angular 表單指令 NgForm 的引用，該指令能夠追蹤表單中每個控制項的值和有效性。原生 <form> 元素沒有 form 屬性，但是

NgForm 指令有。如果 newForm.form.valid 無效，將禁用提交按鈕，並在表單有效時將整個表單控制樹傳遞給元件的 onSubmit 方法。

14.4.2 說明

範本引用變數 (#phone) 與在 *ngFor 中看到的範本輸入變數 (let phone) 不同，範本引用變數的範圍是整個範本，除非在結構指令控制的嵌入式視圖中宣告了引用變數。在嵌入式視圖中宣告的範本引用變數僅對被結構指令嵌入的範本部分可見。

注意：在外部宣告的範本引用變數可以在嵌入式視圖中引用，反之則不能。請勿在同一範本中多次定義相同的變數名稱，執行時期它們的值將不可預測。

14.5 服務

服務最常用的場景是提供資料，資料通常來自伺服器，因此資料服務總是非同步的。只要元件需要，就可以透過依賴注入將資料提供給元件。為了方便演示，本節提供的資料將由本地提供，而非透過伺服器獲取。

14.5.1 定義實體類別

在目錄 lib/src 下創建 Employee 類別，包含 id、name 和 salary 屬性。範例程式如下：

```
//chapter14/service/lib/src/employee.dart
class Employee{
   final int id;
   String name;
   num salary;
   Employee(this.id,this.name,this.salary);
}
```

14.5.2 創建服務

在目錄 lib/src 下創建 employee_service.dart 檔案，匯入 employee.dart 檔案並定義 EmployeeService 類別。在類別上使用 @Injectable() 註釋，表示可被注入。範例程式如下：

```
//chapter14/service/lib/src/employee_service.dart
import 'package:angular/angular.dart';
import 'employee.dart';

@Injectable()
class EmployeeService{
}
```

在類別中定義 getAll() 方法，並用字面量定義一個 Employee 清單。因為在實際使用中服務通常是非同步的，所以可以將返回值使用 Future 封裝，並在 getAll() 方法上使用 async 標記。範例程式如下：

```
//chapter14/service/lib/src/employee_service.dart
import 'package:angular/angular.dart';
import 'employee.dart';

@Injectable()
class EmployeeService{
  Future<List<Employee>> getAll() async{
    var emps = [
      Employee(1, 'Mr. Nice',3000),
      Employee(2, 'Narco',3105),
      Employee(3, 'Bombasto',3988),
      Employee(4, 'Celeritas',3401),
      Employee(5, 'Magneta',9971),
      Employee(6, 'RubberMan',4533),
      Employee(7, 'Dynama',6720),
      Employee(8, 'Dr IQ',4907),
      Employee(9, 'Magma',5278),
      Employee(10, 'Tornado',7800)
    ];
    return emps;
  }
}
```

14.5.3 使用服務

在 AppComponent 元件中透過檔案匯入 EmployeeService 和 Employee 類別。

```
import 'src/employee.dart';
import 'src/employee_service.dart';
```

在元件註釋 @Component 的 providers 參數中標識服務提供者，該參數接收一個清單。通常使用 ClassProvider 方法來載入服務。範例程式如下：

```
@Component(
   selector: 'my-app',
   styleURLs: ['app_component.css'],
   templateURL: 'app_component.html',
   directives: [NgFor],
   providers: [ClassProvider(EmployeeService)],
)
```

增加清單屬性 emps、EmployeeService 類型的私有屬性 _employeeService，並在元件的建構函數中初始化該屬性。範例程式如下：

```
final EmployeeService _employeeService;
List<Employee> emps;
AppComponent(this._employeeService);
```

增加 getEmployees() 方法，方法用於將服務中獲取的資料傳遞給 emps 屬性。由於資料獲取是非同步的，因此方法需要用 async 標記，並且在使用服務獲取資料的敘述前面使用 await 關鍵字。範例程式如下：

```
void getEmployees() async{
   emps = await _employeeService.getAll();
}
```

讓元件實現生命週期函數 ngOnInit() 並呼叫 getEmployees() 方法，該函數用於初始化元件資料。範例程式如下：

```
class AppComponent implements OnInit{
   @override
```

```
    void ngOnInit(){
        getEmployees();
    }
}
```

此時元件 AppComponent 的完整範例程式如下：

```
//chapter14/service/lib/app_component.dart
import 'package:angular/angular.dart';
import 'src/employee.dart';
import 'src/employee_service.dart';
@Component(
    selector: 'my-app',
    styleURLs: ['app_component.css'],
    templateURL: 'app_component.html',
    directives: [NgFor],
    providers: [ClassProvider(EmployeeService)],
)
class AppComponent implements OnInit{
    // 定義一個 EmployeeService 類型的私有變數
    final EmployeeService _employeeService;
    List<Employee> emps;
    // 注入服務
    AppComponent(this._employeeService);
    // 覆寫 ngOnInit 方法
    @override
    void ngOnInit() {
        // 在元件初始化時呼叫 getEmployees 方法
        getEmployees();
    }
    // 使用服務物件 _employeeService 獲取員工清單
    void getEmployees() async{
        emps = await _employeeService.getAll();
    }
}
```

在範本檔案中增加元素區塊，用於疊代 emps 屬性並顯示資料。範本程式如下：

```
//chapter14/service/lib/app_component.html
```

```
<ul class="emps">
   <li *ngFor="let emp of emps">
      <span class="badge">{{emp.id}}</span>
      {{emp.name}}
      <span class="badge salary">{{emp.salary}}</span>
   </li>
</ul>
```

定義元素區塊樣式，在 app_component.css 檔案中增加的樣式程式如下：

```
//chapter14/service/lib/app_component.css
.emps {
   margin: 002em 0;
   list-style-type: none;
   padding: 0;
   width: 15em;
}
.emps li {
   cursor: pointer;
   position: relative;
   left: 0;
   background-color: #EEE;
   margin: .5em;
   padding: .3em 0;
   height: 1.6em;
   border-radius: 4px;
}
.emps li:hover {
   color: #607D8B;
   background-color: #EEE;
   left: .1em;
}
.emps .text {
   position: relative;
   top: -3px;
}
.emps .badge {
   display: inline-block;
   font-size: small;
```

```
    color: white;
    padding: 0.8em 0.7em 00.7em;
    background-color: #607D8B;
    line-height: 1em;
    position: relative;
    left: -1px;
    top: -4px;
    height: 1.8em;
    margin-right: .8em;
    border-radius: 4px 004px;
}
.emps .salary {
    float:right;
    margin-right:0;
    margin-left: .8em;
    border-radius: 04px 4px 0;
}
```

執行結果如圖 14-17 所示。

▲ 圖 14-17 員工清單

14.6 子元件

前面所有的內容都是在範本生成的根元件 AppComponent 中完成的，在本節將介紹如何創建元件，並將其作為根元件 AppComponent 的子元件。

14.6.1 創建元件

通常會將元件及其範本和樣式檔案放在以其名字命名的資料夾中，在 src 目錄下建立 employee 目錄。根據約定與元件相關的 HTML、CSS、dart 檔案需要以 _component 作為檔案名稱尾碼。在 employee 目錄下建立 employee_component.dart 檔案，指定其選擇器為 employee。範例程式如下：

```
//chapter14/child_component/lib/src/employee/employee_component.dart
import 'package:angular/angular.dart';
@Component(
   selector:'employee',
)
class EmployeeComponent{
}
```

這裡就不單獨建立範本檔案了，使用內聯範本，向 @Component 註釋提供 template 參數和值。範例程式如下：

```
//chapter14/child_component/lib/src/employee/employee_component.dart
import 'package:angular/angular.dart';
@Component(
   selector:'employee',
   template: '''
      <div> 員工 ID：5<div>
      <div> 員工名稱：Magneta<div>
      <div> 員工薪資：9971<div>
      ''',
)
class EmployeeComponent{
}
```

這就是一個元件的基本組成部分，在實際使用中可能會複雜一些。

14.6.2 增加到父元件

首先需要在 AppComponent 元件中引入子元件。

```
import 'src/employee/employee_component.dart';
```

因為元件實際上是帶有範本的指令，所以應當將子元件 EmployeeComponent 增加到指令清單 directives 中。一個元件可以包含多個子元件。更新 AppComponent 元件的 @Component 註釋。範例程式如下：

```
//chapter14/child_component/lib/app_component.dart
import 'package:angular/angular.dart';
import 'src/employee/employee_component.dart';

@Component(
   selector: 'my-app',
   styleURLs: ['app_component.css'],
   templateURL: 'app_component.html',
   directives: [coreDirectives,EmployeeComponent],
)
class AppComponent {
}
```

前面在使用內建指令時，都是將每個指令單獨列出，實際上可以使用 coreDirectives 代替所有的內建指令，例如：NgClass、NgIf、NgFor 等。

然後在範本檔案中增加子元件定義的選擇器 employee。範例程式如下：

```
//chapter14/child_component/lib/app_component.html
<employee></employee>
```

執行結果如圖 14-18 所示。

員工ID: 5 員工名稱: Magneta 員工薪資: 9971

▲ 圖 14-18 子元件視圖

14.6.3 輸入輸出屬性

在子元件中範本內容是自訂的，與父元件沒有產生聯繫。在實際開發中子元件的某些資料可能是由父元件提供的，這就需要提供輸入屬性做支援。

1. 輸入屬性

前面的內容主要集中於綁定宣告右側範本運算式和敘述中的元件成員，該位置的成員是資料綁定來源。本節重點介紹綁定的目標，這些目標是綁定宣告左側的指令屬性。這些指令屬性必須宣告為輸入或輸出。

創建 Employee 類別，程式如下：

```
//chapter14/child_component/lib/src/employee.dart
class Employee{
  final int id;
  String name;
  num salary;
  Employee(this.id,this.name,this.salary);
}
```

創建服務 EmployeeService，程式如下：

```
//chapter14/child_component/lib/src/employee_service.dart
import 'package:angular/angular.dart';
import 'employee.dart';

@Injectable()
class EmployeeService{
  Future<List<Employee>> getAll() async{
  var emps = [
        Employee(1, 'Mr. Nice',3000),
        Employee(2, 'Narco',3105),
        Employee(3, 'Bombasto',3988),
        Employee(4, 'Celeritas',3401),
        Employee(5, 'Magneta',9971),
        Employee(6, 'RubberMan',4533),
        Employee(7, 'Dynama',6720),
        Employee(8, 'Dr IQ',4907),
```

```
            Employee(9, 'Magma',5278),
            Employee(10, 'Tornado',7800)
        ];
        return emps;
    }
}
```

在子元件 EmployeeComponent 中匯入 employee.dart 檔案。程式如下：

```
import '../employee.dart';
```

在子元件 EmployeeComponent 中定義一個 Employee 類型的屬性 emp，並使用 @Input 註釋將其標記為輸入屬性。程式如下：

```
@Input()
Employee emp;
```

增加核心指令常數 coreDirectives 到指令清單。使用 NgIf 指令判斷屬性 emp 是否為空，如果不為空，則在範本中使用插值存取屬性 emp 的資料。範例程式如下：

```
//chapter14/child_component/lib/src/employee/employee_component.dart
import 'package:angular/angular.dart';
import '../employee.dart';

@Component(
    selector:'employee',
    template: '''
        <div *ngIf="emp != null">
            <div> 員工 ID：{{emp.id}}</div>
            <div> 員工名稱：{{emp.name}}</div>
            <div> 員工薪資：{{emp.salary}}</div>
        </div>
            ''',
    directives: [coreDirectives],
)
class EmployeeComponent{
    @Input()
    Employee emp;
}
```

綁定的目標是綁定標記 ([]、() 或 [()]) 內的屬性或事件，資料來源位於引號 ("") 或插值符號 ({{}}) 內。綁定目標有了，現在需要綁定資料來源。已知在父元件中有一個清單屬性 emps，因此可以將清單中被選中的項目作為輸入資料來源。在父元件中宣告接收選中項目的 Employee 類型的屬性 selected，宣告透過點擊事件為屬性 selected 設定值的 onSelect() 方法。程式如下：

```
Employee selected;
void onSelect(Employee emp){
   selected = emp;
}
```

根元件 AppComponent 的完整程式如下：

```
//chapter14/child_component/lib/app_component.dart
import 'package:angular/angular.dart';

import 'src/employee.dart';
import 'src/employee_service.dart';
import 'src/employee/employee_component.dart';

@Component(
   selector: 'my-app',
   styleURLs: ['app_component.css'],
   templateURL: 'app_component.html',
   directives: [coreDirectives,EmployeeComponent],
   providers: [ClassProvider(EmployeeService)],
)
class AppComponent implements OnInit{
   // 定義一個 EmployeeService 類型的私有變數
   final EmployeeService _employeeService;
   List<Employee> emps;
   // 注入服務
   AppComponent(this._employeeService);
   // 覆寫 ngOnInit 方法
   @override
   void ngOnInit(){
      // 在元件初始化時呼叫 getEmployees 方法
```

```
        getEmployees();
    }
    // 使用服務物件 _employeeService 獲取員工清單
    void getEmployees() async{
        emps = await _employeeService.getAll();
    }

    // 被選中的員工
    Employee selected;
    // 更新選中的員工
    void onSelect(Employee emp){
        selected = emp;
    }
}
```

更新範本為每個項目綁定點擊事件，並為子元件 employee 增加屬性綁定。程式如下：

```
//chapter14/child_component/lib/app_component.html
<ul class="emps">
    <li *ngFor="let emp of emps"
        (click)="onSelect(emp)" >
        <span class="badge">{{emp.id}}</span>
        {{emp.name}}
        <span class="badge salary">{{emp.salary}}</span>
    </li>
</ul>

<employee [emp]="selected"></employee>
```

更新樣式檔案以獲得更好的互動體驗。樣式程式如下：

```
//chapter14/child_component/lib/app_component.css
.emps {
    margin: 002em 0;
    list-style-type: none;
    padding: 0;
    width: 15em;
}
.emps li {
```

```
    cursor: pointer;
    position: relative;
    left: 0;
    background-color: #EEE;
    margin: .5em;
    padding: .3em 0;
    height: 1.6em;
    border-radius: 4px;
}
.emps li:hover {
    color: #607D8B;
    background-color: #EEE;
    left: .1em;
}
.emps .text {
    position: relative;
    top: -3px;
}
.emps .badge {
    display: inline-block;
    font-size: small;
    color: white;
    padding: 0.8em 0.7em 00.7em;
    background-color: #607D8B;
    line-height: 1em;
    position: relative;
    left: -1px;
    top: -4px;
    height: 1.8em;
    margin-right: .8em;
    border-radius: 4px 004px;
}
.emps .salary {
    float:right;
    margin-right:0;
    margin-left: .8em;
    border-radius: 04px 4px 0;
}
```

執行專案，點擊清單中的任何項目，子元件視圖都會更新為該項目的內容，這裡點擊最後一個項目，執行結果如圖 14-19 所示。

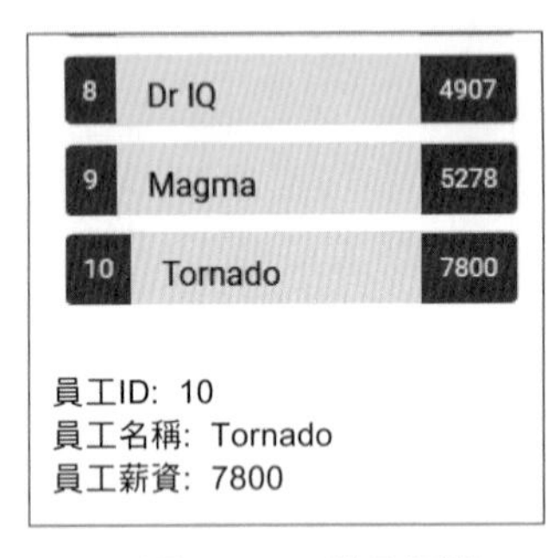

▲ 圖 14-19 輸入屬性

2. 輸出屬性

輸出屬性是事件綁定的目標，表示它會觸發某種事件。指令通常使用 StreamController 物件引發自訂事件，創建 StreamController 物件並將其 stream 作為屬性公開。透過呼叫 StreamController.add(payload) 觸發一個事件，並傳遞攜帶有效酬載的訊息，它可以是任何類型和結構，父指令透過事件綁定到該屬性來監聽事件，並透過 $event 物件存取事件傳遞的資訊。

現在透過範例演示輸出屬性的定義與使用，想法是在子元件中透過手動提供資料並觸發事件。父元件透過與子元件的輸出屬性綁定的敘述接收訊息，並傳給相關屬性以供範本顯示。

在子元件中使用註釋 @Output() 宣告輸出屬性，對外部來説是事件。範例程式如下：

```
//chapter14/child_component/lib/src/employee/employee_component.dart
import 'dart:async';

import 'package:angular/angular.dart';
import '../employee.dart';
```

```
@Component(
   selector:'employee',
   templateURL: 'employee_component.html',
   directives: [coreDirectives],
)
class EmployeeComponent{
   // 宣告輸入屬性
   @Input()
   Employee emp;
   // 宣告 StreamController 物件，該物件可以用於觸發事件
   final _str = StreamController<String>();
   // 宣告輸出屬性 add，對外部來說就是事件
   @Output()
   get add => _str.stream;
   // 用於處理點擊事件的函數
   void addStr(var str){
      // 發送資料到事件
      _str.add(str);
   }
}
```

該元件定義了一個 StreamController 屬性，並透過 add 屬性的 getter 方法公開了該控制器的 stream 物件。

將元件中原有的 HTML 元素放置在範本檔案中並增加新的 HTML 元素，輸入框用於手動輸入資料，按鈕的點擊事件用於將輸入框的值傳遞給 addStr 方法，方法 addStr 呼叫 StreamController 物件的 add 方法以觸發一個事件。範本程式如下：

```
//chapter14/child_component/lib/src/employee/employee_component.html
<div *ngIf="emp != null">
      <div> 員工 ID：{{emp.id}}</div>
      <div> 員工名稱：{{emp.name}}</div>
      <div> 員工薪資：{{emp.salary}}</div>
</div>

<input #input >
<button (click)="addStr(input.value)"> 增加 </button>
```

當使用者點擊增加按鈕時，將呼叫元件的 addStr() 方法以觸發事件並傳遞資料。

在父元件中提供接收資料的屬性 receiver，並提供回應 employee 子元件 add 事件的函數 onAdd。程式如下：

```
//chapter14/child_component/lib/app_component.dart
// 用於接收資料
String receiver;
// 回應 add 事件的處理常式
onAdd(var event){
  receiver = event;
}
```

在父元件範本中增加元素程式如下：

```
//chapter14/child_component/lib/app_component.html
<p *ngIf="receiver != null">{{' 子元件提供的資料：'+receiver}}</p>
<employee [emp]="selected" (add)="onAdd($event)"></employee>
```

在輸入框中隨意輸入資料，點擊「增加」按鈕，add 事件將觸發，Angular 呼叫父元件的 onAdd 方法，並傳遞 $event 物件到其內部設定值給 receiver 屬性，因為控制器 StreamController 的泛型是 String，所以物件 $event 的資料是 String 類型的。執行效果如圖 14-20 所示。

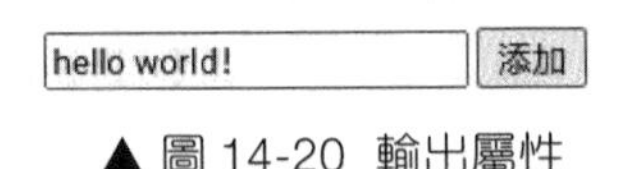

▲ 圖 14-20 輸出屬性

3. 起別名

透過向註釋 @Input 和 @Output 傳遞一個 String 類型的參數，可以為輸入和輸出屬性起別名。範例程式如下：

```
@Input('employee')
Employee emp;

final _str = StreamController<String>();
```

```
@Output('addString')
get add => _str.stream;
void addStr(var str){
   _str.add(str);
}
```

在屬性綁定和事件綁定中使用別名。程式如下：

```
<employee [employee]="selected" (addString)="onAdd($event)"></employee>
```

14.6.4 雙向資料綁定

有時希望既要顯示元件屬性值，又要在使用者進行更改時更新該屬性。對元素來說既要設定元素屬性又要監聽元素更改事件。輸入和輸出屬性為這樣的場景提供了技術支援。

雙向資料綁定結合了屬性綁定和事件綁定，其綁定語法 [(x)] 也結合了屬性綁定語法 [] 和事件綁定語法 ()。

綁定範例程式如下：

```
<element [(x)]="property" ></element>
```

準確來說，雙向資料綁定是屬性綁定和事件綁定的語法糖。與之等值的程式如下：

```
<element [x]="property" (xChange)="property=$event"></element>
```

在命名時屬性名稱可以是符合規範的識別符號，事件名是屬性名稱加尾碼 Change。語法 [()] 很容易被驗證，為元素提供一個可設定的屬性 x，並為其提供名為 xChange 的對應事件。

透過範例演示其工作原理，在 lib/src 目錄下新建 sizer_component.dart 檔案，並創建元件 SizerComponent。程式如下：

```
import 'dart:async';
import 'dart:math';
```

```
import 'package:angular/angular.dart';
const minSize = 8;
const maxSize = minSize * 5;
@Component(
   selector: 'my-sizer',
   template: '''
      <div>
         <button (click)="dec()" [disabled]="size <= minSize">-</button>
         <button (click)="inc()" [disabled]="size >= maxSize">+</button>
         <label [style.font-size.px]="size">FontSize: {{size}}px</label>
      </div>''',
   exports: [minSize, maxSize],
)
class SizerComponent {
   int _size = minSize * 2;
   int get size => _size;
   @Input()
   void set size(/*String|int*/ val) {
      int z = val is int ? val : int.tryParse(val);
      if (z != null) _size = min(maxSize, max(minSize, z));
   }
   final _sizeChange = StreamController<int>();
   @Output()
   Stream<int> get sizeChange => _sizeChange.stream;
   void dec() => resize(-1);
   void inc() => resize(1);
   void resize(int delta) {
      size = size + delta;
      _sizeChange.add(size);
   }
}
```

在 AppComponent 中匯入 sizer_component.dart 檔案，增加 SizerComponent 到指令清單，並提供綁定參數 fontSizePx。程式如下：

```
//chapter14/child_component/lib/app_component.dart
import 'package:angular/angular.dart';

import 'src/employee.dart';
```

```
import 'src/employee_service.dart';
import 'src/employee/employee_component.dart';
import 'src/sizer_component.dart';

@Component(
   selector: 'my-app',
   styleURLs: ['app_component.css'],
   templateURL: 'app_component.html',
   directives: [coreDirectives,EmployeeComponent,SizerComponent],
   providers: [ClassProvider(EmployeeService)],
)
class AppComponent implements OnInit{
   //...
   // 用於雙向資料綁定
   int fontSizePx=10;
}
```

在根元件範本中增加新元素：

```
//chapter14/child_component/lib/app_component.html
<my-sizer [(size)]="fontSizePx" #mySizer></my-sizer>
<div [style.font-size.px]="mySizer.size">Resizable Text</div>
```

點擊加號或減號按鈕可以控制字型大小，執行結果如圖 14-21 所示。

- + FontSize: 19px
Resizable Text

▲ 圖 14-21 雙向資料綁定

14.7 表單

表單是業務類應用程式必備的內容，可以使用表單完成註冊、登入、預定班機及其他資料的輸入任務。

在編寫表單時，對輸入資料做驗證以有效指導使用者和良好的輸入體驗是非常重要的。本節將介紹透過 angular_forms 套件對表單元素進行雙向資料綁定、更改追蹤、驗證和錯誤處理。

Angular 表單由 angular_forms 套件提供指令和樣式，它在 HTML 基礎資料表單提供進階功能，將套件增加到依賴項並執行 pub get 命令：

```
//chapter14/forms/pubspec.yaml
dependencies:
  angular: ^5.3.0
  angular_components: ^0.13.0
  angular_forms: ^2.1.2
```

14.7.1 建立資料模型

當使用者輸入表單資料時，需要捕捉更改資訊以更新模型的實例。模型通常是表單中各個輸入項的集合，這裡模型 Employee 包含了 id、age、name 和 department 4 個欄位。範例程式如下：

```
//chapter14/forms/lib/src/employee.dart
class Employee{
  int id,age;
  String name,department;
  Employee(this.id,this.name,this.age,this.department);
  // 覆寫 toString 方法
  String toString(){
    return '$id: $name ($age) 部門：$department';
  }
}
```

14.7.2 建立表單

表單分為兩個部分：以 HTML 為基礎的範本和處理資料及使用者互動的元件類別。

在元件中增加程式如下：

```
//chapter14/forms/lib/app_component.dart
import 'dart:convert';

import 'package:angular/angular.dart';
import 'package:angular_forms/angular_forms.dart';
import 'src/employee.dart';

@Component(
  selector: 'my-app',
  styleURLs: ['app_component.css'],
  templateURL: 'app_component.html',
  directives: [coreDirectives,formDirectives],
)
class AppComponent {
  // 表單中下拉清單需要的選項清單
  static const List<String> deps = ['設計部','技術部','財務部','行政部'];
  // 初始化表單模型資料
  Employee model = Employee(1,'姚環',23,deps[0]);
}
```

元件內容主要匯入了模型 Employee，並且為 model 和 deps 提供了模擬資料。submitted 用於記錄表單提交狀態，onSubmit 方法用於處理表單的提交請求，這裡並沒有透過 onSubmit 方法將表單資料提交到伺服器，只是將資料編碼為 json 並儲存在屬性 data 中。

在範本檔案中編寫表單程式如下：

```
//chapter14/forms/lib/app_component.html
<div class="container">
  <h1>員工表單</h1>
  <form>
    <div class="form-group">
      <label for="name">名字 *</label>
      <input type="text" class="form-control" id="name" required>
    </div>
    <div class="form-group">
      <label for="age">年齡</label>
```

```
            <input type="number" class="form-control" id="age">
        </div>
        <div class="form-group">
            <label for="department">所屬部門 *</label>
            <select class="form-control" id="department" required>
                <option *ngFor="let dep of deps" [value]="dep">{{dep}}</option>
            </select>
        </div>
        <div class="row">
            <div class="col-auto">
                <button type="submit" class="btn btn-primary">提交</button>
            </div>
            <small class="col text-right">* 必需</small>
        </div>
    </form>
</div>
```

表單列出了 3 個表單元件，它們分別展示模型 Employee 的 name、age 和 department 欄位。使用者可以在輸入框中提供內容，或在下拉清單中選擇項目。

控制項 input 具有 required 屬性，表示必填。在表單中還會發現用於控制結構的元素上設定有 form-group、form-control 樣式類別，以及為提交按鈕設定的一些樣式類別。樣式類別 form-group 是將其包裹的元素作為一組，form-control 樣式類別應用於基礎表單控制項，label 元素中的 for 屬性的值是與其連結的表單控制項的 id 名。這些樣式類別是在 Bootstrap 框架中定義的，在項目的 index.html 檔案中的 <head> 元素中引用以下樣式檔案。引用程式如下：

```
<link rel="stylesheet"
href="https://maxcdn.bootstrapcdn.com/bootstrap/4.0.0-beta/css/bootstrap.min.css"
integrity="sha384-/Y6pD6FV/Vv2HJnA6t+vslU6fwYXjCFtcEpHbNJ0lyAFsXTsjBbfaDjzALeQ
sN6M" crossorigin="anonymous">
```

執行專案結果如圖 14-22 所示。

員工表單

名字*

年齡

所屬部門*

設計部

Submit

*必備

▲ 圖 14-22 表單

14.7.3 表單指令

此時並沒有將表單控制項與 Employee 模型資料產生聯繫，這種聯繫指模型資料能夠在對應表單控制項中顯示，還要能透過輸入控制項更新模型資料。實現這樣的功能首先想到的是雙向資料綁定，而 angular_forms 套件提供了指令 NgModel，該指令使得表單控制項與模型實現雙向資料綁定，其語法如下：

```
<element [(ngModel)]="expression"></element>
```

NgModel 指令會創建 NgControl 實例，並將其綁定到表單控制項元素。表單 NgControl 實例追蹤控制項的值、使用者互動和驗證狀態，並使視圖與模型保持同步。NgModel 指令旨在用作獨立值，如果希望將其作為 Angular 表單系統的一部分，則必須使用 NgControl 指令為其指定一個名字，該名字可以是整個表單欄位任何唯一的值。NgControl 指令的作用是將控制項註冊到 Angular 表單系統，在內部 Angular 創建 NgFormControl 實例，每個 NgFormControl 都以分配給 ngControl 指令的名稱註冊。

在表單控制項增加 NgModel 和 NgControl 指令，修改後的部分範本程式如下：

```
//chapter14/forms/lib/app_component.html
<div class="form-group">
        <label for="name"> 名字  *</label>
        <input type="text" class="form-control" id="name" required
          [(ngModel)]="model.name" ngControl="name">
      </div>
      <div class="form-group">
        <label for="age"> 年齡 </label>
        <input type="number" class="form-control" id="age"
          [(ngModel)]="model.age" ngControl="age">
      </div>
      <div class="form-group">
        <label for="department"> 所屬部門  *</label>
        <select class="form-control" id="department" required
            [(ngModel)]="model.department" ngControl="department">
          <option *ngFor="let dep of deps" [value]="dep">{{dep}}</option>
        </select>
</div>
```

現在執行專案，結果如圖 14-23 所示。

▲ 圖 14-23 表單指令

NgControl 指令使得 Angular 表單的每個控制項都能夠追蹤自己的狀態，並透過以下屬性使得狀態可供檢查。

(1) dirty 和 pristine：指示控制項的值是否已更改。

(2) touched 和 untouched：表示控制項是否已被存取。
(3) valid：反映控制項值的有效性。

其中控制項的 valid 屬性最常用，為了提供良好的視覺回饋，使用 Bootstrap 表單樣式類別 is-valid 和 is-invalid 標識控制項值是否有效。

在名字的 input 元素上增加範本引用變數，變數名稱叫 name。範本引用變數 name 透過語法 #name="ngForm" 綁定到與輸入元素連結的 NgModel。指令的 exportAs 屬性是可以在範本中使用的名稱，用於將該指令分配給變數。因為 ngModel 指令的 exportAs 屬性值為 ngForm，所以將範本引用變數 name 的值設定為 ngForm。

使用引用變數 name 和 CSS 類別綁定有條件地分配適當的表單有效性 CSS 類別。範本程式如下：

```
<div class="form-group">
   <label for="name">名字 *</label>
   <input type="text" class="form-control" id="name" required
      [(ngModel)]="model.name" ngControl="name"
      #name="ngForm"
      [class.is-valid]="name.valid"
      [class.is-invalid]="!name.valid">
</div>
```

14.7.4 提交表單

此時填寫完符合要求的資料後提交表單，在表單底部的提交按鈕不會執行任何操作，但由於其類型為 submit，因此會觸發表單提交事件。

為了使表單可以正常使用，將元件的 onSubmit 方法與表單的 ngSubmit 事件綁定。程式如下：

```
<form (ngSubmit)="onSubmit()" #empForm="ngForm">
```

Angular 會自動創建一個 NgForm 指令並附加在 form 元素上，NgForm 指令的 exportAs 屬性值為 ngForm，因此可以將其設定值給範本引用變數。NgForm 指令透過附加功能補充了 form 元素，它包含使用 ngModel 和 ngControl 指令為表單元素創建的控制項，並監視它們的屬性，包括有效性。

透過引用變數 empForm 將表單的整體有效性綁定到提交按鈕的 disabled 屬性。程式如下：

```
<button [disabled]="!empForm.form.valid" type="submit" class="btn btn-
primary"> 提交 </button>
```

Angular 表單是某些層次結構中控制項的集合，可以透過 NgForm 指令的屬性 value 存取整數個表單的資料，屬性 value 是一個反映表單資料結構的 JSON 物件。增加 onSubmit 方法並提供一個接收表單值的屬性 data。範例程式如下：

```
//chapter14/forms/lib/app_component.dart
import 'dart:convert';

import 'package:angular/angular.dart';
import 'package:angular_forms/angular_forms.dart';
import 'src/employee.dart';

@Component(
   selector: 'my-app',
   styleURLs: ['app_component.css'],
   templateURL: 'app_component.html',
   directives: [coreDirectives,formDirectives],
)
class AppComponent {
   // 表單中下拉清單需要的選項清單
   static const List<String> deps = [' 設計部 ',' 技術部 ',' 財務部 ',' 行政部 '];
   // 初始化表單模型資料
   Employee model = Employee(1,' 姚環 ',23,deps[0]);
   // 用於儲存表單資料
   String data;
```

```
    // 用於更新表單資料
    void onSubmit(val){
        // 接收表單中的資料並編碼為 json
        data = json.encode(val);
    }
}
```

修改表單上 ngSubmit 事件綁定的資訊，並在表單外透過插值顯示 data 屬性的值。範本程式如下：

```
//chapter14/forms/lib/app_component.html
<div class="container">
    <h1> 員工表單 </h1>
    <form (ngSubmit)="onSubmit(empForm.value)" #empForm="ngForm">
        <div class="form-group">
            <label for="name"> 名字  *</label>
            <input type="text" class="form-control" id="name" required
                [(ngModel)]="model.name" ngControl="name"
                #name="ngForm"
                [class.is-valid]="name.valid"
                [class.is-invalid]="!name.valid">
            </div>
            <div class="form-group">
                <label for="age"> 年齡 </label>
                <input type="number" class="form-control" id="age"
                    [(ngModel)]="model.age" ngControl="age">
            </div>
            <div class="form-group">
                <label for="department"> 所屬部門  *</label>
                <select class="form-control" id="department" required
                        [(ngModel)]="model.department" ngControl="department">
                    <option *ngFor="let dep of deps" [value]="dep">{{dep}}</option>
                </select>
            </div>
            <div class="row">
                <div class="col-auto">
                    <button [disabled]="!empForm.form.valid" type="submit"
class="btn btn-primary"> 提交 </button>
                </div>
```

```
            <small class="col text-right">*  必需 </small>
        </div>
    </form>
    <p>{{data}}</p>
</div>
```

執行專案，點擊 Submit 按鈕，執行效果如圖 14-24 所示。

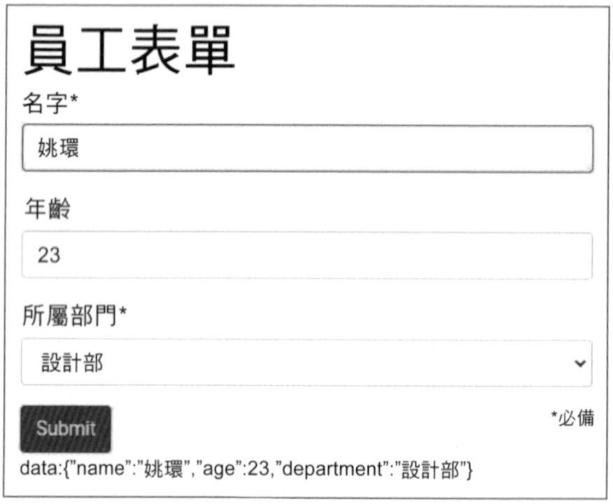

▲ 圖 14-24 表單資料處理

14.8 Angular 架構回顧

至此已經學習了 Angular 框架的基礎知識，現在來回顧它的各個組成部分：範本、指令、註釋、元件、服務、資料綁定、依賴注入和提供者。

(1) 範本：通常由 HTML 元素組成，它告訴 Angular 如何呈現元件。範本與普通 HTML 又有一些區別，例如：可在元素上使用屬性綁定、事件綁定；可以將元素與 NgIf 或 NgFor 等指令配合使用；甚至可以包含自訂元素。其中自訂元素是在 Angular 元件的註釋中指定的自訂選擇器名。

(2) 指令：範本是動態的，當 Angular 繪製它們時，將根據指定指令轉換 DOM。指令包含元件、屬性指令和結構指令。

(3) 註釋：註釋告訴 Angular 如何處理一個類別或屬性。註釋 @Component 將一個類別標識為元件，註釋 @Directive 將一個類別標識為指令，註釋 @Injectable 將一個類別標識為服務，註釋 @Pipe 將一個類別標識為管道，註釋 @Input 將一個屬性標識為輸入屬性，註釋 @Output 將一個屬性標識為輸出屬性。

(4) 元件：元件用於控制其範本呈現的視圖，在元件類別中定義應用程式邏輯，該類別透過屬性和方法的 API 與視圖進行互動。

(5) 服務：可以涵蓋的內容很多，主要包括提供應用程式需要的功能和資料。

(6) 資料綁定：一種協調範本和元件的機制，將綁定標記增加到 HTML 範本，以告訴 Angular 如何連接範本和元件。資料綁定包含插值、屬性綁定、事件綁定和雙向資料綁定。

(7) 依賴注入：依賴注入是一種為類別的新實例提供所需的完整依賴關係的方法，大多數依賴項是服務。最常見的場景是使用依賴注入為元件提供所需的服務，Angular 透過查看建構函數參數的類型來判斷元件需要哪些服務。注入器維護其已創建的服務實例的容器，在 Angular 創建元件時，會先在注入器詢問元件所需的服務，如果請求的服務實例不在容器中，則注入程式將創建一個服務並將其增加到容器中，然後再將服務返回到 Angular。解析並返回所有請求的服務後，Angular 可以使用這些服務作為參數來呼叫元件的建構函數。

(8) 提供者：在使用依賴注入前，必須使用注入器註冊一個服務的提供者，提供者可以創建或返回服務，並且通常是服務類別本身。可以向元件註冊提供者，也可以在啟動應用程式時透過根注入器註冊。註冊提供者的常見方法是在 @Component 註釋的 providers 參數值中指定。當元件需要某個服務時，如果注入器中沒有對應服務，注入器將使用註冊的提供者創建該服務。

Chapter 15

Angular 進階

15.1 屬性指令

一個屬性指令常用於更改 DOM 元素的外觀或行為。

Angular 中有 3 種指令：

(1) 元件：帶有範本的指令。
(2) 結構指令：透過增加和刪除 DOM 元素更改 DOM 佈局。
(3) 屬性指令：更改元素、元件或其他指令的外觀或行為。

屬性指令有 2 種類型：

(1) 以類別為基礎：使用類別實現功能齊全的屬性指令。
(2) 函數式：使用頂層函數實現的無狀態屬性指令。

15.1.1 以類別為基礎的屬性指令

創建以類別為基礎的屬性指令需要先編寫帶有 @Directive 註釋的控制類別，並且需向註釋提供屬性選擇器。控制器類別的建構函數接收一個

HTML 元素作為參數，並透過操作該元素實現所需的指令行為。開始前先創建新專案，只留下根元件 AppComponent。

1. 編寫指令程式

在專案 lib/src 目錄下創建 fontsize_directive.dart 檔案，並編寫 Fontsize Directive 指令。程式如下：

```
//chapter15/attribute_directive/lib/src/fontsize_directive.dart
import 'dart:html';
import 'package:angular/angular.dart';
// 指令註釋
//selector 為屬性選擇器
@Directive(selector: '[myFontsize]')
class FontsizeDirective {
  // 指令的建構函數
  // 注入宿主元素物件
  FontsizeDirective(Element el) {
    // 修改元素的樣式：字型大小
    el.style.fontSize = '24px';
  }
}
```

@Directive 註釋中的 CSS 選擇器用於標識在範本中與指令連結的 HTML 元素。CSS 選擇器使用中括號包裹識別符號來表示屬性選擇器，這裡的屬性選擇器是 [myFontsize]。Angular 會將指令行為應用於範本中所有具有 myFontsize 屬性的元素。

儘管屬性命名為 fontsize 比 myFontsize 更簡潔，但對於自訂屬性採用字首實際上是最好的做法，這裡的字首是 my。這樣可以避免與標準 HTML 屬性衝突，也可以減少與第三方命名衝突的機率。字首不能使用 ng，該字首僅為 Angular 內部使用，使用該字首可能導致錯誤。

@Directive 註釋後邊是指令的控制器類別，名為 FontsizeDirective。Angular 為每個匹配的元素創建指令的控制器類別的新實例，並將 HTML

元素 (Element) 注入控制器類別的建構函數。在函數本體中透過操作元素物件控制元素的外觀或行為。

2. 應用屬性指令

要使用指令 FontsizeDirective，需在範本中為元素提供 myFontsize 屬性。此範本將指令作為屬性應用於段落 (<p>) 元素，在 Angular 術語中 p 元素是屬性 myFontsize 的宿主。

在根元件範本中增加以下內容：

```
//chapter15/attribute_directive/lib/app_component.html
<p myFontsize>Hello World!</p>
```

現在，在根元件 AppComponent 中匯入指令檔案，並將 FontsizeDirective 指令增加到指令清單中。這樣當 Angular 在根元件範本中遇到 myFontsize 屬性時就會辨識指令。程式如下：

```
//chapter15/attribute_directive/lib/app_component.dart
import 'package:angular/angular.dart';
import 'src/fontsize_directive.dart';

@Component(
   selector: 'my-app',
   styleURLs: ['app_component.css'],
   templateURL: 'app_component.html',
   directives: [FontsizeDirective],
)
class AppComponent {
}
```

執行應用程式，在瀏覽器中 myFontsize 屬性將段落文字的字型大小置為 24px，執行結果如圖 15-1 所示。

AngularDart App 2

Hello World!

▲ 圖 15-1 以類別為基礎的屬性指令

Angular 在 p 元素上找到了 myFontsize 屬性，它創建了指令 FontsizeDirective 的新實例，並對 p 元素的引用注入指令的建構函數中，該建構函數將 p 元素的字型大小修改為 24px。

3. 回應使用者事件

現在 myFontsize 僅使用固定值設定元素的字型大小，為讓其更加動態，可以監聽使用者行為。例如：將滑鼠移入或移除元素，並透過修改字型大小來做出回應。

宣告 _el 私有屬性並在建構函數中使用宿主元素物件初始化。除此之外還需要記錄元素初始的字型大小，使用 _size 屬性完成此工作。程式如下：

```
Element _el;
String _size;

FontsizeDirective(Element el){
   _el = el;
   _size = _el.style.fontSize;
}
```

增加一個輔助方法，該方法設定宿主元素的字型大小。程式如下：

```
_fontsize([String size]){
   _el.style.fontSize = size ?? _size;
}
```

增加兩個事件處理常式，使它們在滑鼠進入或離開時做出回應，每個事件處理常式都由 @HostListener 註釋修飾。在處理常式體中使用輔助方法控制宿主元素的字型大小。範例程式如下：

```
@HostListener('mouseenter')
void onMouseEnter() {
   _fontsize('36px');
}

@HostListener('mouseleave')
void onMouseLeave() {
```

```
    _fontsize();
  }
  _fontsize([String size]){
    _el.style.fontSize = size ?? _size;
  }
```

@HostListener 註釋用於監聽指令或元件宿主元素的事件，它接收一個 DOM 事件名作為參數。這裡的宿主元素是 p，事件名分別是 mouseenter 和 mouseleave。

完整的指令程式如下：

```
//chapter15/attribute_directive/lib/src/fontsize_directive.dart
import 'dart:html';
import 'package:angular/angular.dart';
// 指令註釋
//selector 為屬性選擇器
@Directive(selector: '[myFontsize]')
class FontsizeDirective {
  // 快取宿主元素物件
  Element _el;
  // 快取宿主元素初始字型大小
  String _size;

  // 指令的建構函數
  // 注入宿主元素物件
  FontsizeDirective(Element el) {
    _el = el;
    _size = _el.style.fontSize;
  }
  // 輔助方法
  _fontsize([String size]){
    // 若提供參數 size 則採用該值，否則使用初始字型大小
    _el.style.fontSize = size ?? _size;
  }
  // 監聽宿主元素的滑鼠移入事件
  @HostListener('mouseenter')
  void onMouseEnter() {
    // 這裡使用固定字型大小
```

```
        _fontsize('24px');
    }

    // 監聽宿主元素的滑鼠移出事件
    @HostListener('mouseleave')
    void onMouseLeave() {
        _fontsize();
    }
}
```

刷新瀏覽器，執行程式，此時元素中的字型大小為其預設值，當滑鼠移到元素上時，元素的字型大小變為 24px。

4. 為指令設定值

目前字型大小都是在指令中強制寫入的，在本節將使用指令動態設定字型大小。

首先在指令類別增加輸入屬性 fontsize。範例程式如下：

```
// 輸入屬性
@Input()
String fontsize;
```

使用註釋 @Input 標記 fontsize 屬性，使 fontsize 屬性可用於屬性綁定。若屬性沒有註釋 @Input 標記，則 Angular 會拒絕綁定。

修改滑鼠移入事件的處理常式的邏輯程式。範例程式如下：

```
// 監聽宿主元素的滑鼠移入事件
@HostListener('mouseenter')
void onMouseEnter() {
    // 若 fontsize 不為空則使用該值，否則使用 24px
    _fontsize(fontsize ?? '24px');
}
```

在元素上應用 myFontsize 屬性指令時，可以使用輸入屬性 fontsize 自訂字型大小。最新範本程式如下：

```
//chapter15/attribute_directive/lib/app_component.html
<p myFontsize fontsize="13px">Hello World!</p>
<p myFontsize [fontsize]="'17px'">Hello World!</p>
```

刷新瀏覽器，將滑鼠移入 p 元素，p 元素內的字型大小將根據提供的值變化。

5. 綁定別名

現在已經可以動態地設定字型大小，但是此時需要提供兩個屬性。幸運的是，可以根據需要為指令屬性加上別名，此時可以使用屬性指令的名字作為輸入屬性的別名。範例程式如下：

```
@Input('myFontsize')
String fontsize;
```

在指令內部，該屬性是 fontsize，在指令外部它是 myFontsize。此時就將預想的屬性名稱和綁定語法結合在了一起，最新範本程式如下：

```
//chapter15/attribute_directive/lib/app_component.html
<p [myFontsize]="'19px'">Hello World!</p>
```

指令 FontsizeDirective 的最新版本程式如下：

```
//chapter15/attribute_directive/lib/src/fontsize_directive.dart
import 'dart:html';
import 'package:angular/angular.dart';
// 指令註釋
//selector 為屬性選擇器
@Directive(selector: '[myFontsize]')
class FontsizeDirective {
    // 快取宿主元素物件
    Element _el;
    // 快取宿主元素初始字型大小
    String _size;

    // 輸入屬性
    @Input('myFontsize')
    String fontsize;
```

```
    // 指令的建構函數
    // 傳入元素物件
    FontsizeDirective(Element el) {
      _el = el;
      _size = _el.style.fontSize;
    }
    // 輔助方法
    _fontsize([String size]){
      // 若提供參數 size 則採用該值，否則使用初始字型大小
      _el.style.fontSize = size ?? _size;
    }
    // 監聽宿主元素的滑鼠移入事件
    @HostListener('mouseenter')
    void onMouseEnter() {
    // 若 fontsize 不為空則使用該值，否則使用 24px
    _fontsize(fontsize ?? '24px');
    }

    // 監聽宿主元素的滑鼠移出事件
    @HostListener('mouseleave')
    void onMouseLeave() {
      _fontsize();
    }
}
```

15.1.2 函數式指令

函數式指令是呈現一次的無狀態指令，可以使用註釋 @Directive 修飾一個頂層函數來創建函數式指令。

創建函數式屬性指令，程式如下：

```
//chapter15/attribute_directive/lib/src/auto_id_directive.dart
import 'dart:html';
import 'package:angular/angular.dart';
// 計數變數
int _id = 0;
// 在頂層函數使用屬性指令註釋
```

```
// 注入宿主元素，並注入宿主屬性 auto-id 的值
@Directive(selector:'[auto-id]')
void autoIdDirective(Element el,@Attribute('auto-id') String prefix){
   // 以 auto-id 屬性的值為字首加計數變數 _id 的值
   // 為宿主元素的 id 屬性設定值
   el.id = '$prefix${_id++}';
}
```

指令的選擇器是 [auto-id]，表示指令將應用於帶有屬性 auto-id 的元素。與以類別為基礎的建構函數一樣第一個參數是宿主元素物件。@Attribute 註釋用於注入宿主元素中指定屬性的屬性值，這裡注入宿主元素 auto-id 屬性的值。在函數內部將屬性 auto-id 的值作為字首，附加計數變數 _id 的值作為宿主元素 id 屬性的值。

編寫功能指令時，需遵循以下規則：

(1) 使函數返回類型為 void。
(2) 在 @Directive 註釋中，僅使用 selector 參數。

在元件中匯入函數式指令檔案，並將指令增加到指令清單中。範例程式如下：

```
//chapter15/attribute_directive/lib/app_component.dart
import 'package:angular/angular.dart';
import 'src/fontsize_directive.dart';
import 'src/auto_id_directive.dart';

@Component(
   selector: 'my-app',
   styleURLs: ['app_component.css'],
   templateURL: 'app_component.html',
   directives: [FontsizeDirective,autoIdDirective],
)
class AppComponent {
}
```

在根元件範本中增加程式如下：

```
//chapter15/attribute_directive/lib/app_component.html
<div #d1 auto-id="div-">Auto-ID:{{d1.id}}</div>
<div #d2 auto-id="div-">Auto-ID:{{d2.id}}</div>
<p #p1 auto-id="p-">Auto-ID:{{p1.id}}</p>
<div #d3 auto-id="div-">Auto-ID:{{d3.id}}</div>
<div #d4 auto-id="div-">Auto-ID:{{d4.id}}</div>
<p #p2 auto-id="p-">Auto-ID:{{p2.id}}</p>
```

儘管功能指令是無狀態的，但它們可能受全域狀態影響，執行效果如圖 15-2 所示。

Auto-ID:div-0
Auto-ID:div-1

Auto-ID:p-2

Auto-ID:div-3
Auto-ID:div-4

Auto-ID:p-5

▲ 圖 15-2 函數式指令

15.2 元件樣式

Angular 應用程式採用標準 CSS 設定範本元素的樣式，即可以在應用程式中使用 CSS 樣式表、選擇器、規則和媒體查詢。

Angular 將元件樣式與元件綁定在一起，即元件樣式只對元件範本內的元素有效，這樣可以避免多個元件的樣式相互干擾，為樣式的模組化設計提供便利。

前面已經介紹了在元件中使用 styles 參數提供內聯樣式，使用 styleURLs 參數提供樣式表檔案。在這裡主要介紹一些特殊選擇器，它們可以用於設定父元件、子元件甚至全域樣式。

15.2.1 :host

使用 :host 偽類別選取器，可以在元件的宿主元素中應用目標樣式，而非在元件範本內定位元素。宿主元素是指在定義元件時提供的選擇器，:host 選擇器是存取宿主元素的唯一途徑，因為它不是元件範本的一部分，該宿主元素位於父元件的範本中。

元件定義程式如下：

```
//chapter15/component_styles/lib/src/child/child_component.dart
import 'package:angular/angular.dart';

@Component(
   selector: 'child',
   styleURLs: ['child_component.css'],
   templateURL: 'child_component.html',
   directives: [],
)
class ChildComponent{
}
```

在範本中增加元素程式如下：

```
//chapter15/component_styles/lib/src/child/child_component.html
<p>child 元件 </p>
```

定義宿主元素樣式程式如下：

```
//chapter15/component_styles/lib/src/child/child_component.css
// 匹配宿主元素
:host{
   // 宿主元素以區塊元素顯示
   display: block;
   margin:6px;
   // 在宿主元素增加 1 像素實線黑色邊框
   border: 1px solid black;
   // 將宿主元素設定為綠色
   background-color:green;
}
```

在根元件 AppComponent 中引入子元件，並將其增加到指令清單。程式如下：

```
//chapter15/component_styles/lib/app_component.dart
import 'package:angular/angular.dart';
import 'src/child/child_component.dart';

@Component(
   selector: 'my-app',
   styleURLs: ['app_component.css'],
   templateURL: 'app_component.html',
   directives: [ChildComponent],
)
class AppComponent {
}
```

在根元件範本中增加元素程式如下：

```
//chapter15/component_styles/lib/app_component.html
<div>AppComponent 中 </div>
<child></child>
```

執行結果如圖 15-3 所示。

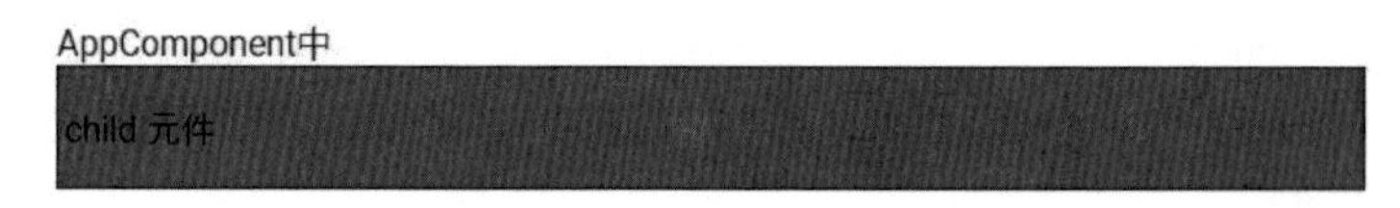

▲ 圖 15-3 宿主元素樣式

15.2.2 :host()

使用形式 :host() 是在 :host 的基礎上提供另一個選擇器，使用該選擇器篩選宿主元素，滿足條件的宿主元素將被匹配。

以下選擇器仍然將宿主元素作為目標，篩選條件是 CSS 類別中帶有 class 名 active 的宿主元素。程式如下：

```
//chapter15/component_styles/lib/src/child/child_component.css
// 匹配帶有 class 名 active 的宿主元素
```

```
:host(.active) {
  // 將宿主元素邊框寬度設定為 3 像素
  border-width: 3px;
}
```

在根元件範本中增加一個 CSS 類別中帶有 class 名 active 的 child 元素。範本程式如下：

```
//chapter15/component_styles/lib/app_component.html
<div>AppComponent 中 </div>
<child></child>
<child class="active"></child>
```

執行結果如圖 15-4 所示。

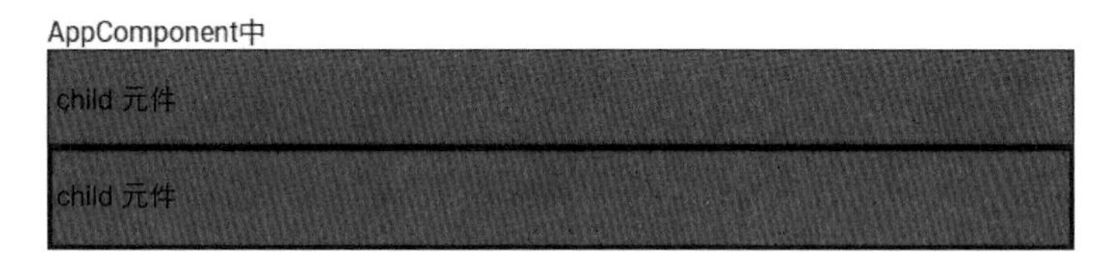

▲ 圖 15-4 樣式應用於部分宿主元素

15.2.3 :host-context()

有時需要根據元件視圖外部的某些條件更改元件範本內元素的外觀。偽類別選取器 :host-context() 工作方式與 :host() 的形式相同，:host-context() 接收一個選擇器作為參數，在元件宿主元素的所有父項目中尋找該選擇器，直到文件根，若宿主元素的任意父項目滿足該選擇器的尋找條件，則該宿主元素被匹配。:host-context() 選擇器需與另一個選擇器結合使用，當找到滿足條件的父項目後，再在元件範本中匹配第二個選擇器所指定的元素。

以下範例僅在宿主元素的任意父項目具有 CSS 類別 theme 的情況下，將背景顏色樣式應用於元件範本內的所有 h2 元素。程式如下：

```
//chapter15/component_styles/lib/src/child/child_component.css
```

```
/* 匹配父項目帶有 CSS 類別 theme 的宿主元素，再匹配滿足條件的宿主元素下的 h2 元素
*/
:host-context(.theme) h2{
/* 設定 h2 元素的背景顏色 */
background-color: #eef;
}
```

更新元件元素程式如下：

```
//chapter15/component_styles/lib/src/child/child_component.html
<p>child 元件 </p>
<h2>child h2</h2>
```

在根元件範本中增加元素程式如下：

```
<div class="theme">
   <h2>AppComponent .theme h2</h2>
   <child></child>
</div>
```

為了便於瞭解，為父項目擁有類別 theme 的 child 元素增加同級元素 h2，這樣就可以比較樣式應用的範圍。

執行結果如圖 15-5 所示。

▲ 圖 15-5 樣式應用於部分宿主元素的子元素

15.2.4 ::ng-deep

元件樣式通常僅適用於元件自己範本中的 HTML，使用 ::ng-deep 選擇器可將樣式向下強制透過子元件樹應用於所有子元件範本中滿足條件的元素。

::ng-deep 選擇器適用於巢狀結構元件的任何深度，並且適用於該元件的視圖子級和內容子級。在根元件樣式檔案中增加樣式程式如下：

```
//chapter15/component_styles/lib/app_component.css
:host ::ng-deep h2{
   font-style: italic;
}
```

該樣式將應用於根元件 AppComponent 範本中的所有 h2 元素，以及所有子元件範本中的所有 h2 元素，執行結果如圖 15-6 所示。

AppComponent .theme h2

▲ 圖 15-6 樣式應用於子元件中的元素

15.2.5 樣式匯入

可以在一個 CSS 檔案中匯入另一個 CSS 檔案，規則適用於標準 CSS @import 規則。創建外部 CSS 檔案 external.css，並增加樣式程式如下：

```
//chapter15/component_styles/lib/src/child/external.css
p{
   /* 字型顏色：白色 */
   color: white;
   /* 背景顏色：木色 */
   background-color: bURLywood;
}
```

在元件 ChildComponent 樣式檔案中採用相對路徑匯入該檔案的 URL：

```
//chapter15/component_styles/lib/src/child/child_component.css
@import 'external.css';
```

刷新瀏覽器，執行結果如圖 15-7 所示。

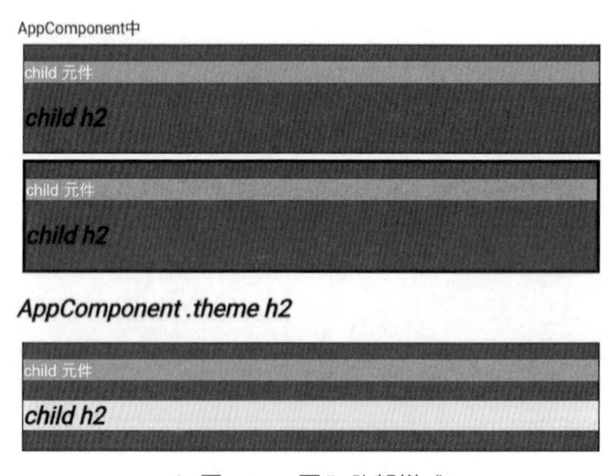

▲ 圖 15-7 匯入外部樣式

15.2.6 視圖封裝

預設情況下元件樣式被封裝，僅影響元件範本中的 HTML 元素或自訂元素。可以使用特殊的選擇器來影響元件視圖之外的元素，也可以完全禁用元件的視圖封裝。

禁用視圖封裝會使元件的樣式變為全域樣式，為此需將元件的中繼資料 encapsulation 參數設定為 ViewEncapsulation.None。程式如下：

```
@Component(
    //・・・
    encapsulation: ViewEncapsulation.None,
)
```

ViewEncapsulation 列舉可以具有兩個值：

(1) Emulated：預設值，Angular 透過前置處理 CSS 來模擬 shadow DOM 的行為，以便有效地將 CSS 範圍限定在元件的視圖中。

(2) None：Angular 不進行視圖封裝，相反它使元件的樣式變為全域樣式。前面討論的範圍規則、隔離和保護不適用，這本質上與將元件的樣式貼上到專案 web 目錄下的 styles.css 檔案中相同。

15.3 依賴注入

Angular 具有分層的依賴注入系統：實際上有一個與應用程式的元件樹平行的注入器樹，可以在該元件樹的任何等級重新設定注入器。

15.3.1 注入器樹

Angular 應用程式是一棵元件樹，每個元件實例都有其自己的注入器。注入器樹與元件樹平行。

一個元件的注入器可能是元件樹中更進階別的祖先注入器的代理，這可以提高效率並節省資源。幾乎察覺不到它們之間的差異，需要關注的是每個元件都有自己的注射器。

在前面的例子中，根部是 AppComponent 元件，它包含一些子元件。其中之一是 EmployeeListComponent，EmployeeListComponent 也可以擁有其他元件。總之，Angular 從根元件開始逐漸延伸到所有元件，如圖 15-8 所示。

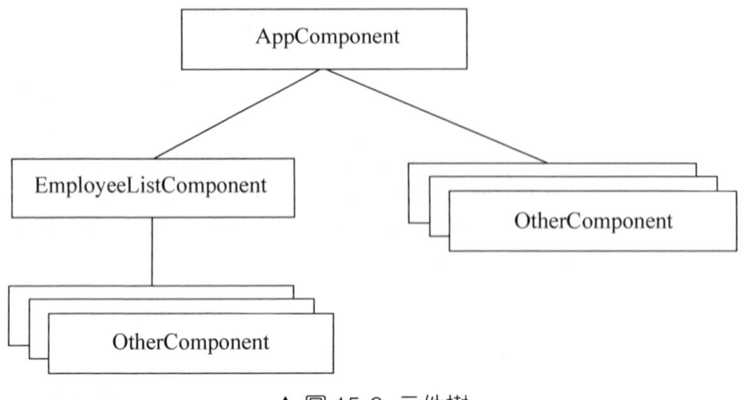

▲ 圖 15-8 元件樹

當元件請求依賴時，Angular 會嘗試透過在該元件自己的注入器中註冊的提供程式來滿足該依賴。如果元件的注入器缺少提供者，它將把請求傳遞到其父元件的注入器，如果該注入器仍然無法滿足請求，則繼續向上傳遞。請求一直上浮，直到 Angular 找到可以處理該請求的注入器或耗盡注入器為止。如果注入器樹中沒有符合請求的依賴，則 Angular 會拋出錯誤。

也可以限制上浮，使用 viewProviders 參數代替 Providers，依賴項注入提供者清單將僅限於此元件及其子元件的範本。

15.3.2 服務隔離

雖然可以在根元件的注入器中提供所有服務，以便在所有元件中都可以透過注入來使用服務。出於架構原因，更好的做法是對服務的存取限制為它所屬的應用程式定義域。

在專案 lib/src 目錄下創建 employee 目錄，並創建資料模型 Employee，程式如下：

```
//chapter15/dependency-injection/lib/src/employee/employee.dart
class Employee{
   final int id;
   String name;
```

```
    num salary;
    Employee(this.id,this.name,this.salary);
}
```

創建服務 EmployeeService，它的 getAll 方法能夠返回使用字面量建構的 Employee 類型的清單 emps，程式如下：

```
//chapter15/dependency-injection/lib/src/employee/employee_service.dart
import 'package:angular/angular.dart';
import 'employee.dart';

@Injectable()
class EmployeeService{
    var emps = [
        Employee(5, 'Magneta',9971),
        Employee(6, 'RubberMan',4533),
        Employee(7, 'Dynama',6720),
        Employee(8, 'Dr IQ',4907),
        Employee(9, 'Magma',5278),
        Employee(10, 'Tornado',7800)
    ];
    List<Employee> getAll(){
        return emps;
    }
}
```

創建元件 EmployeeListComponent，它從服務 EmployeeService 獲取 emps，並在範本中展示 emps 清單。

這裡沒有在 AppComponent 元件的提供者清單中增加 EmployeeService，而是在 EmployeeListComponent 的提供者清單中增加 EmployeeService，程式如下：

```
//chapter15/dependency-injection/lib/src/employee/employee_list_component.dart
import 'package:angular/angular.dart';

import 'employee.dart';
import 'employee_service.dart';
```

```
@Component(
   selector:'employee-list',
   templateURL: 'employee_list_component.html',
   directives: [coreDirectives],
   providers: [ClassProvider(EmployeeService)],
)
class EmployeeListComponent{
   final EmployeeService _employeeService;
   List<Employee> emps;
   // 注入服務 EmployeeService
   EmployeeListComponent(this._employeeService){
      emps = _employeeService.getAll();
   }
}
```

在範本增加元素程式如下：

```
//chapter15/dependency-injection/lib/src/employee/employee_list_component.html
<div>
   <h3> 員工清單 </h3>
   <table *ngIf="emps != null">
      <tr>
         <th> 員工 ID</th>
         <th> 員工名稱 </th>
         <th> 員工薪資 </th>
      </tr>
      <!-- 根據 emps 清單疊代員工 -->
      <tr *ngFor="let emp of emps">
         <td>{{emp.id}}</td>
         <td>{{emp.name}}</td>
         <td>{{emp.salary}}</td>
      </tr>
   </table>
</div>
```

透過在 EmployeeListComponent 中繼資料中提供 EmployeeService，並且僅在 EmployeeListComponent 及其子元件樹中提供該服務。這表示服務 EmployeeService 只在能使用到它的地方被提供，而根元件 AppComponent 下的其他元件無法存取它。

15.3.3 多個編輯階段

許多應用程式允許使用者同時處理多個打開的任務。例如在薪資分配應用程式中，會計人員可能同時打開多個員工的薪資分配表。

要打開員工的薪資分配表，會計人員需點擊員工名稱，以便打開一個用於編輯該薪資分配表的元件。每個選定的員工薪資分配表都會在其自己的元件實例中被打開，並且可以同時打開多個薪資分配表。

每個薪資分配表元件都具有以下特徵：

(1) 薪資分配表元件是其自身的薪資分配表的編輯階段。
(2) 可以更改薪資分配表而不會影響其他元件的資料。
(3) 可以保存更改後的薪資或取消。

為模型類別增加工廠命名建構函數 Employee.copy，用於複製原物件。範例程式如下：

```
//chapter15/dependency-injection/lib/src/employee/employee.dart
class Employee{
   final int id;
   String name;
   num salary;
   Employee(this.id,this.name,this.salary);
   // 返回原物件的備份
   factory Employee.copy(Employee e) =>
   Employee(e.id,e.name,e.salary);
}
```

在服務 EmployeeService 增加 saveSalary 方法，用於更新清單 emps 中匹配員工的薪資。程式如下：

```
//chapter15/dependency-injection/lib/src/employee/employee_service.dart
import 'package:angular/angular.dart';
import 'employee.dart';
@Injectable()
class EmployeeService{
```

```
  var emps = [
    Employee(5, 'Magneta',9971),
    Employee(6, 'RubberMan',4533),
    Employee(7, 'Dynama',6720),
    Employee(8, 'Dr IQ',4907),
    Employee(9, 'Magma',5278),
    Employee(10, 'Tornado',7800)
  ];
  List<Employee> getAll(){
    return emps;
  }
  // 保存匹配員工的薪資資訊
  void saveSalary(Employee emp){
    // 判斷 emps 清單是否包含傳入的物件
    // 如果包含就返回該物件，否則返回空
    var employee = emps.firstWhere((e){
      return e.id == emp.id;
    },orElse: ()=> null);

    if(employee != null ){
      // 修改對應員工的薪資
      employee.salary = emp.salary;
    }
  }
}
```

增加服務 EmployeeSalaryService，它快取單一 Employee，追蹤該物件的更改，並可以保存或恢復它。它還委派給應用程式範圍內的單例 EmployeeService，並透過注入獲得。程式如下：

```
//chapter15/dependency-injection/lib/src/employee/employee_salary_service.dart
import 'employee.dart';
import 'employee_service.dart';

class EmployeeSalaryService{
  final EmployeeService _employeeService;
  Employee _currentEMP, _originalEMP;

  // 注入應用程式範圍內的單例 EmployeeService
```

```
  EmployeeSalaryService(this._employeeService);

  //employee 的 setter 方法
  void set employee(Employee emp) {
    // 使用 _originalEMP 快取原物件
    _originalEMP = emp;
    // 複製原物件到 _currentEMP
    _currentEMP = Employee.copy(emp);
  }

  //employee 的 getter 方法返回最新員工資訊
  Employee get employee => _currentEMP;

  // 恢復薪資資訊
  void restoreSalary(){
    // 恢復原物件
    employee = _originalEMP;
  }

  // 保存薪資資訊
  void saveSalary(){
    // 修改單例 EmployeeService 中的資料
    _employeeService.saveSalary(_currentEMP);
  }
}
```

增加元件 EmployeeSalaryComponent，它負責展示薪資分配表，並提供與服務 EmployeeSalaryService 的互動。程式如下：

```
//chapter15/dependency-injection/lib/src/employee/employee_salary_component.dart
import 'dart:async';

import 'package:angular/angular.dart';
import 'package:angular_forms/angular_forms.dart';

import 'employee.dart';
import 'employee_salary_service.dart';

@Component(
```

```
    selector:'employee-salary',
    templateURL: 'employee_salary_component.html',
    styleURLs: ['employee_salary_component.css'],
    directives: [coreDirectives,formDirectives],
    providers: [ClassProvider(EmployeeSalaryService)],
)
class EmployeeSalaryComponent{
    final EmployeeSalaryService _employeeSalaryService;
    String message = '';
    // 注入服務 EmployeeSalaryService
    EmployeeSalaryComponent(this._employeeSalaryService);
    // 宣告輸入物件
    @Input()
    void set employee(Employee emp){
        _employeeSalaryService.employee = emp;
    }
    Employee get employee => _employeeSalaryService.employee;

    // 保存資料
    Future<void> onSaved() async {
        await _employeeSalaryService.saveSalary();
        await flashMessage(' 已保存 ');
    }

    // 取消更改，還原物件資訊
    Future<void> onCanceled() async {
        _employeeSalaryService.restoreSalary();
        await flashMessage(' 已取消 ');
    }

    final _close = StreamController<Null>();
    // 宣告輸出物件
    @Output()
    Stream<Null> get close => _close.stream;
    // 關閉薪資分配表
    // 因為只需觸發事件而不需要傳遞資料，因此向 add 傳遞 null 物件
    void onClose() => _close.add(null);

    // 刷新訊息
```

```
    void flashMessage(String msg) async {
        message = msg;
        await Future.delayed(Duration(milliseconds: 500));
        message = '';
    }
}
```

增加薪資分配表對應的範本程式如下：

```
//chapter15/dependency-injection/lib/src/employee/employee_salary_component.html
<div class="salary">
    <div class="msg" [class.canceled]="message==='Canceled'">{{message}}</div>
    <fieldset>
        <span id="name">{{employee.name}}</span>
        <label id="id">ID: {{employee.id}}</label>
    </fieldset>
    <fieldset>
        <label>
            薪資：<input type="number" [(ngModel)]="employee.salary" class="num">
        </label>
    </fieldset>
    <fieldset>
        <button (click)="onSaved()">保存</button>
        <button (click)="onCanceled()">取消</button>
        <button (click)="onClose()">關閉</button>
    </fieldset>
</div>
```

將元件 EmployeeSalaryComponent 增加到元件 EmployeeListComponent 的指令清單，增加儲存處於編輯狀態的員工清單 selectedEmployees，並提供在該清單增加和移除員工資訊的方法 showSalary 與 closeSalary。程式如下：

```
//chapter15/dependency-injection/lib/src/employee/employee_list_component.dart
import 'package:angular/angular.dart';

import 'employee.dart';
import 'employee_service.dart';
import 'employee_salary_component.dart';
```

```
@Component(
  selector:'employee-list',
  templateURL: 'employee_list_component.html',
  directives: [coreDirectives,EmployeeSalaryComponent],
  providers: [ClassProvider(EmployeeService)],
)
class EmployeeListComponent{
  final EmployeeService _employeeService;
  List<Employee> emps;
  // 注入服務 EmployeeService
  EmployeeListComponent(this._employeeService){
    emps = _employeeService.getAll();
  }
  // 儲存當前處於編輯狀態的員工清單
  final List<Employee> selectedEmployees = [];

  // 透過將單一員工增加到 selectedEmployees 清單以顯示相關薪資分配表
  void showSalary(Employee emp){
    // 判斷 selectedEmployees 清單是否已存在該員工
    if(!selectedEmployees.any((e)=> e.id == emp.id)){
      selectedEmployees.add(emp);
    }
  }
  // 透過移除 selectedEmployees 清單中的對應元素關閉薪資分配表
  void closeSalary(int index){
    selectedEmployees.removeAt(index);
  }
}
```

更新元件 EmployeeListComponent 的範本，在員工清單的每個項目增加點擊事件並綁定到 showSalary 方法，使用清單 selectedEmployees 疊代多個薪資分配表，向每個 EmployeeSalaryComponent 元件實例注入當前疊代員工資訊，並將實例的 close 事件綁定到 showSalary 方法。程式如下：

```
//chapter15/dependency-injection/lib/src/employee/employee_list_component.html
<div>
  <h3> 員工清單 </h3>
  <table *ngIf="emps != null">
    <tr>
```

```
            <th> 員工 ID</th>
            <th> 員工名稱 </th>
            <th> 員工薪資 </th>
        </tr>
            <!-- 根據 emps 清單疊代員工 -->
            <!-- 增加點擊事件綁定到 showSalary 方法 -->
            <tr *ngFor="let emp of emps" (click)="showSalary(emp)">
                <td>{{emp.id}}</td>
                <td>{{emp.name}}</td>
                <td>{{emp.salary}}</td>
            </tr>
        </table>
    </div>
<!-- 根據 selectedEmployees 清單疊代多個子元件實例 -->
<employee-salary *ngFor="let selected of selectedEmployees;let i = index"
        [employee]="selected" (close)="closeSalary(i)">
</employee-salary>
```

任意點擊員工清單中的兩個項目，執行效果如圖 15-9 所示。

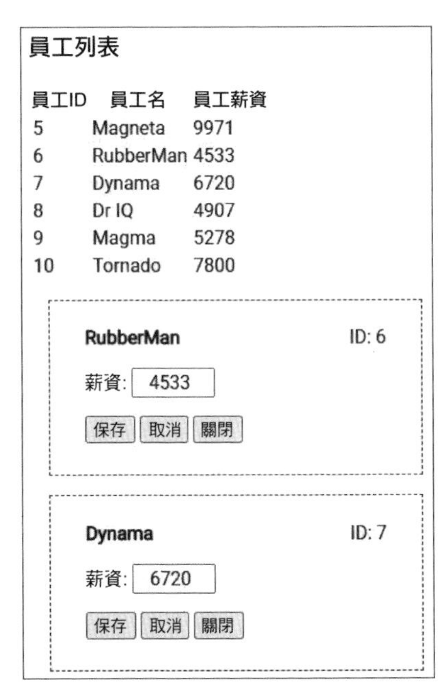

▲ 圖 15-9 多編輯階段

15.4 生命週期掛鉤

Angular 會創建和繪製元件及其子級，在資料綁定屬性更改時對元件進行檢查，並在元件從 DOM 中刪除之前將其銷毀。

Angular 提供了一組生命週期掛鉤函數，它們為元件或指令提供了關鍵時刻的可視性及發生時的操作能力。

指令具有相同的生命週期掛鉤集，減去特定於元件內容和視圖的掛鉤 , 如圖 15-10 所示。

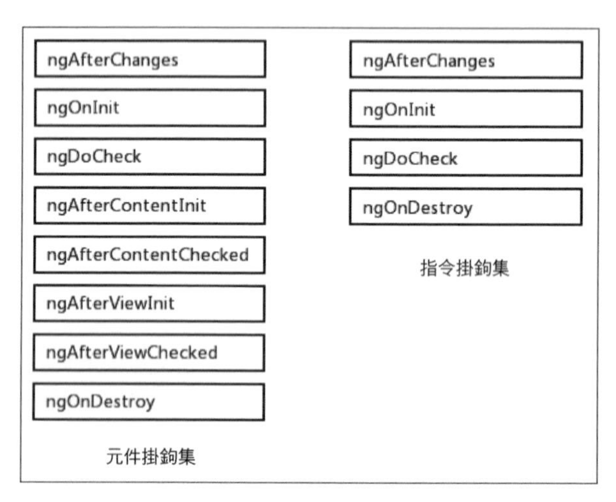

▲ 圖 15-10 元件和指令掛鉤集

15.4.1 元件生命週期掛鉤

指令和元件實例都具有生命週期，因為由 Angular 負責創建、更新和銷毀它們。透過實現 Angular 核心函數庫中一個或多個生命週期掛鉤介面，開發人員可以利用生命週期中的關鍵時刻完成一些操作。

每個介面都有一個掛鉤函數，以 ng 為字首，其名稱為介面名稱。舉例來說，OnInit 介面具有一個名為 ngOnInit 的掛鉤函數，Angular 在創建元件後不久會呼叫該方法。範例程式如下：

```
//chapter15/lifecycle-hooks/lib/app_component.dart
import 'package:angular/angular.dart';
import 'package:angular_forms/angular_forms.dart';

@Component(
   selector: 'my-app',
   template:'''<p *ngFor="let msg of msgs">{{msg}}</p>''',
   directives: [coreDirectives,formDirectives],
)
class AppComponent implements OnInit{
   List<String> msgs = [];
   AppComponent(){
      msgs.add('AppComponent 建構函數 ');
   }
   @override
   void ngOnInit(){
      msgs.add('AppComponent ngOnInit');
   }
}
```

執行結果如圖 15-11 所示。

AppComponent 构造函数

AppComponent ngOnInit

▲ 圖 15-11　初始化週期函數

沒有指令或元件將實現所有生命週期掛鉤函數，並且某些掛鉤僅對元件有意義。Angular 僅在實現了指令或元件掛鉤函數的情況下才呼叫它。

15.4.2 生命週期序列

透過呼叫建構函數創建元件或指令後，Angular 在特定時刻按以下順序呼叫生命週期掛鉤函數，如表 15-1 所示。

表 15-1 生命週期函數

掛鉤函數	目的和時間
ngAfterChanges	當 Angular 設定或更新資料綁定的輸入屬性時回應。 在 ngOnInit 之前及當一個或多個資料綁定輸入屬性更改時呼叫
ngOnInit	在 Angular 第一次顯示資料綁定屬性並設定指令或元件的輸入屬性後，初始化指令或元件
	在 ngOnChanges 第一次呼叫之後呼叫一次
ngDoCheck	檢測 Angular 無法或無法自行檢測到的變化並採取措施。 在每次更改檢測執行期間呼叫，緊接在 ngOnChanges 和 ngOnInit 之後
ngAfterContentInit	在 Angular 將外部內容投射到元件的視圖中後做出回應。 在第一次 ngDoCheck 呼叫之後呼叫一次。 僅元件掛鉤
ngAfterContentChecked	Angular 檢查投影到元件中的內容後回應。 在 ngAfterContentInit 和每個後續的 ngDoCheck 呼叫之後呼叫。 僅元件掛鉤
ngAfterViewInit	在 Angular 初始化元件的視圖和子視圖之後回應。 在第一次 ngAfterContentChecked 呼叫之後呼叫一次。 僅元件掛鉤
ngAfterViewChecked	在 Angular 檢查元件的視圖和子視圖之後回應。 在 ngAfterViewInit 和隨後的每個 ngAfterContentChecked 之後呼叫。 僅元件掛鉤
ngOnDestroy	在 Angular 銷毀指令 / 元件之前進行清理。取消訂閱可觀察物件並分離事件處理常式，以避免記憶體洩漏。 在 Angular 銷毀指令 / 元件之前呼叫

15.4.3 其他生命週期掛鉤

除了這些元件掛鉤之外，其他 Angular 子系統可能還有自己的生命週期掛鉤。舉例來說，路由器還具有自己的路由器生命週期掛鉤，使得開發人員可以利用路由導覽中的特定時刻。路由的生命週期掛鉤 routerOnActivate

等於元件的生命週期掛鉤 ngOnInit，兩者都有字首以避免衝突，並且都在初始化元件時正確執行。

第三方函數庫也可以實現自己的掛鉤，以使開發人員可以更進一步地控制這些函數庫的使用方式。

15.4.4 生命週期練習

該範例透過在根目錄 AppComponent 的控制下呈現演示了元件的各個生命週期掛鉤。它遵循一種通用模式：父元件充當子元件的測試平台，子元件實現多個生命週期掛鉤方法。

首先定義一個日誌服務，其作用是列印各個生命週期傳遞的訊息。範例程式如下：

```
//chapter15/lifecycle-hooks/lib/src/log_service.dart
import 'package:angular/angular.dart';
@Injectable()
class LogService{
  // 快取日誌
  List<String> logs = [];
  // 增加日誌
  void log(String msg){
    logs.add(msg);
  }
  // 安排視圖刷新以確保顯示及時
  tick() => Future(() {});
}
```

然後定義元件 HooksComponent，注入日誌服務 LogService，這裡不要使用 providers 參數提供服務，而要使用父元件 AppComponent 注入器中的服務。使其實現所有生命週期掛鉤函數，並在實現方法中向日誌服務提供訊息。範例程式如下：

```
//chapter15/lifecycle-hooks/lib/src/hooks_component.dart
import 'package:angular/angular.dart';
```

```
import 'package:lifecycle_hooks/src/log_service.dart';
import 'package:angular_forms/angular_forms.dart';

@Component(
  selector: 'hooks',
  template:'''
  <p> 透過修改屬性，觸發 ngDoCheck、AfterContentChecked 和 AfterViewChecked</p>
  <input [(ngModel)]="name">''',
  directives: [coreDirectives,formDirectives],
)
class HooksComponent implements
    AfterChanges,
    OnInit,
    DoCheck,
    AfterContentInit,
    AfterContentChecked,
    AfterViewInit,
    AfterViewChecked,
    OnDestroy{
  final LogService _log;
  // 輸入屬性
  @Input()
  String name;
  // 記錄與視圖相關的掛鉤函數的執行次數
  int _afterContentCheckedCounter = 1;
  int _afterViewCheckedCounter = 1;
  int _afterChangesCounter = 1;
  int _doCheckCounter = 1;
  // 建構函數，注入服務 LogService
  HooksComponent(this._log){
    _log.log('HooksComponent 建構函數 ');
  }

  @override
  void ngAfterChanges() {
    _log.log('HooksComponent ngAfterChanges (${_afterChangesCounter++})');
  }
  @overri
  de
```

```
  void ngAfterContentChecked() {
    _log.log('HooksComponent ngAfterContentChecked(${_afterContentCheckedCo
unter++})');
  }

  @override
  void ngAfterContentInit() {
    _log.log('HooksComponent ngAfterContentInit');
  }

  @override
  void ngAfterViewChecked() {
    _log.log('HooksComponent ngAfterViewChecked(${_
afterViewCheckedCounter++})');
  }

  @override
  void ngAfterViewInit() {
    _log.log('HooksComponent ngAfterViewInit');
  }

  @override
    void ngDoCheck() {
  _log.log('HooksComponent ngDoCheck(${_doCheckCounter++})');
  }

  @override
  void ngOnDestroy() {
    _log.log('HooksComponent ngOnDestroy');
  }

  @override
  void ngOnInit() {
    _log.log('HooksComponent ngOnInit');
  }
}
```

在 AppComponent 中注入服務 LogService，並將 HooksComponent 增加到指令清單。增加 toggleChild 函數用於創建或銷毀 HooksComponent 元件

實例。範例程式如下：

```
//chapter15/lifecycle-hooks/lib/app_component.dart
import 'package:angular/angular.dart';
import 'package:angular_forms/angular_forms.dart';

import 'src/hooks_component.dart';
import 'src/log_service.dart';

@Component(
   selector: 'my-app',
   styleURLs: ['app_component.css'],
   templateURL: 'app_component.html',
   directives: [coreDirectives,formDirectives,HooksComponent],
   providers: [ClassProvider(LogService)],
)
class AppComponent{
   final LogService _log;
   String name = 'lei';
   AppComponent(this._log);
   //logs 用於指向服務 LogService 中的 logs
   List<String> get logs => _log.logs;
   // 元件創建和銷毀控制變數
   bool isShow = false;
   // 元件創建或銷毀控制方法
   toggleChild(){
      isShow = !isShow;
      // 用於觸發資料更新以便更新視圖
      _log.tick();
   }
}
```

在範本增加元素程式如下：

```
//chapter15/lifecycle-hooks/lib/app_component.html
<button (click)="toggleChild()">{{isShow ? ' 銷毀 ':' 創建 '}}hooks 實例 </
button>
<div *ngIf="isShow">
   <p> 透過修改屬性 name，更改傳入 HooksComponent 實例的輸入屬性 name</p>
   <p> 觸發元件 HooksComponent 的鉤子函數 ngAfterChanges</p>
```

```
    <input [(ngModel)]="name">
</div>

<hr>

<hooks *ngIf="isShow" [name]="name"></hooks>

<h4>-- 生命週期掛鉤函數執行日誌 --</h4>
<div *ngFor="let msg of logs;let i = index">{{msg}}</div>
```

執行結果如圖 15-12 所示。

銷毀hooks實例

透過修改屬性name，更改傳入HooksComponent實例的輸入屬性name

觸發元件HooksComponent的鉤子函數ngAfterChanges

lei2

透過修改屬性，觸發ngDoCheck、AfterContentChecked和AfterViewChecked

lei2

– 生命週期掛鉤函數執行日誌 –

HooksComponent 建構函數
HooksComponent ngAfterChanges (1)
HooksComponent ngOnInit
HooksComponent ngDoCheck(1)
HooksComponent ngAfterContentInit
HooksComponent ngAfterContentChecked(1)
HooksComponent ngAfterViewInit
HooksComponent ngAfterViewChecked(1)
HooksComponent ngDoCheck(2)
HooksComponent ngAfterContentChecked(2)
HooksComponent ngAfterViewChecked(2)
HooksComponent ngAfterChanges (2)
HooksComponent ngDoCheck(3)
HooksComponent ngAfterContentChecked(3)
HooksComponent ngAfterViewChecked(3)

▲ 圖 15-12 生命週期函數實例

應當充分使用提供的操作，控制屬性更改和元件狀態變化，仔細分析以便更準確了解各個生命週期掛鉤的使用與呼叫時機。

15.5 管道

每個應用程式都會獲取資料，通常將其原始 toString 值直接傳遞到視圖，但在某些時候會導致不好的使用者體驗。舉例來說，在大多數使用情況下，使用者喜歡以簡單的格式 (如 1988 年 4 月 15 日) 查看日期，而非原始的字串格式 (Fri 1988 年 4 月 15 日 00:00:00 GMT-0700，太平洋日光節約時間)。

有些值可以透過一些小改動而變得易於閱讀。應用程式中可能對某些值進行許多相同的轉換，幾乎可以將它們視為樣式。實際上，可以像對待樣式一樣在 HTML 範本中應用它們。

本節介紹管道，這是一種可以在 HTML 中對顯示值進行轉換的方法。

15.5.1 使用管道

使用管道前需要提供 pipes 參數，常數 commonPipes 包含了所有內建管道。

```
pipes:[commonPipes],
```

在範本中使用格式範例程式如下：

```
{{expr | pipeName}}
```

管道將資料作為輸入，並將其轉為所需的輸出。在插值運算式中，將運算式 expr 的值透過管道運算子 (|) 傳遞到右側的管道函數，所有管道都以這種方式工作。

Angular 內建大量管道：

(1) DatePipe：日期管道，用於格式化日期。

(2) UpperCasePipe：將字串轉為大寫。

(3) LowerCasePipe：將字串轉為小寫。

(4) CurrencyPipe：將數字轉為本地貨幣的表示形式。

(5) PercentPipe：將數字轉為百分比的表示形式。

在元件中定義日期和字串屬性，程式如下：

```
//chapter15/pipes/lib/app_component.dart
import 'package:angular/angular.dart';
@Component(
   selector: 'my-app',
   styleURLs: ['app_component.css'],
   templateURL: 'app_component.html',
   pipes: [commonPipes],
)
class AppComponent {
   DateTime birthday = DateTime(2020,6,25);
   String str = 'Lower Upper';
   int number = 99;
}
```

在範本中增加元素，程式如下：

```
//chapter15/pipes/lib/app_component.html
<h6>原日期格式：{{birthday}}</h6>
<p>date:{{birthday | date}}</p>
<h6> 原字串：{{str}}</h6>
<p>uppercase:{{str | uppercase}}</p>
<p>lowercase:{{str | lowercase}}</p>
<h6> 原數字：{{number}}</h6>
<p>currency:{{number | currency}}</p>
<p>percent:{{number | percent}}</p>
```

執行結果如圖 15-13 所示。

原日期格式: 2020-06-25 00:00:00.000

date:Jun 25, 2020

原字串: Lower Upper

uppercase:LOWER UPPER

lowercase:lower upper

原數字: 99

currency:USD99

percent:9,900%

▲ 圖 15-13 內建管道

15.5.2 參數化管道

管道可以接收任意數量的可選參數來微調其輸出。要將參數增加到管道，需在管道名稱後加上冒號 (:)，然後輸入參數值，例如：currency:'EUR'。如果管道接收多個參數，則需用冒號分隔值，例如：slice:1:5。

日期管道的參數組成元素的使用格式如表 15-2 所示。

表 15-2 日期參數格式

構成元素	數字表示	兩位數表示
年	y(2020)	yy(20)
月	M(6)	MM(06)
日	d(9)	dd(09)
小時 (12)	h(1 PM)	hh(01 PM)
小時 (24)	H(13)	HH(13)
分鐘	m(3)	mm(03)
秒	s(9)	ss(09)

參數值可以是任何有效的範本運算式，例如：字串或元件屬性。在元件增加日期格式切換方法 toggleFormat，它控制元件的 format 屬性在格式 yMMdd 和格式 MM/dd/yy 之間切換。範例程式如下：

```
//chapter15/pipes/lib/app_component.dart
// 切換控制變數
bool toggle = true;
// 格式化字串
get format => toggle ? 'yMMdd' : 'MM/dd/yy';
// 切換控制方法
void toggleFormat() {
  toggle = !toggle;
}
```

可以直接將字串 MM/dd/yy 用作日期管道的參數，也可以將管道的 format 參數綁定到元件的 format 屬性。在範本增加一個按鈕，並將其點擊事件綁定到元件的 toggleFormat() 方法。在範本增加程式如下：

```
//chapter15/pipes/lib/app_component.html
<p>birthday:{{ birthday | date:"MM/dd/yy" }}</p>
<p>birthday:{{ birthday | date:format }}</p>
<button (click)="toggleFormat()"> 切換格式 </button>
```

刷新瀏覽器，點擊「切換格式」按鈕，日期在 "20200625" 和 "06/25/20" 之間交替顯示。

15.5.3 管道鏈

可以將管道以潛在有用的組合方式連接在一起，即可以同時使用多個管道。可以將 birthday 連接到 DatePipe 並連接到 UpperCasePipe。範例程式如下：

```
{{birthday | date | uppercase}}
```

此範例連接了與上述相同的管道，並向 date 管道傳遞了一個參數。範例程式如下：

```
{{birthday | date:'fullDate' | uppercase}}
```

在範本增加元素程式如下：

```
//chapter15/pipes/lib/app_component.html
<p>birthday:{{ birthday | date}}</p>
<p>birthday:{{ birthday | date | uppercase}}</p>

<p>birthday:{{ birthday | date:'fullDate'}}</p>
<p>birthday:{{ birthday | date:'fullDate' | uppercase}}</p>
```

15.5.4 自訂管道

可以編寫自訂管道，管道是使用 @Pipe 註釋修飾的類別，@Pipe 註釋的參數是管道名。管道類別必須實現 PipeTransform 介面的 transform 方法，該方法可以接收多個參數，第一個參數是透過管道符傳入的值，其他參數是在使用管道時提供的參數。方法內部執行轉換操作，並返回轉換後的值。參數和返回值可以是任何類型。

這是一個名為 RoundAreaPipe 的自訂管道，它的作用是根據圓的半徑計算圓的面積。範例程式如下：

```
//chapter15/pipes/lib/src/round_area_pipe.dart
import 'package:angular/angular.dart';

@Pipe('roundArea')
class RoundAreaPipe extends PipeTransform{
   // 半徑作為參數並返回圓的面積
   num transform(num r){
      var pi = 3.14;
      return pi*r*r;
   }
}
```

使用管道時，首先在元件中匯入該管道檔案，然後在 pipes 清單中列出該管道。範例程式如下：

```
//chapter15/pipes/lib/app_component.dart
import 'package:angular/angular.dart';
import 'src/round_area_pipe.dart';

@Component(
   selector: 'my-app',
   styleURLs: ['app_component.css'],
   templateURL: 'app_component.html',
   directives: [coreDirectives,formDirectives],
   pipes: [commonPipes,RoundAreaPipe],
)
class AppComponent {
}
```

隨後就可以在範本中增加元素，程式如下：

```
//chapter15/pipes/lib/app_component.html
<p> 半徑為 9 公分的圓的面積為 {{9 | roundArea}} 平方公分 </p>
```

15.5.5 管道和變更檢測

Angular 透過在每個 DOM 事件 (每次擊鍵、滑鼠移動、計時器滴答和伺服器回應) 之後執行的更改檢測處理程序來尋找對資料綁定值的更改，這可能佔用大量資源，Angular 盡可能地降低影響。使用管道時，Angular 選擇一種更簡單、更快速的變更檢測演算法。

在元件中增加泛型為字串的 list 清單，在 reset 方法中為該清單初始化值。在建構函數中呼叫 reset 方法，增加方法 addStr 用於在清單 list 增加字串。範例程式如下：

```
//chapter15/pipes/lib/app_component.dart
List<String> list;
AppComponent(){
   reset();
}
void addStr(String str){
   list.add(str);
}
void reset(){
   list = ['Bombasto','RubberMan','Magneta','Magma'];
}
```

使用預設的主動更改檢測策略來監視和更新 list 清單中每個字串的顯示。在範本中增加元素程式如下：

```
//chapter15/pipes/lib/app_component.html
<input type="text" #box
   (keyup.enter)="addStr(box.value); box.value=''"
   placeholder=" 請輸入字串 ">
<button (click)="reset()"> 重置清單 </button>
<h6> 未使用管道 </h6>
<div *ngFor="let str of list">
   {{str}}
</div>
```

然後定義管道 PrefixPipe，它的作用是根據提供的字首 prefix，返回 list

清單中所有滿足條件的字串組成的新清單。範例程式如下：

```
//chapter15/pipes/lib/src/prefix_pipe.dart
import 'package:angular/angular.dart';

@Pipe('prefix')
class PrefixPipe extends PipeTransform{
  // 第一個參數是字串類型的清單，第二個參數表示字串字首
  List<String> transform(List<String> list,String prefix){
    // 返回符合字首 prefix 的字串的新清單
    return list.where((str)=> str.startsWith(prefix)).toList();
  }
}
```

然後將管道 PrefixPipe 增加到元件的 pipes 清單中，並在範本中增加元素，程式如下：

```
//chapter15/pipes/lib/app_component.html
<h6> 使用管道 prefix</h6>
<div *ngFor="let str of (list | prefix:'M')">
  {{str}}
</div>
```

刷新瀏覽器，輸入字串並確認，沒有使用管道的範本循環會在增加字串後立即更新顯示，使用管道的範本循環則不會更新顯示。

15.5.6 純與不純

管道有兩類：純管道和不純的管道。預設情況下，管道是純管道。到目前為止，所使用的每個管道都是純的。透過將管道的 pure 標示設定為 false，可以使其變為不純的管道。

可以像這樣使 PrefixPipe 變為不純的管道，範例程式如下：

```
@Pipe('prefix', pure:false)
```

先來了解純與不純之間的區別。

1. 純的管道

Angular 僅在檢測到輸入值的純更改時才執行純管道。在 Angular 中，純更改僅由物件引用的更改引起。

Angular 忽略複合物件內的更改。例如更改 List 或 Map 物件中的資料，則不會呼叫純管道。這似乎很嚴格，但速度很快。物件引用檢查很快，比深入檢查符合物件內部的差異快得多，因此 Angular 可以快速確定是否可以跳過管道執行和視圖更新。

因此，當可以使用預設變更檢測策略時，最好使用純管道。如果不能，則可以使用不純的管道。

2. 不純的管道

Angular 在每個元件更改檢測週期內執行不純管道。每次擊鍵或移動滑鼠時，都會頻繁呼叫不純的管道。考慮到這一點，應格外小心地使用不純的管道。它會消耗大量運算資源，長時間執行的管道可能會破壞使用者體驗。

定義一個不純的管道 PrefixImpurePipe，在這裡直接使用 PrefixImpurePipe 繼承 PrefixPipe 就可以了，完整範例程式如下：

```
//chapter15/pipes/lib/src/prefix_pipe.dart
@Pipe('prefixImpure',pure: false)
class PrefixImpurePipe extends PrefixPipe{}
```

從繼承 PrefixPipe 來證明內部沒有做任何更改，唯一的區別是 pipe 中繼資料中的 pure 標示。然後將 PrefixImpurePipe 增加到元件的 pipes 清單，在範本中增加元素，程式如下：

```
//chapter15/pipes/lib/app_component.html
<h6> 使用不純的管道 prefixImpure</h6>
<div *ngFor="let str of (list | prefixImpure:'M')">
   {{str}}
</div>
```

此時在 list 清單中增加字串，不純的管道 prefixImpure 就會檢查更改，帶有 M 開頭的字串將被顯示在視圖中。

3. 非同步管道

AsyncPipe 是不純的管道，可以使用 Future 或 Stream 作為輸入，並且會自動訂閱輸入。AsyncPipe 也是有狀態的。管道維護對輸入 Stream 的訂閱，並不斷地從串流中傳遞值到管道。

在元件中增加流量控制器 st，增加一個接收 st 的 Stram 物件的變數 message，在方法 addMes 中由流量控制器 st 發送資料到 message。範例程式如下：

```
//chapter15/pipes/lib/app_component.dart
//Stream 控制器
StreamController<String> st = StreamController<String>();
// 返回 st 的串流物件
Stream<String> get message => st.stream;
// 觸發事件並傳遞資料
addMes(String str){
   st.add(str);
}
```

在範本中透過輸入控制項的 enter 事件呼叫 addMes 方法，並在插值中使用非同步管道。在範本中增加元素程式如下：

```
//chapter15/pipes/lib/app_component.html
<h6>Async 管道 </h6>
<input type="text" #mes
   (keyup.enter)="addMes(mes.value); mes.value=''"
   placeholder=" 請輸入訊息並確認 ">
<p>message:{{message | async}}</p>
```

每當在輸入控制項中輸入資料並確認，message 資訊就會更新。

15.6 路由

當使用者執行應用程式任務時，Angular 路由器可以從一個視圖導覽到下一個視圖。本指南涵蓋了路由器的主要功能，並透過可即時執行的小應用程式的演變來說明它們。

15.6.1 路由基礎

1. 增加依賴項

路由器功能位於 angular_router 包中，將套件增加到 pubspec 依賴項，並執行 pub get 命令：

```
//chapter15/router/pubspec.yaml
dependencies:
   angular: ^5.3.0
   angular_components: ^0.13.0
   angular_router: ^2.0.0-alpha+22
```

在任何需要使用路由器功能的 Dart 檔案中，匯入路由器函數庫，範例程式如下：

```
import 'package:angular_router/angular_router.dart';
```

2. 增加全域路由提供者

在應用程式的啟動函數中指定 routerProvidersHash，使 Angular 知道應用程式使用了路由功能。範例程式如下：

```
//chapter15/router/web/main.dart
import 'package:angular/angular.dart';
import 'package:angular_router/angular_router.dart';
import 'package:router/app_component.template.dart' as ng;

import 'main.template.dart' as self;

const useHashLS = false;
```

```
@GenerateInjector(
   // 在生產環境使用 routerProviders
   routerProvidersHash,
)
final InjectorFactory injector = self.injector$Injector;

void main() {
   runApp(ng.AppComponentNgFactory, createInjector: injector);
}
```

預設的路由定位策略 LocationStrategy 採用路徑定位策略 PathLocation Strategy，所以在生產環境中使用 routerProviders。在開發環境中使用 routerProvidersHash，因為 webdev serve 不支援深層次的連結，即無法透過連結直接存取首頁以外的頁面。

3. 設定 base href

增加 <base href> 元素到應用程式的 index.html 檔案。當引用 CSS 檔案、指令稿和圖型時，瀏覽器使用 href 的值作為相對 URL 的字首。在頁面間導覽時，路由也會使用 href 的值作為相對 URL 的字首。

在開發環境中動態設定 <base> 元素，以便可以在開發過程中使用任何官方推薦的工具來執行和測試應用程式。範例程式如下：

```
//chapter15/router/web/index.html
<!DOCTYPE html>
<html>
   <head>
      <title>router</title>
      <meta charset="utf-8">
      <meta name="viewport" content="width=device-width, initial-scale=1">
      <link rel="stylesheet" href="styles.css">
      <link rel="icon" type="image/png" href="favicon.png">
      <script>
         // 警告：不要在生產環境中動態設定 <base href>
         (function () {
            var m = document.location.pathname.match(/^(\/[-\w]+)+\/web($|\/)/);
            document.write('<base href="' + (m ? m[0]: '/') + '" />');
```

```
        }());
      </script>
        <script defer src="main.dart.js"></script>
  </head>
  <body>
    <my-app>Loading...</my-app>
  </body>
</html>
```

在生產環境中，將 script 指令稿替換為 base 元素，其中 href 設定為應用程式的根路徑。如果路徑為空，則使用 "/"。範例程式如下：

```
<head>
  <base href="/">
</head>
```

創建任務和員工清單元件，TaskListComponent 元件程式如下：

```
//chapter15/router/lib/src/task/task_list_component.dart
import 'package:angular/angular.dart';

@Component(
  selector: 'task-list',
  template: '''
    <h2> 任務清單 </h2>
    <p> 具體任務 </p>
  ''',
)
class TaskListComponent{}
```

EmployeeListComponent 元件程式如下：

```
//chapter15/router/lib/src/employee/employee_list_component.dart
import 'package:angular/angular.dart';

@Component(
  selector: 'employee-list',
  template: '''
    <h2> 員工清單 </h2>
    <p> 員工資訊 </p>
  ''',
```

```
)
class EmployeeListComponent{}
```

上述兩個元件範本中沒有實質性內容。

4. 路由

路由告訴路由器使用者點擊連結或直接貼上 URL 到瀏覽器時顯示哪些視圖。

1) 路由路徑

RoutePath 類別的建構函數宣告程式如下：

```
RoutePath({
   String path,
   this.parent,
   this.useAsDefault = false,
   this.additionalData,
}) : this.path = URL.trimSlashes(path);
```

參數 path 表示路徑。parent 是 RoutePath 類型，表示父路由路徑。useAs Default 表示是否用作預設路由。additionalData 表示可以是任意類型的附加資料。

通常會將所有的路由路徑定義並封裝在一個檔案中。為每個應用視圖定義一個路由路徑，並將這些路由路徑作為類別變數封裝在類別 RoutePaths 中。範例程式如下：

```
//chapter15/router/lib/src/route_paths.dart
import 'package:angular_router/angular_router.dart';

class RoutePaths{
   static final tasks = RoutePath(path:'tasks');
   static final employees = RoutePath(path:'employees');
}
```

透過在單獨的檔案中定義路由路徑，可以避免導覽結構複雜的應用程式中路由定義之間的循環依賴性。

2) 路由定義

路由器根據路由定義清單協調應用導覽，路由定義將路由路徑與元件相連結，元件負責處理到路徑的導覽及相關檢視的繪製。

路由定義由 RouteDefinition 類別負責，其工廠建構函數等於以下程式：

```
factory RouteDefinition({
   String path,
ComponentFactory<Object> component,
   bool useAsDefault,
   dynamic additionalData,
   RoutePath routePath})
   : this.path = Url.trimSlashes(path ?? routePath?.path),
   this.useAsDefault = useAsDefault ?? routePath?.useAsDefault ?? false,
   this.additionalData = additionalData ?? routePath?.additionalData;
```

參數 path 表示路徑。component 表示與路徑連結的元件實現，ComponentFactory 是指用 @Component 註釋的類別 T 背後的支持實現。useAsDefault 表示是否用作預設路由，additionalData 表示可以是任意類型的附加資料。routePath 表示定義的路徑路由。若提供了 routePath 參數，且 path、useAsDefault 或 additionalData 未指定，它們的值將被 routePath 中的資料覆寫。

定義一組路由，並將它們作為類別變數封裝在類別 Routes 中。範例程式如下：

```
//chapter15/router/lib/src/routes.dart
import 'package:angular_router/angular_router.dart';

import 'employee/employee_list_component.template.dart' as employee_list_template;
import 'task/task_list_component.template.dart' as task_list_template;
import 'route_paths.dart';
export 'route_paths.dart';

class Routes{
   // 導覽到任務清單元件的路由
   static final tasks = RouteDefinition(
```

```
      routePath: RoutePaths.tasks,
      component: task_list_template.TaskListComponentNgFactory,
   );

   // 導覽到員工清單元件的路由
   static final employees = RouteDefinition(
      routePath: RoutePaths.employees,
      component: employee_list_template.EmployeeListComponentNgFactory,
   );
   // 返回已定義路由清單
   static final all = <RouteDefinition>[
      tasks,
      employees,
   ];
}
```

將 AppComponent 作為路由元件，匯入路由定義檔案，在指令清單中增加路由器指令常數 routerDirectives。出現了新的 @Component 註釋參數 exports，它的值是一個清單，在該清單中定義的識別符號可以在範本中引用，其使用方式和元件屬性一樣。這裡將靜態類別 RoutePaths 和 Routes 匯出到範本，使得範本中可以引用它們。範例程式如下：

```
//chapter15/router/lib/app_component.dart
import 'package:angular/angular.dart';
import 'package:angular_router/angular_router.dart';

import 'src/routes.dart';

@Component(
   selector: 'my-app',
   styleURLs: ['app_component.css'],
   templateURL: 'app_component.html',
   directives: [routerDirectives],
   exports: [RoutePaths,Routes],
)
class AppComponent{}
```

5. 導覽

路由包中定義了 3 個指令用於與導覽相關的操作：

RouterLink 指令需要在錨標籤 a 上使用。將路由路徑透過屬性綁定到 routerLink，該指令會為錨標籤增加 href 屬性並設值。可以直接為 routerLink 屬性指定連結，也可以綁定到路由路徑。範例程式如下：

```
<!-- 指定明確的連結 -->
<a routerLink="/employees">員工清單</a>
<!-- 指定路由路徑 -->
<a [routerLink]="RoutePaths.employees.toURL()">員工清單</a>
```

RouterLinkActive 與 RouterLink 指令配合使用，當連接的路由處於活動狀態時，將指定 CSS 類別增加到綁定的元素。可以將其和 RouterLink 指令綁定在同一個元素，也可以綁定到 RouterLink 指令綁定的元素的父項目上。範例程式如下：

```
<!-- 與 RouterLink 指令綁定在同一元素上 -->
<a routerLink="/employees" routerLinkActive="active-route">員工清單</a>
<!-- 綁定在 RouterLink 指令綁定元素的父項目上 -->
<div routerLinkActive="active-route">
   <a routerLink="/employees">員工清單</a>
</div>
```

RouterOutlet 指令作用物件是 router-outlet 元素。在 DOM 中，路由器透過在 router-outlet 元素之後插入視圖元素作為同級來顯示視圖。使用時需要將需要展示的路由綁定到輸入屬性 routes 上。

```
<router-outlet [routes]="Routes.all"></router-outlet>
```

更新 AppComponent 範本，使其具有兩個路由的導覽列和控制視圖的 router-outlet 元素。範例程式如下：

```
//chapter15/router/lib/app_component.html
<h1>Angular 路由</h1>
<nav>
   <a [routerLink]="RoutePaths.tasks.toURL()"
```

```
      [routerLinkActive]="'active-route'"> 任務清單 </a>
   <a [routerLink]="RoutePaths.employees.toURL()"
      [routerLinkActive]="'active-route'"> 員工清單 </a>
</nav>
<router-outlet [routes]="Routes.all"></router-outlet>
```

在樣式檔案中增加 CSS 類別，程式如下：

```
//chapter15/router/lib/app_component.css
/* 導覽連結樣式 */
nav a {
   padding: 5px 10px;
   text-decoration: none;
   margin-right: 10px;
   margin-top: 10px;
   display: inline-block;
   background-color: #eee;
   border-radius: 4px;
}
nav a:visited, a:link {
   color: #607D8B;
}
nav a:hover {
   color: #039be5;
   background-color: #CFD8DC;
}
nav a.active {
   color: #039be5;
}
/* 路由處於活躍狀態的連結樣式 */
.active-route {color: #039be5}
```

路由器將顯示視圖作為 <router-outlet> 元素的同級元素插入 DOM 中，其中顯示視圖是當前活躍路由相連結的元件的範本，且該路由必須存在於綁定到 <router-outlet> 元素輸入屬性 routes 的路由集合中。這裡 Routes.all 包含所有已定義的路由。

15.6.2 常用設定

1. 設定預設路由

啟動應用程式後，瀏覽器的初始 URL 與任何已設定的路由都不匹配，這表示應用啟動後不會顯示任何元件，只會顯示導覽列。使用者必須點擊導覽列中的連結，才能觸發對應元件視圖的顯示。

如果希望應用啟動後就顯示某個元件，可以在路由定義中將參數 useAsDefault 的值設定為 true。範例程式如下：

```
static final employees = RouteDefinition(
   routePath: RoutePaths.employees,
   component: employee_list_template.EmployeeListComponentNgFactory,
   useAsDefault: true
);
```

刷新瀏覽器，應用程式顯示員工清單，URL 路徑為 "/"。

2. 重新導向路由

可以透過匹配路徑重新導向到另一個路由，重新導向路由的建構函數程式如下：

```
factory RouteDefinition.redirect({
   String path,
   String redirectTo,
   bool useAsDefault,
   additionalData,
   RoutePath routePath,
})
```

redirectTo 參數的值為 URL，表示將要跳躍的位址。刪除路由定義中的設定參數 useAsDefault，增加重新導向路由，該路由會匹配到 URL 路徑 "/"，然後跳躍到 "/#/employees"。範例程式如下：

```
static final all = <RouteDefinition>[
   tasks,
```

```
   employees,
   RouteDefinition.redirect(
      path: '',
      redirectTo: RoutePaths.employees.toURL(),
   )
];
```

3. 萬用字元路由

目前已經創建了到 "/#/tasks" 和 "/#/employees" 的路由，如果向路由器提供了其他的路徑，則可以透過萬用字元路由匹配未找到的路由。

首先創建元件 NotFoundComponent，用於攔截未定義的路由。範例程式如下：

```
//chapter15/router/lib/src/not_found_component.dart
import 'package:angular/angular.dart';

@Component(
   selector: 'not-found',
   template: '<h2> 頁面未找到 </h2>',
)
class NotFoundComponent {}
```

完整的路由定義程式如下：

```
//chapter15/router/lib/src/routes.dart
import 'package:angular_router/angular_router.dart';

import 'employee/employee_list_component.template.dart' as employee_list_
template;
import 'task/task_list_component.template.dart' as task_list_template;
import 'not_found_component.template.dart' as not_found_template;
import 'route_paths.dart';
export 'route_paths.dart';

class Routes{
   static final tasks = RouteDefinition(
      routePath: RoutePaths.tasks,
```

```
      component: task_list_template.TaskListComponentNgFactory,
   );

   static final employees = RouteDefinition(
      routePath: RoutePaths.employees,
      component: employee_list_template.EmployeeListComponentNgFactory,
   );

   static final all = <RouteDefinition>[
      tasks,
      employees,
      RouteDefinition.redirect(
         path: '',
         redirectTo: RoutePaths.employees.toURL(),
      ),
      RouteDefinition(
         path: '.+',
         component: not_found_template.NotFoundComponentNgFactory,
      )
   ];
}
```

正規表示法 ".+" 匹配所有不可為空路徑。因為路由器會根據路由定義清單 all 中的順序依次匹配路由，因此應當將萬用字元路由放在路由定義清單 all 中的最後一個。

15.6.3 函數導覽

本節實現的功能是透過點擊員工清單中的項目跳躍到所對應員工資訊頁，跳躍是透過與項目點擊事件綁定的函數完成的。

首先創建 Employee 類別，程式如下：

```
//chapter15/router/lib/src/employee/employee.dart
class Employee{
   final int id;
   String name;
```

```
    num salary;
    Employee(this.id,this.name,this.salary);
}
```

創建提供資料服務的類別 EmployeeService，程式如下：

```
//chapter15/router/lib/src/employee/employee_service.dart
import 'employee.dart';
class EmployeeService{
    // 定義員工清單
    var emps = [
        Employee(5, 'Magneta',9971),
        Employee(6, 'RubberMan',4533),
        Employee(7, 'Dynama',6720),
        Employee(8, 'Dr IQ',4907),
        Employee(9, 'Magma',5278),
        Employee(10, 'Tornado',7800)
    ];
    // 獲取所有員工資訊
    List<Employee> getAll(){
        return emps;
    }
    // 根據 id 獲取單一員工資訊
    Employee getById(int id){
        return emps.firstWhere((emp)=> emp.id == id);
    }
}
```

將服務 EmployeeService 增加到元件 EmployeeListComponent 的提供者清單。

```
providers:[ClassProvider(EmployeeService)],
```

增加路由路徑 employee，程式如下：

```
//chapter15/router/lib/src/route_paths.dart
import 'package:angular_router/angular_router.dart';
// 定義常數 idParam，它的值為 id
const idParam = 'id';
class RoutePaths{
```

```
    // 定義任務清單頁的路由路徑物件 tasks
    static final tasks = RoutePath(path:'tasks');
    // 定義員工清單頁的路由路徑物件 employees
    static final employees = RoutePath(path:'employees');
    // 定義員工資訊頁的路由路徑物件 employee
    // 並需在路徑中攜帶有參數 id
    static final employee = RoutePath(path: '${employees.path}/:$idParam');
}
// 解析參數清單中參數 id 的值
int getId(Map<String,String> parameters){
    final id = parameters[idParam];
    return id == null ? null : int.parse(id);
}
```

在路由路徑物件 employee 的 path 參數值中，"${employees.path}" 代表引用路由路徑物件 employees 的路徑；"/" 代表分隔符號；":$idParam" 表示一個參數，參數以冒號 (:) 開始，從程式可以看出 $idParam 的值為 id，此處 id 是一個預留位置。例如：查看 id 為 5 的員工資訊，實際 URL 中其路徑是 "/employees/5"。

如果需要傳遞多個參數，則每個參數都應以冒號開始加上預留位置，在參數間必須使用分割符號，且參數的實際值不能包含與分割符號一樣的字串序列。常用的分隔符號是斜線 (/) 或逗點 (,)。例如：範例程式中包含 id 和 other 兩個參數，它們之間使用逗點作為分隔符號。

```
RoutePath(path: '${employees.path}/:$idParam,:other');
```

方法 getId 是一個幫助函數，用於解析參數清單中參數 id 的值。

增加元件 EmployeeComponent，用於展示員工資訊。範例程式如下：

```
//chapter15/router/lib/src/employee/employee_component.dart
import 'package:angular/angular.dart';
import 'package:angular_router/angular_router.dart';

import '../route_paths.dart';
import 'employee.dart';
import 'employee_service.dart';
```

```
@Component(
   selector: 'employee',
   templateURL: 'employee_component.html',
   styleURLs: ['employee_component.css'],
   directives: [coreDirectives],
   providers: [ClassProvider(EmployeeService)],
)
// 實現 OnActivate 介面
class EmployeeComponent implements OnActivate {
   final EmployeeService _employeeService;
   final Router _router;
   Employee emp;
   // 注入服務 EmployeeService 和路由器 Router
   EmployeeComponent(this._employeeService,this._router);

   // 實現路由器的生命週期掛鉤函數 onActivate
   @override
   void onActivate(_, RouterState current){
      // 解析當前路由中參數 id 的值
      var id = getId(current.parameters);
      // 如果 id 值存在，則獲取對應 id 的員工資訊
      if (id != null) emp = _employeeService.getById(id);
   }

   // 導覽到員工清單
   Future<NavigationResult> goBack(){
      // 使用路由器物件導覽到指定路由
      _router.navigate(RoutePaths.employees.toURL());
   }
}
```

在提供者清單中增加並注入服務 EmployeeService。注入了路由器物件 Router，該物件的 navigate 方法用於導覽到新的路由，它接收一個必選參數 path 和一個可選參數 navigationParams。

```
Future<NavigationResult> navigate(
   String path, [
   NavigationParams navigationParams,
]);
```

方法 navigate 會導覽到與參數 path 匹配的路由，物件 navigationParams 用於向路由指定可選參數清單。可以向 NavigationParams 的建構函數提供參數 queryParameters 以建構新範例。範例程式如下：

```
Future<NavigationResult> goBack() => _router.navigate(
   RoutePaths.heroes.toURL(),
      NavigationParams(queryParameters: {idParam: '${emp.id}'}));
```

查詢參數 queryParameters 是一個 Map 集合，可以提供多個參數。以 id 為 5 的員工為例，其最終跳躍路徑等於 "/employees?id=5"。查詢參數是以問號 ? 啟動的，參數間使用符號 & 分隔。

將 OnActivate 介面增加到元件的實現介面清單中，實現路由器的生命週期函數 onActivate，當元件透過路由啟動時會觸發該生命週期函數。它會接收兩個參數，範例程式如下：

```
void onActivate(RouterState previous, RouterState current);
```

兩個參數都是表示路由器狀態的 RouterState 物件，該物件包含路由的路徑 path、路由定義清單、參數集合 parameters、查詢參數集合 queryParameters。第一個 RouterState 代表上一個路由的狀態，第二個 RouterState 代表當前路由的狀態，因為用不到上一個路由的狀態，因此在程式中使用預留位置 "_" 代替。

更新元件 EmployeeComponent 的範本，範例程式如下：

```
//chapter15/router/lib/src/employee/employee_component.html
<div *ngIf="emp!= null">
   <h2>{{emp.name}}</h2>
   <div>
      <label>id: </label>{{emp.id}}</div>
   <div>
      <label> 薪資：</label>{{emp.salary}}
   </div>
   <button (click)="goBack()"> 返回 </button>
</div>
```

更新元件 EmployeeComponent 的樣式，程式如下：

```
//chapter15/router/lib/src/employee/employee_component.css
label {
   display: inline-block;
   width: 3em;
   margin: .5em 0;
   color: #607D8B;
   font-weight: bold;
}

button {
   margin-top: 20px;
   font-family: Arial;
   background-color: #eee;
   border: none;
   padding: 5px 10px;
   border-radius: 4px;
   cursor: pointer; cursor: hand;
}
button:hover {
   background-color: #cfd8dc;
}
```

匯入元件 EmployeeComponent 的實現類別，並增加路由定義 employee。程式如下：

```
//chapter15/router/lib/src/routes.dart
import 'employee/employee_component.template.dart' as employee_template;
//・・・
class Routes {
   //・・・
   static final employee = RouteDefinition(
   routePath: RoutePaths.employee,
   component: employee_template.EmployeeComponentNgFactory,
);

static final all = <RouteDefinition>[
      //・・・
      employee,
```

```
      //・・・
   ];
}
```

更新元件 EmployeeListComponent 的範本程式如下：

```
//chapter15/router/lib/src/employee/employee_list_component.html
<ul *ngIf="emps != null" class="list">
   <li *ngFor="let emp of emps"
      (click)="onSelect(emp)">
      <span class="badge">{{emp.id}}</span>
      <span class="text">{{emp.name}}</span>
   </li>
</ul>
```

更新元件 EmployeeListComponent 的程式如下：

```
//chapter15/router/lib/src/employee/employee_list_component.dart
import 'package:angular/angular.dart';
import 'package:angular_router/angular_router.dart';

import '../route_paths.dart';
import 'employee_service.dart';
import 'employee.dart';

@Component(
   selector: 'employee-list',
   templateURL: 'employee_list_component.html',
   styleURLs: ['employee_list_component.css'],
   directives: [coreDirectives],
   providers: [ClassProvider(EmployeeService)],
)
class EmployeeListComponent{
   final EmployeeService _employeeService;
   final Router _router;
   List<Employee> emps;
   EmployeeListComponent(this._employeeService,this._router){
      // 初始化員工清單 emps
      emps = _employeeService.getAll();
   }
```

```
    // 根據指定 id 建構導覽到員工資訊頁的路由
    String _empURL(int id) =>
        RoutePaths.employee.toURL(parameters: {idParam: '$id'});

    // 導覽到指定路由
    Future<NavigationResult> _gotoDetail(int id) =>
        _router.navigate(_empURL(id));

    // 回應員工清單項目的點擊事件
    void onSelect(Employee emp) {
        _gotoDetail(emp.id);
    }
}
```

元件注入了服務 EmployeeService 和路由器 Router，並在建構函數中透過服務的 getAll 方法對員工清單 emps 進行了初始化。

當使用者點擊員工清單中的項目時會執行 onSelect 方法，該方法又會呼叫 gotoDetail 方法，該方法的功能是透過路由器的 navigate 方法導覽到員工資訊頁，其參數透過 empURL 方法獲得，empURL 方法內部透過 RoutePath 物件的 toURL 方法建構 URL，並指定參數值。

更新元件 EmployeeListComponent 的樣式，程式如下：

```
//chapter15/router/lib/src/employee/employee_list_component.css
.list {
    margin: 002em 0;
    list-style-type: none;
    padding: 0;
    width: 15em;
}
.list li {
    cursor: pointer;
    position: relative;
    left: 0;
    background-color: #EEE;
    margin: .5em;
    padding: .3em 0;
```

```
    height: 1.6em;
    border-radius: 4px;
}
.list li:hover {
    color: #607D8B;
    background-color: #EEE;
    left: .1em;
}
.list .text {
    position: relative;
    top: -3px;
}
.list .badge {
    display: inline-block;
    font-size: small;
    color: white;
    padding: 0.8em 0.7em 00.7em;
    background-color: #607D8B;
    line-height: 1em;
    position: relative;
    left: -1px;
    top: -4px;
    height: 1.8em;
    margin-right: .8em;
    border-radius: 4px 004px;
}
```

保存更改並刷新瀏覽器，點擊員工清單中的項目，觀察視圖更改和路由變化。

15.6.4 子路由

前面已經介紹了與路由相關的所有重要內容，在本節中將創建一個與應用程式根元件無關的路由元件，它將管理與任務相關的路由。

將 TaskListComponent 作為路由元件，它擁有自己的路由集 routes 和 router-outlet，就像 AppComponent 一樣。透過該路由元件導覽到 TaskHome

Component 和 TaskComponent。TaskComponent 元件接收一個參數並展示相關任務的資訊。TaskHomeComponent 元件則單純作為子路由的預設路由。

創建服務 TaskService，用於提供與任務相關的資料和功能。程式如下：

```
//chapter15/router/lib/src/task/task_service.dart
class Task{
   int id;
   String task;
   Task(this.id,this.task);
}

class TaskService{
   var tasks = [
      Task(5, '業務分析'),
      Task(6, '產品分析'),
      Task(7, '結構設計'),
      Task(8, '視覺設計')
   ];
   List<Task> getAll(){
      return tasks;
   }
   Task getById(int id){
      return tasks.firstWhere((task)=> task.id == id);
   }
}
```

創建元件 TaskHomeComponent，它將用作子路由的預設路由。程式如下：

```
//chapter15/router/lib/src/task/task_home_component.dart
import 'package:angular/angular.dart';

@Component(
   selector: 'task-home',
   template: '''
   <div>
      任務中心
```

```
    </div>
    ''',
    directives: [coreDirectives],
)
class TaskHomeComponent{}
```

創建元件 TaskComponent，它將用於展示單一任務資訊。程式如下：

```
//chapter15/router/lib/src/task/task_component.dart
import 'package:angular/angular.dart';
import 'package:angular_router/angular_router.dart';

import 'task_service.dart';
import 'route_paths.dart';

@Component(
    selector: 'task',
    template: '''
        <div *ngIf="task!= null">
            <div>
                <label>id: </label>{{task.id}}</div>
            <div>
                <label> 任務： </label>{{task.task}}
            </div>
            <button (click)="goBack()"> 返回 </button>
        </div>
    ''',
    styleURLs: ['task_component.css'],
    directives: [coreDirectives],
)
class TaskComponent implements OnActivate{
    final TaskService _taskService;
    final Router _router;
    Task task;
    // 注入服務 TaskService 和 Router
    TaskComponent(this._taskService,this._router);
    // 實現路由器生命週期函數 onActivate
    @override
    void onActivate(RouterState previous, RouterState current) {
        var id = getId(current.parameters);
```

```
      task = _taskService.getById(id);
   }
   // 跳躍到子路由的預設路由
   Future<NavigationResult> goBack(){
      _router.navigate(RoutePaths.home.toURL());
   }
}
```

樣式程式如下：

```
//chapter15/router/lib/src/task/task_component.css
label {
   display: inline-block;
   width: 3em;
   margin: .5em 0;
   color: #607D8B;
   font-weight: bold;
}

button {
   margin-top: 20px;
   font-family: Arial;
   background-color: #eee;
   border: none;
   padding: 5px 10px;
   border-radius: 4px;
   cursor: pointer; cursor: hand;
}

button:hover {
   background-color: #cfd8dc;
}
```

定義子路由的路由路徑。程式如下：

```
//chapter15/router/lib/src/task/route_paths.dart
import 'package:angular_router/angular_router.dart';
import '../route_paths.dart' as _parent;
export '../route_paths.dart' show idParam,getId;
class RoutePaths{
```

```
    // 定義子路由的預設路由
    static final home = RoutePath(
      path:'',
      parent: _parent.RoutePaths.tasks,
    );
    // 定義子路由下導覽到任務資訊頁的路由
    static final task = RoutePath(
      path:':${_parent.idParam}',
      parent: _parent.RoutePaths.tasks,
    );
}
```

該檔案匯入了父路由定義的路由路徑檔案，其別名為 _parent。將常數 idParam 和方法 getId 匯出，使得引入當前路由路徑檔案的函數庫可以使用。

在路由路徑定義中出現了新的參數 parent，它表示當前路徑的父路由，該參數的值引用當前路由元件 TaskListComponent 在父路由中定義的路由路徑 tasks。

增加子路由的路由定義。程式如下：

```
//chapter15/router/lib/src/task/routes.dart
import 'package:angular_router/angular_router.dart';

import 'task_component.template.dart' as task_template;
import 'task_home_component.template.dart' as task_home_template;
import 'route_paths.dart';
export 'route_paths.dart';

class Routes{
    // 導覽到子路由下 TaskHomeComponent 元件的路由定義
    // 並設定為子路由下的預設路由
    static final home = RouteDefinition(
      routePath: RoutePaths.home,
      component: task_home_template.TaskHomeComponentNgFactory,
      useAsDefault: true,
    );
```

```
    // 導覽到子路由下 TaskComponent 元件的路由定義
    static final task = RouteDefinition(
      routePath: RoutePaths.task,
      component: task_template.TaskComponentNgFactory,
    );
    // 引用所有已定義路由
    static final all = <RouteDefinition>[
      home,
      task,
    ];
}
```

更新 TaskListComponent 元件。程式如下：

```
//chapter15/router/lib/src/task/task_list_component.dart
import 'package:angular/angular.dart';
import 'package:angular_router/angular_router.dart';

import 'task_service.dart';
import 'routes.dart';
@Component(
    selector: 'task-list',
    template: '''
    <ul *ngIf="tasks != null" class="list">
      <li *ngFor="let task of tasks"
        (click)="onSelect(task)">
        <span class="badge">{{task.id}}</span>
        <span class="text">{{task.task}}</span>
      </li>
    </ul>
    <router-outlet [routes]="Routes.all"></router-outlet>
    ''',
    styleURLs: ['task_list_component.css'],
    directives: [coreDirectives,routerDirectives],
    providers: [ClassProvider(TaskService)],
    exports: [RoutePaths,Routes],
)
class TaskListComponent {
    final TaskService _taskService;
```

```
  final Router _router;
  List<Task> tasks;
  TaskListComponent(this._taskService,this._router){
  // 初始化任務清單 tasks
  tasks = _taskService.getAll();
  }
  // 回應任務清單項目的點擊事件
  void onSelect(Task task) {
    _gotoDetail(task.id);
  }
  // 根據指定 id 建構導覽到任務資訊頁的路由
  String _empURL(int id) =>
  RoutePaths.task.toURL(parameters: {idParam: '$id'});
  // 導覽到指定路由
  Future<NavigationResult> _gotoDetail(int id) =>
    _router.navigate(_empURL(id));
}
```

更新元件的樣式，程式如下：

```
//chapter15/router/lib/src/task/task_list_component.css
.list {
  margin: 002em 0;
  list-style-type: none;
  padding: 0;
  width: 15em;
}
.list li {
  cursor: pointer;
  position: relative;
  left: 0;
  background-color: #EEE;
  margin: .5em;
  padding: .3em 0;
  height: 1.6em;
  border-radius: 4px;
}
.list li:hover {
  color: #607D8B;
```

```
    background-color: #EEE;
    left: .1em;
}
.list .text {
    position: relative;
    top: -3px;
}
.list .badge {
    display: inline-block;
    font-size: small;
    color: white;
    padding: 0.8em 0.7em 00.7em;
    background-color: #607D8B;
    line-height: 1em;
    position: relative;
    left: -1px;
    top: -4px;
    height: 1.8em;
    margin-right: .8em;
    border-radius: 4px 004px;
}
```

保存檔案並刷新瀏覽器，進入任務清單，點擊其中的任意項目，觀察子路由變化情況。本節的重點應該關注在路由路徑定義檔案上，其他部分與 15.6.3 節的內容相似。通常不會在開發中使用子路由，它的適用場景是重用路由元件，即組合多個現有路由器。

15.6.5 生命週期函數

預設情況下，任何使用者都可以在應用程式中任意進行導覽，對於完整的應用程式這樣做並不總是正確的，例如：

(1) 使用者無權導覽到目標群元件。

(2) 使用者必須先經過許可權驗證。

(3) 在顯示目標群元件前初始化一些資料。

(4) 使用者可能在離開元件前有表單資料未保存。

路由提供了一系列生命週期函數，以幫助開發者處理這些情況 , 如表 15-3 所示。

表 15-3 路由生命週期函數

掛鉤函數	目的和時間
onActivate	當路由啟動元件時呼叫，常用於資料初始化。例如：從網路請求中獲取資料
onDeactivate	當路由停用元件時呼叫
canNavigate	控制當前路由是否可跳躍
canReuse	設定當前元件實例是否可重複使用。注意：父元件可重複使用，子元件實例才可重複使用
canDeactivate	允許有條件地停用路由
canActivate	允許有條件地啟用路由

所有的生命週期函數都包含在名稱相同介面中，對應的介面名稱分別為 OnActivate、OnDeactivate、CanNavigate、CanReuse、CanDeactivate、CanActivate。

首先在元件 TaskListComponent 上使用 mixin 類別 CanReuse，其作用是保證 TaskHomeComponent 元件實例能被重用。程式如下：

```
//chapter15/router/lib/src/task/task_list_component.dart
class TaskListComponent with CanReuse{
//...
}
```

在元件中 TaskHomeComponent 實現所有路由週期函數對應的介面，並實現週期函數。範例程式如下：

```
//chapter15/router/lib/src/task/task_home_component.dart
import 'package:angular/angular.dart';
import 'package:angular_router/angular_router.dart';

@Component(
  selector: 'task-home',
```

```
    template: '''
    <div>
      任務中心
    </div>
    ''',
    directives: [coreDirectives],
)
class TaskHomeComponent with CanReuse implements CanActivate,CanDeactivate,Can
Navigate,OnActivate,OnDeactivate {

    // 在路由啟動元件時呼叫
    @override
    void onActivate(RouterState previous, RouterState current) {
      // 常用於初始化資料
      print('TaskHomeComponent: onActivate: ${previous?.toURL()} ->
${current?.toURL()}');
    }

    // 在路由停用元件之前呼叫
    @override
    void onDeactivate(RouterState current, RouterState next) {
      // 常用於停用元件實例前處理必須完成的任務
      print('TaskHomeComponent: onDeactivate: ${current?.toURL()} -> ${next?.
toURL()}');
    }

    // 允許有條件地啟動新路由
    @override
    Future<bool> canActivate(RouterState current, RouterState next) async{
      print('TaskHomeComponent: canActivate: ${current?.toURL()} -> ${next?.
toURL()}');
      // 在實際使用中需根據條件控制是否允許啟動當前路由
      // 這裡沒有增加控制條件，直接返回 true，表示允許
      return true;
    }

    // 允許有條件地停用路由
    @override
    Future<bool> canDeactivate(RouterState current, RouterState next) async{
```

```
        print('TaskHomeComponent: canDeactivate: ${current?.toURL()} -> ${next?.
toURL()}');
        // 在實際使用中需根據條件控制是否允許停用當前路由
        // 這裡沒有增加控制條件，直接返回 true，表示允許
        return true;
    }

    // 允許有條件地阻止導覽
    @override
    Future<bool> canNavigate() async{
        print('TaskHomeComponent: canNavigate');
        // 在實際使用中需根據條件控制是否允許導覽
        // 這裡沒有增加控制條件，直接返回 true，表示允許
        return true;
    }

    // 允許重用現有的元件實例
    @override
    Future<bool> canReuse(RouterState current, RouterState next) async{
        print('TaskHomeComponent: canReuse: ${current?.toURL()} -> ${next?.
toURL()}');
        // 在實際使用中需根據條件控制是否允許重用此元件實例
        // 這裡沒有增加控制條件，直接返回 true，表示允許
        return true;
    }
}
```

刷新瀏覽器，打開主控台。點擊導覽中的任務清單，再點擊清單中任意一個項目，再點擊「返回」按鈕。執行結果如圖 15-14 所示。

```
TaskHomeComponent: canActivate: /employees -> /tasks
TaskHomeComponent: onActivate: /employees -> /tasks
TaskHomeComponent: canNavigate
TaskHomeComponent: canDeactivate: /tasks -> /tasks/7
TaskHomeComponent: onDeactivate: /tasks -> /tasks/7
TaskHomeComponent: canReuse: /tasks -> /tasks/7
TaskHomeComponent: canActivate: /tasks/7 -> /tasks
TaskHomeComponent: onActivate: /tasks/7 -> /tasks
>
```

▲ 圖 15-14 路由生命週期函數

15.7 結構指令

前面已經介紹了結構指令的使用，這裡將介紹結構指令的更多細節。

15.7.1 星號字首

在使用結構指令時會注意到指令前面會帶上星號 (*)，本節將介紹它的具體用途。

1. NgIf

使用 *ngIf 根據範本運算式判斷是否顯示元素及內容。程式如下：

```
<div *ngIf="name != null">名字：{{name}}</div>
```

星號實際上是語法糖，在內部處理時，Angular 會將其分解為 <template> 元素，並將宿主元素 <div> 包裹在 <template> 元素中。程式如下：

```
<template [ngIf]="name != null">
   <div>名字：{{name}}</div>
</template>
```

此時 *ngIf 指令轉移到 <template> 元素上，並且成為其屬性指令 [ngIf]。其餘部分將按原結構包裹在 template 元素中。

2. NgFor

*ngFor 完整功能的使用範例程式如下：

```
<div *ngFor="let emp of emps;let i=index;
let odd=odd;trackBy:trackById" [class.odd]="odd">
   Id: {{emp.id}} Name：{{emp.name}} Salary：{{emp.salary}}
</div>
```

Angular 會將其分解為 <template> 元素，程式如下：

```
<template ngFor let-emp [ngForOf]="emps" let-i="index" let-odd="odd"
```

```
      [ngForTrackBy]="trackById">
   <div [class.odd]="odd">Id: {{emp.id}} Name：{{emp.name}} Salary：{{emp.
salary}}</div>
</template>
```

這顯然比 NgIf 使用起來更加複雜，需要說明的是在 *ngFor 後邊字串之外的部分仍然會停留在宿主元素中，此範例 [ngClass]="odd" 就停留在 <div> 中。

Angular 微語法可以在一個字串中簡潔有效地設定指令，而微語法解析器將字串轉為元素 <template> 上的屬性：

(1) let 關鍵字宣告在範本中引用的範本輸入變數。此範例中輸入變數是 emp、i 和 odd。解析器將 let emp、let i 和 let odd 轉為名為 let-emp、let-i 和 let-odd 的範本輸入變數。

(2) 微語法解析器將 of 和 trackby 改名為 Of 和 TrackBy，並且使用 ngFor 作為字首，最終名稱為 ngForOf 和 ngForTrackBy。它們是 NgFor 兩個輸入屬性的名稱，這樣，指令就可以知道清單是 emps，而追蹤函數是 trackById。

(3) 當 NgFor 指令遍歷清單時，它會設定和重置自身上下文物件的屬性。這些屬性包括 index 和 odd，以及一個特殊屬性 $implicit。

(4) 將 let-i 和 let-odd 變數定義為 let i=index 和 let odd=odd。Angular 將它們設定為上下文的 index 和 odd 屬性的當前值。

(5) 未向 let-hero 指定內容屬性，因為它的來源是隱式的。Angular 將 let-hero 設定為上下文的 $implicit 屬性的值，$implicit 屬性的值由 NgFor 使用當前疊代的 emp 為其進行初始化。

使用 let 關鍵字宣告範本輸入變數，變數的作用範圍僅限於重複範本的單一實例。可以在範本包含的其他結構指令中使用相同的變數名稱。

宿主元素可以使用多個屬性指令，但是只能使用一個結構指令。因為結構指令會對宿主元素及其後代進行複雜處理，當多個指令對同一宿主元素進行處理時，哪一個應該優先成為一個難題。因此正常的做法是將多個結構指令應用於不同的宿主元素上。

3. NgSwitch

從前面的知識已經了解到 NgSwitch 實際上是一組協作指令：NgSwitch、NgSwitchCase、NgSwitchDefault。

先來看一個範例，程式如下：

```
<div [ngSwitch]="color">
   <span *ngSwitchCase="'Red'"> 紅色 </span>
   <span *ngSwitchCase="'Green'"> 綠色 </span>
   <span *ngSwitchCase="'Black'"> 黑色 </span>
   <span *ngSwitchDefault > 藍色 </span>
</div>
```

NgSwitch 是一個屬性指令，用於控制其他兩個協作指令的行為。因此使用 [ngSwitch] 而非 *ngSwitch。

NgSwitchCase 和 NgSwitchDefault 是結構性指令，因此使用形式為 *ngSwitchCase 和 *ngSwitchDefault。與其他結構指令一樣，它們也可簡化為 <template> 元素的形式。程式如下：

```
<div [ngSwitch]="color">
   <template [ngSwitchCase]=="'Red'">
      <span> 紅色 </span>
   </template>
   <template [ngSwitchCase]=="'Green'">
      <span> 綠色 </span>
   </template>
   <template [ngSwitchCase]=="'Black'">
      <span> 黑色 </span>
   </template>
   <template ngSwitchDefault>
      <span> 藍色 </span>
   </template>
</div>
```

在應用結構指令時可以手動指定為 <template> 元素的形式，但推薦使用星號的模式。

15.7.2 自訂結構指令

在本節編寫名叫 RepeatDirective 的結構指令，該指令的作用是根據提供的次數重複範本內容。

指令是使用註釋 @Directive 宣告的類別。建立指令的基本結構，程式如下：

```
import 'package:angular/angular.dart';

@Directive(selector: '[myRepeat]')
class RepeatDirective{
}
```

指令的選擇器通常是中括號包裹的屬性選擇器 [myRepeat]，指令屬性名稱採用小駝峰命名法，並且帶有字首。避免使用 ng 字首，該字首是 Angular 佔有的。為了編碼規範指令類別名稱以 Directive 結尾。

1. 嵌入式範本

TemplateRef 表示能夠用於實例化嵌入式視圖的嵌入式範本。

可透過兩種方式存取 TemplateRef：第一種，將指令放置在 <template> 元素上或使用指令時以 * 開頭，並透過 TemplateRef 權杖將此嵌入式視圖的 TemplateRef 物件注入指令的建構函數中。第二種，透過 Query 從元件或指令中尋找 TemplateRef。

2. 視圖容器

ViewContainerRef 表示可以附加一個或多個視圖的容器。該容器可以包含兩種視圖：第一種，透過 createComponent 方法實例化一個元件創建的宿主視圖。第二種，透過 createEmbeddedView 方法實例化嵌入式範本 TemplateRef 創建的嵌入式視圖。

要存取元素的 ViewContainerRef，可以將 ViewContainerRef 注入指令的建構函數中，或透過 ViewChild 查詢得到。

ViewContainerRef 類別包含以下一些重要方法：

(1) clear()：銷毀該容器中的所有視圖。
(2) createEmbeddedView(TemplateRef templateRef) → EmbeddedViewRef：以 TemplateRef 為基礎實例化一個嵌入式視圖，並將其附加到此容器中。
(3) insertEmbeddedView(TemplateRef templateRef, int index) → EmbeddedViewRef：以 TemplateRef 為基礎實例化一個嵌入式視圖，並將其插入此容器中的指定索引 index 處。

使用 TemplateRef 可以獲取 Angular 生成的 <template> 元素中的內容，使用 ViewContainerRef 可以存取宿主元素的視圖容器。

透過建構函數將 TemplateRef 和 ViewContainerRef 注入指令類別的私有變數中。程式如下：

```
// 嵌入式範本
TemplateRef _templateRef;
// 視圖容器
ViewContainerRef _viewContainer;
// 注入視圖容器和嵌入式範本
RepeatDirective(this._templateRef, this._viewContainer);
```

該指令還需要接收一個值，宣告一個輸入屬性 myRepeat。因為不需要獲取它的值，所以只提供了該屬性的 setter 方法。程式如下：

```
@Input()
set myRepeat(int times){
   if(times>0){
      for(int i = 0; i < times; i++){
         // 向視圖容器插入嵌入式範本
         _viewContainer.insertEmbeddedView(templateRef,i);
      }
   }
}
```

times 表示需要重複嵌入式範本的次數，在迴圈本體中將嵌入式範本 TemplateRef 不斷附加到視圖容器 ViewContainerRef 中。

完整的指令程式如下：

```
//chapter15/structural_directive/lib/src/repeat_directive.dart
import 'package:angular/angular.dart';

@Directive(selector: '[myRepeat]')
class RepeatDirective{
  // 嵌入式範本
  TemplateRef _templateRef;
  // 視圖容器
  ViewContainerRef _viewContainer;
  // 注入視圖容器和嵌入式範本
  RepeatDirective(this._templateRef, this._viewContainer);
  // 宣告輸入屬性並接收一個 int 類型的參數
  // 該參數表示在視圖容器中重複嵌入式範本的次數
  @Input()
  set myRepeat(int times){
    if(times>0){
      for(int i = 0; i < times; i++){
        // 向視圖容器插入嵌入式範本
        _viewContainer.insertEmbeddedView(_templateRef, i);
      }
    }
  }
}
```

將此指令增加到 AppComponent 的指令清單中。程式如下：

```
//chapter15/structural_directive/lib/app_component.dart
import 'package:angular/angular.dart';
import 'src/repeat_directive.dart';
@Component(
  selector: 'my-app',
  styleURLs: ['app_component.css'],
  templateURL: 'app_component.html',
  directives: [RepeatDirective],
)
```

```
class AppComponent {
}
```

然後在範本中使用指令，程式如下：

```
//chapter15/structural_directive/lib/app_component.html
<p *myRepeat="9">Repeat( 重複指令 )</p>
```

執行結果如圖 15-15 所示。

```
Repeat (重複指令)
Repeat (重複指令)
Repeat (重複指令)
Repeat (重複指令)
Repeat (重複指令)
Repeat (重複指令)
Repeat (重複指令)
Repeat (重複指令)
Repeat (重複指令)
```

▲ 圖 15-15 結構指令

15.8 HTTP 連接

前端應用通常需要透過 HTTP 與服務端通訊。基礎請求可以使用 dart:html 包中的 HttpRequest 類別。若需要做複雜操作，推薦使用 http 包中的方法和類別。

15.8.1 http 套件

http 套件包含一組進階函數和類別，可輕鬆使用 HTTP 資源。它是多平台的，並且支持行動裝置、桌面和瀏覽器。

常用函數：

(1) delete(dynamic URL, {Map<String, String> headers})：將具有指定標頭的 HTTP DELETE 請求發送到指定 URL，該 URL 可以是 Uri 或 String。
(2) get(dynamic URL, {Map<String, String> headers})：將具有指定標頭的 HTTP GET 請求發送到指定 URL，該 URL 可以是 Uri 或 String。
(3) post(dynamic URL, {Map<String, String> headers, dynamic body, Encoding encoding})：將具有指定標頭和正文的 HTTP POST 請求發送到指定的 URL，該 URL 可以是 Uri 或 String。
(4) put(dynamic URL, {Map<String, String> headers, dynamic body, Encoding encoding})：將具有指定標頭和正文的 HTTP PUT 請求發送到指定的 URL，該 URL 可以是 Uri 或 String。

對於 put 和 post 方法請求的正文可以是 String、List<int> 或 Map<String, String>：①如果是 String，將使用 encoding 對請求進行編碼，請求的 content-type 預設是 text/plain；②如果是 List，它將用作請求正文的位元組清單；③如果是 Map，將使用 encoding 為表單欄位編碼，請求的 content-type 被設定為 application/x-www-form-URLencoded，並且不可更改。encoding 預設為 UTF-8。

上述函數會返回一個由 Future 封裝的 Response 物件，http 包中 Response 類別常用屬性如下：

(1) body：將回應正文作為字串返回。
(2) bodyBytes：將回應正文作為位元組陣列返回。

15.8.2 資料轉換

在用戶端和服務端傳輸資料時，通常會對模型資料進行轉碼和解碼。最常用的資料傳輸格式是 JSON，json_string 套件使得資料在 JSON 和 Dart

物件間相互轉換變得容易。該套件並不是 JSON 資料和 Dart 物件間相互轉換最好的套件，但它最能夠展現 JSON 資料和 Dart 物件相互轉換的原理，並且具有極佳的自訂特徵。

json_string 套件定義了一個 Mixin。

Jsonable：使用它並實現 toJson 方法可以將 Dart 物件編碼為某些有效的 JSON 物件。除此之外還需要為 Dart 物件定義類方法 fromJson，該方法作為將 JSON 資料轉為 Dart 物件的解碼器。

json_string 套件定義了一個類別 JsonString，它具有以下建構函數：

(1) JsonString(String source, {bool enableCache: false})：如果 source 是有效的 JSON，則構造一個 JsonString 物件。如果可選的 enableCache 參數設定為 true，那麼將提供解碼值的快取。
(2) JsonString.encode(Object value, {dynamic encoder(Object object)})：創建一個將 value 轉為有效 JSON 的 JsonString。
(3) JsonString.orNull(String source, {bool enableCache: false})：如果 source 是有效的 JSON，則構造一個 JsonString 物件；如果不是，則返回 null。

類別 JsonString 具有以下類別方法：

(1) encodeObject<T extends Object>(T value, {JsonObjectEncoder<T> encoder})：創建一個將值轉為有效 JSON 物件的 JsonString。如果 T 實現 Jsonable 介面，則在轉換期間使用 toJson 方法返回的結果。如果沒有，則必須提供編碼器函數。
(2) encodeObjectList<T extends Object>(List<T> list, {JsonObjectEncoder <T> encoder})：創建一個將 list 轉為有效 JSON 清單的 JsonString。如果 T 實現 Jsonable 介面，則在轉換期間使用 toJson 方法返回的結果。如果沒有，則必須提供編碼器函數。

類別 JsonString 具有以下實例方法：

(1) decodeAsObject<T extends Object>(JsonObjectDecoder<T> decoder)：將 JSON 資料解碼為 T 的實例。

(2) decodeAsObjectList<T extends Object>(JsonObjectDecoder<T> decoder)：將 JSON 資料解碼為 List<T> 的實例。

15.8.3 服務端

創建命令列應用程式，專案名稱為 http_server。

更新依賴項並執行 pub get 命令，程式如下：

```
//chapter15/http_server/pubspec.yaml
name: http_server
description: A simple command-line application.
#version: 1.0.0
#homepage: https://www.example.com

environment:
  sdk: '>=2.7.0 <3.0.0'

dependencies:
  shelf: ^0.7.7
  shelf_router: ^0.7.2
  json_string: ^2.0.1

dev_dependencies:
  shelf_router_generator: ^0.7.0+1
  build_runner: ^1.3.1
  pedantic: ^1.8.0
```

首先建立模型 Employee，它包含欄位 id、name 和 salary。使用 mixin 類別 Jsonable，並實現 toJson 方法用於將 Employee 實例編碼為 JSON。提供 fromJson 方法用於將 JSON 物件解碼為 Employee 實例。程式如下：

```
//chapter15/http_server/bin/employee.dart
```

```
// 也用作 chapter15/http_client/lib/src/employee.dart
import 'package:json_string/json_string.dart';
// 使用 mixin 類別 Jsonable
class Employee with Jsonable{
   int id;
   String name;
   num salary;
   Employee({this.id,this.name,this.salary});

   // 實現 toJson 方法，將模型轉為 json 編碼器
   @override
   Map<String,dynamic> toJson(){
      // 返回 json 物件
      return {
         'id':id,
         'name':name,
         'salary':salary
      };
   }

   // 類別方法 fromJson 將 json 轉為模型解碼器
   static Employee fromJson(Map<String,dynamic> json){
      // 返回 Employee 實例
      return Employee(
         id:json['id'],
         name:json['name'],
         salary:json['salary']
      );
   }
}
```

因為目前沒有連接資料庫，所以以強制寫入的方式實例化一個 Employee 類型的清單。提供 getAll 方法用於返回整個清單，提供 getById 方法用於根據 id 返回對應 Employee 實例。程式如下：

```
//chapter15/http_server/bin/employee_service.dart
import 'employee.dart';
class EmployeeService{
   var emps = [
```

```
        Employee(id:5,name:'Magneta',salary:9971),
        Employee(id:6,name:'RubberMan',salary:4533),
        Employee(id:7,name:'Dynama',salary:6720),
        Employee(id:8,name:'Dr IQ',salary:4907),
        Employee(id:9,name:'Magma',salary:5278),
        Employee(id:10,name:'Tornado',salary:7800)
    ];
    List<Employee> getAll(){
        // 返回整個 emps 清單
        return emps;
    }
    Employee getById(int id){
        // 根據 id 從 emps 清單中返回單一 Employee 實例
        return emps.firstWhere((emp)=> emp.id == id,orElse: (){
            // 當 emps 清單中不存在 id 相匹配的實例時返回 null
            return null;
        });
    }
}
```

定義路由器，此部分內容一方面是回顧路由器方面的知識，另一方面是使用 json_string 套件提供的功能轉換 Dart 物件和 JSON 資料。程式如下：

```
//chapter15/http_server/bin/routers.dart
import 'dart:async' show Future;
import 'dart:convert';
import 'package:shelf/shelf.dart';
import 'package:shelf_router/shelf_router.dart';
import 'package:json_string/json_string.dart';
import 'employee_service.dart';
import 'employee.dart';
// 將生成的檔案作為本函數庫的一部分
// 該檔案名稱作為路由生成器生成檔案的依據
// 當未使用 library 指令時，單一檔案預設為一個函數庫
part 'routers.g.dart';
```

```
class Emp{
  // 回應路由 '/emp/' 下帶路徑參數的 GET 請求
  @Route.get('/emp/<empId>')
  Future<Response> _getById(Request request)async{
    // 解析查詢參數 empId
    var empId = int.parse(params(request, 'empId'));
    // 根據 id 返回對應實例
    var emp = EmployeeService().getById(empId);
    // 狀態碼為 200 成功回應並返回 Json 資料
    // 使用 jsonEncode 方法將 Employee 實例編碼為 Json 資料並返回
    //jsonEncode 方法是由函數庫 dart:convert 提供的
    return Response.ok(jsonEncode(emp));
  }
  // 回應路由 '/emps' 下的 GET 請求
  @Route.get('/emps')
  Future<Response> _getAll(Request request)async{
    // 返回整個清單
    var emps = EmployeeService().getAll();
    // 狀態碼為 200 成功回應
    // 使用 jsonEncode 將 emps 清單編碼為 Json 資料並返回
    return Response.ok(jsonEncode(emps));
  }
  // 回應路由 '/emp' 下的 POST 請求
  @Route.post('/emp')
  Future<Response> _create(Request request)async{
    var emp;
    //request.read() 方法將請求主體轉化為 Stream<List<int>>
    //utf8.decoder.bind() 方法對 Stream<List<int>> 進行解碼
    //join() 方法用於將可能的多個資料區塊組合在一起
    await utf8.decoder.bind(request.read()).join().then((content){
      // 將 Json 內容解碼為 Employee 實例
      emp = JsonString(content).decodeAsObject(Employee.fromJson);
    });
    return Response.ok(jsonEncode(emp));
  }

  @Route.all('/<ignored|.*>')
```

```
  Future<Response> _notFound(Request request) async{
    return Response.notFound(' 頁面未找到 ');
  }
  // 生成的 _$AppRouter 函數可用於獲取此物件的 handler
  // 用於返回此路由器的處理常式
  Handler get handler => _$EmpRouter(this).handler;
}
```

然後在編輯器中的命令列輸入並執行以下命令：

```
pub run build_runner build
```

成功生成 routers.g.dart 檔案後，增加跨域中介軟體，用於回應跨域請求。程式如下：

```
//chapter15/http_server/bin/main.dart
// 創建中介軟體並提供 corsHeaders 設定
final cors = createCorsHeadersMiddleware(
  corsHeaders:{
    'Access-Control-Allow-Origin': '*',
    'Access-Control-Expose-Headers': 'Authorization, Content-Type',
    'Access-Control-Allow-Headers': 'Authorization, Origin, X-Requested-
With, Content-Type, Accept',
    'Access-Control-Allow-Methods': 'GET, POST, PUT, PATCH, DELETE'
  }
);

Middleware createCorsHeadersMiddleware({Map<String, String> corsHeaders}) {
  // 未提供 corsHeaders 時，使用預設設定
  corsHeaders ??= {'Access-Control-Allow-Origin': '*'};

  // 請求處理常式
  Response handleOptionsRequest(Request request) {
    if (request.method == 'OPTIONS') {
      return Response.ok(null, headers: corsHeaders);
    } else {
      return null;
    }
  }
```

```
  // 回應處理常式，將 corsHeaders 設定應用於回應
  Response addCorsHeaders(Response response) => response.change(headers: 
corsHeaders);
  // 創建中介軟體並返回
  return createMiddleware(requestHandler: handleOptionsRequest, 
responseHandler: addCorsHeaders);
}
```

實例化路由器物件；透過 Pipeline 物件組合中介軟體 logRequests 和 cors 及路由器實例提供的處理常式；使用介面卡 serve 啟動服務。程式如下：

```
//chapter15/http_server/bin/main.dart
import 'package:shelf/shelf.dart';
import 'package:shelf/shelf_io.dart' as io;
import 'routers.dart';

void main()async{
  // 路由器實例
  final service = Emp();

  // 透過 Pipeline 物件組合中介軟體和單一處理常式
  var handler = Pipeline()
    // 增加預設中介軟體，請求日誌
    .addMiddleware(logRequests())
    // 增加跨域中介軟體
    .addMiddleware(cors)
    // 增加路由器實例提供的處理常式
    .addHandler(service.handler);

  // 透過 shelf_io.serve 方法啟動一個 HttpServer
  var server = await io.serve(handler, 'localhost', 1024);
  print(' 服務位址：http://${server.address.host}:${server.port}');
}
```

啟動指令稿，主控台輸出以下資訊表示服務啟動成功。

服務位址：http://localhost:1024

15.8.4 用戶端

創建 AngularDartWebApp 範本生成的專案，專案命為 http_client。刪除專案 lib/src 目錄下的所有檔案，並刪除根元件中的匯入資訊和指令。

更新依賴項並執行 pub get 命令，程式如下：

```
//chapter15/http_client/pubspec.yaml
name: http_client
description: A web app that uses AngularDart Components
#version: 1.0.0
#homepage: https://www.example.com

environment:
  sdk: '>=2.7.0 <3.0.0'

dependencies:
  angular: ^5.3.0
  angular_components: ^0.13.0
  json_string: ^2.0.1
  http: ^0.12.2

dev_dependencies:
  angular_test: ^2.3.0
  build_runner: ^1.6.0
  build_test: ^0.10.8
  build_web_compilers: ^2.3.0
  pedantic: ^1.8.0
  test: ^1.6.0
```

將服務端的 employee.dart 檔案複製到當前專案的 lib/src 目錄下。創建服務，使用 http 套件下的方法向服務端發出請求，並透過 Response 物件的 body 屬性獲取回應正文的字串。使用 json_string 中的 JsonString 建構函數建構 JsonString 實例，使用 JsonString 實例的 decodeAsObject 或 decodeAsObjectList 方法將 Json 資料解碼為 Dart 模型類別或模型類別組成的清單。模型類別有時會被稱為實體類別，在本專案中指類別 Employee。程式如下：

```
//chapter15/http_client/lib/src/service.dart
import 'dart:convert';
import 'package:angular/angular.dart';
import 'package:http/http.dart' as http;
import 'package:json_string/json_string.dart';

import 'employee.dart';
// 服務的位址和通訊埠
const _URL = 'http://localhost:1024';

@Injectable()
class EmployeeService{
  // 根據 id 獲取員工資訊
  Future<Employee> getById(int id)async{
    var emp;
    // 使用 http 套件下的 get 方法發起 GET 請求
    await http.get('$_URL/emp/$id').then((response){
      if(response.statusCode == 200){
        //response.body 將回應正文作為 String 返回
        // 服務端返回的是 Json 資料
        // 可以使用 JsonString 建構函數建構 JsonString 物件並解碼為 Employee
        實例
        emp = JsonString(response.body).decodeAsObject(Employee.fromJson);
      }
    });
    return emp;
  }
  // 獲取所有員工資訊
  Future<List<Employee>> getAll()async{
    var emps;
  await http.get('$_URL/emps').then((response){
    if(response.statusCode == 200){
      emps = JsonString(response.body).decodeAsObjectList(Employee.fromJson);
    }
  });
  return emps;
}
// 增加員工
Future<Employee> post(Employee emp)async{
```

```
    var new_emp;
    // 使用 http 套件下的 post 方法發起 POST 請求
    // 透過 body 參數傳遞 json 資料
    await http.post('$_URL/emp',body: jsonEncode(emp)).then((response){
      if(response.statusCode == 200){
        new_emp = JsonString(response.body).decodeAsObject(Employee.fromJson);
        }
      });
      return new_emp;
    }
}
```

在根元件中注入服務，並將服務類別 EmployeeService 增加到 providers 參數清單中。在參數 directives 中增加常用指令和表單指令清單 formDirectives。所有需要向服務端發出請求的操作都透過注入的服務代理完成。程式如下：

```
//chapter15/http_client/lib/app_component.dart
import 'package:angular/angular.dart';
import 'package:angular_forms/angular_forms.dart';
import 'src/service.dart';
import 'src/employee.dart';

@Component(
    selector: 'my-app',
    styleURLs: ['app_component.css'],
    templateURL: 'app_component.html',
    providers: [EmployeeService],
    directives: [NgFor,NgIf,formDirectives],
)
class AppComponent implements OnInit{
    // 服務實例
    final EmployeeService _employeeService;
    // 快取員工清單
    List<Employee> emps = List<Employee>();
    // 快取透過 id 獲取的員工資訊
    Employee emp_get;
    // 快取增加員工的資訊
    Employee emp_add = Employee();
```

```
    // 注入服務
    AppComponent(this._employeeService);
    @override
    void ngOnInit()async{
        // 從服務端獲取員工清單
        var all = await _employeeService.getAll();
        // 在員工清單 emps 中增加服務端獲取的員工資訊
        emps.addAll(all);
    }
    void getById(String id)async{
        // 將透過 id 獲取的員工設定值給 emp_get
        emp_get = await _employeeService.getById(int.parse(id));
    }
    void add()async{
        // 向服務端發出新增請求
        var emp= await _employeeService.post(emp_add);
        // 將新增員工增加到員工清單 emps 中
        emps.add(emp);
    }
}
```

在根元件範本中增加疊代員工清單、新增員工資訊表單及透過 id 獲取員工資訊的元素區塊。程式如下：

```
//chapter15/http_client/lib/app_component.html
<h1>HTTP 用戶端 </h1>
<h6> 員工清單 </h6>
<table *ngIf="emps != null">
    <tr>
        <th> 編號 </th>
        <th> 名字 </th>
        <th> 薪資 </th>
    </tr>
    <tr *ngFor="let emp of emps">
        <td>{{emp.id}}</td>
        <td>{{emp.name}}</td>
        <td>{{emp.salary}}</td>
    </tr>
</table>
```

```
<h6> 透過 id 獲取單一員工資訊 </h6>
<input type="number" #input placeholder=" 輸入清單中存在的單一 id">
<button (click)="getById(input.value)"> 獲取單一員工資訊 </button>
<div *ngIf="emp_get != null">
   <span>{{emp_get.id}}</span>
   <span>{{emp_get.name}}</span>
   <span>{{emp_get.salary}}</span>
</div>

<h6> 增加員工 </h6>
<form action="#">
   <input type="number" [(ngModel)]="emp_add.id" placeholder=" 員工編號 ">
   <input type="text" [(ngModel)]="emp_add.name" placeholder=" 員工名稱字 ">
   <input type="number" [(ngModel)]="emp_add.salary" placeholder=" 員工薪資 ">
   <button (click)="add()"> 增加員工 </button>
</form>
```

執行專案 web 目錄下的 index.html 檔案。執行成功後增加員工和獲取當前員工資訊的結果如圖 15-16 所示。

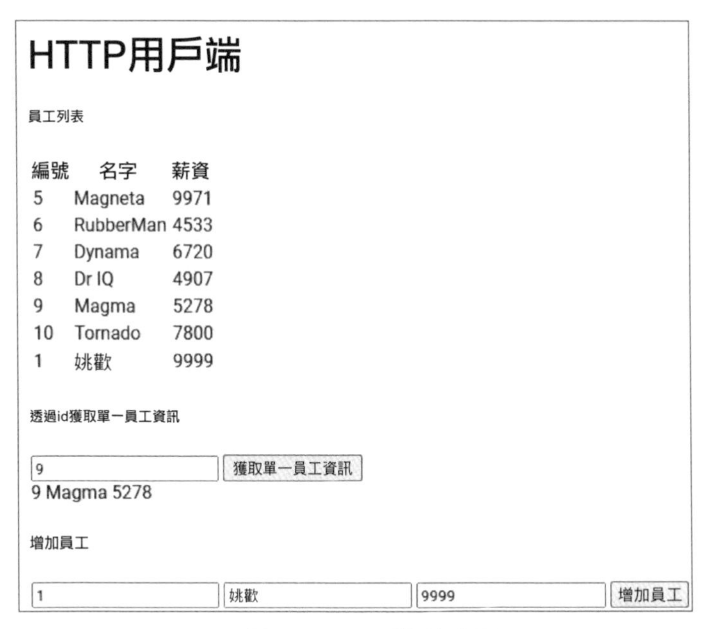

▲ 圖 15-16 用戶端請求頁

15.9 部署專案

部署 AngularDart 專案，首先需要將應用程式編譯為 JavaScript。使用 webdev 工具建構應用程式，將應用程式編譯為 JavaScript 並生成部署所有資源。dart2js 編譯器支援 tree-shaking，它只匯入所需的函數庫並忽略未使用的類別、函數、方法等。因此在使用 dart2js 建構應用時，可以得到一個相當小的 JavaScript 檔案。

透過一些額外的工作，可以使可部署的應用程式更小、更快、更可靠。

15.9.1 webdev 工具

在使用 webdev 工具前需要全域安裝它，以便在任何地方都可以直接使用該工具。編輯器會在合適的時候自動安裝，若是手動安裝則可在命令列中執行以下命令：

```
pub global activate webdev
```

使用 webdev 時，它還依賴於 build_runner 和 build_web_compilers 軟體套件。使用範本創建的專案將自動增加依賴，否則可在開發依賴中增加依賴項並執行 pub get 命令：

```
dev_dependencies:
  build_runner: ^1.6.0
  build_web_compilers: ^2.3.0
```

在開發應用程式時在命令列中使用 webdev serve 命令，該命令會在開發過程中持續建構 Web 應用程式。它預設使用 webdevc 編譯器編譯應用程式並發佈在位址 localhost:8080 上，如果使用編輯器執行專案，通訊埠將由編輯器隨機指定。

```
pub global run webdev serve
```

開發編譯器 webdevc 僅支援 Chrome 瀏覽器，如果需要在其他瀏覽器中也能正常執行，則可使用 dart2js 編譯器。增加 --release 標示以便使用 dart2js：

```
pub global run webdev serve --release
```

也可以為專案指定通訊埠為 80，使用範例如下：

```
pub global run webdev serve web:80
```

使用 webdev build 命令創建應用程式的可部署版本，預設使用 dart2js 編譯器。

```
pub global run webdev build
```

命令執行完畢後，會在專案根目錄下生成一個 build 資料夾，其中包含所有編譯後的檔案。dart2js 編譯器將應用程式編譯為 JavaScript，並將結果保存在 build/web/main.dart.js 檔案中，透過 build 檔案下的 index.html 首頁檔案即可存取部署完成的應用程式。

15.9.2 dart2js 選項

在編譯為 JavaScript 檔案時，可以向編譯器指定選項以便控制編譯最佳化程度。

(1) --minify：生成壓縮的輸出，預設壓縮。
(2) -O0：禁用所有最佳化。禁用內聯、禁用類型推斷、禁用 rti 最佳化。
(3) -O1：啟用預設最佳化。
(4) -O2：啟用遵循語義的最佳化並且對所有程式適用，它會更改類型的字串表示形式。在 minify 的基礎上對執行時期類型返回的字串表示形式要求降低。
(5) -O3：僅在不拋出 Error 子類型的程式上啟用最佳化並遵循語言語義。在 -O2 的基礎上省略隱式檢查。

(6) -O4：在 -O3 的基礎上最佳化程度更進一步，但容易受使用者輸入中邊界測試的影響。

需要使用設定檔以便應用 dart2js 選項，在專案根目錄下創建 build.yaml 檔案。以 15.9.1 節的專案 http_client 為例，增加程式如下：

```
//chapter15/http_client/build.yaml
targets:
  $default:
    builders:
      build_web_compilers|entrypoint:
        # 這是編譯的進入點的位置
        generate_for:
          - test/**.browser_test.dart
          - web/**.dart
        options:
          compiler: dart2js
          #dart2js 指定參數清單，也可以忽略
          dart2js_args:
            - -O2
```

需遵循以下做法，以便幫助 dart2js 更進一步地進行類型推斷，從而可以生成更小和更快的 JavaScript 程式：

(1) 不要使用 Function.apply() 方法。
(2) 不要覆寫 noSuchMethod() 方法。
(3) 避免將變數設定為 null。
(4) 傳遞給每個函數或方法的參數類型應保持一致。

然後在命令列執行以下命令：

```
pub global run webdev build
```

專案建構完成後的目錄結構如圖 15-17 所示。

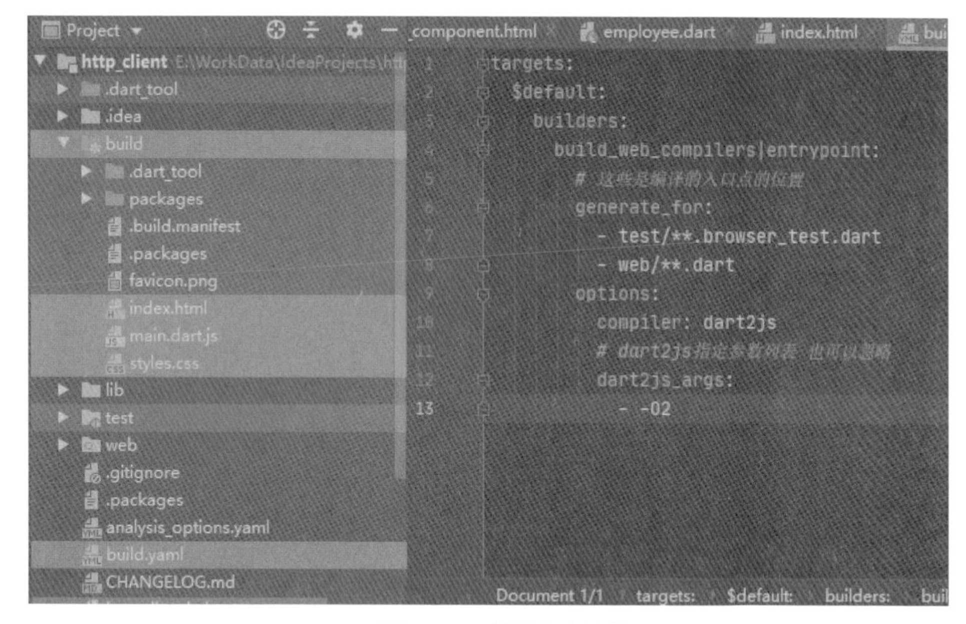

▲ 圖 15-17 部署專案結構

Chapter 16

材質化元件

本章介紹以 Angular 框架為基礎的材質化元件庫，該函數庫中定義了常用的網頁元件。可以直接將它們應用於專案，它們支援複雜的任務邏輯、高效且穩定。

在傳統元件所使用的 Dart、HTML 和 CSS 語言的基礎上，該函數庫還使用了一種名叫 Sass 的語言。Sass 是一款強化 CSS 的輔助工具，它在 CSS 語法的基礎上增加了變數 (variables)、巢狀結構 (nested rules)、混合 (mixins)、匯入 (inline imports) 等進階功能，這些拓展令 CSS 更加強大與優雅。使用 Sass 有助更進一步地組織管理樣式檔案，以及更高效率地開發專案。

在材質化元件庫中 Sass 採用的語法格式是 SCSS(Sassy CSS)，這種格式僅在 CSS3 語法的基礎上進行拓展，所有 CSS3 語法在 SCSS 中都是通用的，同時加入 Sass 的特色功能。這種格式以 .scss 作為檔案副檔名。

在專案中使用材質化元件庫前，首先需要增加依賴項並執行 pub get 命令：

```
name: project_name
description: A web app that uses AngularDart Components
#version: 1.0.0
#homepage: https://www.example.com

environment:
  sdk: '>=2.7.0 <3.0.0'

dependencies:
  angular: ^5.3.0
  angular_components: ^0.13.0

dev_dependencies:
  sass_builder: ^2.1.3
  angular_test: ^2.3.0
  build_runner: ^1.6.0
  build_test: ^0.10.8
  build_web_compilers: ^2.3.0
  pedantic: ^1.8.0
  test: ^1.6.0
```

依賴項 angular_components 就是材質化元件庫，可以在需要的地方匯入其中的或多個元件。開發依賴項 sass_builder 使用 build 套件和 Sass 的 Dart 實現轉換 Sass 檔案，該工具會將 *.scss 檔案編譯為 *.css 檔案。本章創建的所有專案都需要增加這兩個函數庫，後續將不再贅述。

元件都具有預設樣式，它們適用於大部分使用場景。如果需要修改元件的預設樣式，最好具備 Sass 語言基礎。本章與 Sass 相關的知識僅涉及樣式表中 mixin 的使用，因此跟隨本章後續對 Sass 的介紹也可以滿足正常使用。

在 Sass 中使用指令 @mixin 定義一個可以在整個樣式表中重複使用的樣式，其定義格式程式如下：

```
@mixin mixin-name($arg1,$arg2:value2,$arg3:value3){
  // 樣式內容
}
```

從程式中可以看出其定義格式類似於方法。mixin 名是普通識別符號，括號中是需要向 mixin 傳遞的參數。識別符號前加 $ 表示變數，在大括號 ({}) 中可以引用。可以像 Dart 中的函數一樣為參數提供預設值，例如：變數 $arg2 和 $arg3 都有預設值。

在 Sass 中使用 @include 指令將 mixin 引入文件中，定義和使用 mixin 的範例程式如下：

```
// 定義 mixin bordered
// 它帶有兩個參數
@mixin bordered($width, $color) {
   //$width 表示邊框的寬度
   //$color 表示邊框的顏色
   border: $width solid $color;
}

// 使用 mixin 並傳入參數
.myArticle {
   @include bordered(1px, blue);
}
.myNotes {
   @include bordered(2px, green );
}
```

上述內容存放在 .scss 檔案中，它會被轉換成名稱相同 .css 檔案，並且其內容將被轉為以下 CSS 內容：

```
.myArticle {
   border: 1px solid blue;
}

.myNotes {
   border: 2px solid green;
}
```

元件庫中幾乎每個元件都提供 mixin 集合，它們包含在對應元件目錄下的 _mixins.scss 檔案中。匯入該檔案，並透過提供的 mixin 可以控制元件的樣式細節。

對於元件的定義與各個細節在前兩章中已經介紹完畢，本章將不再詳細介紹。本章會先列出元件可用的 mixin、屬性、輸入和輸出屬性，並將重點放在元件的使用上。

16.1 圖示

材質化風格的圖示，它的元件名稱是 MaterialIconComponent，它在範本中的選擇器是 <material-icon> 元素。可用圖示參考網址：https://material.io/resources/icons。

1. 屬性

(1) size String：圖示的大小。值包括 x-small、small、medium、large 和 x-large，分別對應 12px、13px、16px、18px 和 20px。如果未指定 size，則預設為 24px。
(2) flip：是否應該翻轉圖示以適應 RTL 的語言。方向為 RTL 時，並非所有圖示都應翻轉。一般來説處理空間關係的圖示應翻轉。代表時間關係、物理物件或產品徽章的圖示不應翻轉。
(3) light：是否應減少圖示的不透明度。
(4) baseline：圖示是否需要與基準線對齊。

2. 輸入屬性

icon dynamic：該元件應顯示的 Icon 模型或圖示識別符號。

3. Sass mixin

(1) material-icon-size($size)：更改圖示的大小。
(2) svg-icon($svg-icon)：將 SVG 圖片或資料用作圖示而非字型圖示。
(3) svg-icon-size($size)：設定 material-icon 的內部 <i> 元素的大小，進而控制 SVG 圖示大小。

創建 Angular 專案 icon_demo，增加依賴項並啟動專案。在根元件中匯入材質化元件庫，將 MaterialIconComponent 增加到指令清單，屬性 iconColor 用於樣式綁定以便控制圖示的顏色，屬性 iconModel 用於與圖示元件的輸入屬性 icon 綁定。範例程式如下：

```
//chapter16/icon_demo/lib/app_component.dart
import 'package:angular/angular.dart';
// 匯入元件庫
import 'package:angular_components/angular_components.dart';
// 匯入模型類別
import 'package:angular_components/model/ui/icon.dart';

@Component(
  selector: 'my-app',
  styleURLs: ['app_component.css'],
  templateURL: 'app_component.html',
  directives: [MaterialIconComponent],
)
class AppComponent {
  // 自訂圖示顏色
  String iconColor = 'blue';
  // 新建圖示模型
  Icon iconModel = Icon('edit');
}
```

範本程式如下：

```
//chapter16/icon_demo/lib/app_component.html
<h6> 普通圖示 </h6>
<p><material-icon icon="eco"></material-icon> 預設 24px</p>

<h6> 圖示大小 </h6>
<p><material-icon size="x-small" icon="eco"></material-icon> x-small 12px</p>
<p><material-icon size="small" icon="eco"></material-icon> small 13px</p>
<p><material-icon size="medium" icon="eco"></material-icon> medium 16px</p>
<p><material-icon size="large" icon="eco"></material-icon> large 18px</p>
<p><material-icon size="x-large" icon="eco"></material-icon> x-large 20px</p>

<h6> 圖示 light</h6>
```

```
<p><material-icon light icon="eco"></material-icon> 減少圖示的不透明度 </p>

<h6> 圖示 baseline</h6>
<p> 圖示 <material-icon baseline icon="eco"></material-icon> 與基準線對齊 </p>
<p> 圖示 <material-icon icon="eco"></material-icon> 未與基準線對齊 </p>

<h6> 自訂圖示顏色 </h6>
<p><material-icon [style.color]="'green'" icon="eco"></material-icon></p>
<p><material-icon [style.color]="'orange'" icon="eco"></material-icon></p>
<p><material-icon [style.color]="iconColor" icon="eco"></material-icon></p>

<h6>Icon 模型 </h6>
<p><material-icon [icon]="iconModel"></material-icon></p>

<h6>mixin 自訂圖示大小 </h6>
<p><material-icon class="custom-size" icon="eco"></material-icon></p>

<h6>mixin 自訂圖示 (SVG)</h6>
<material-icon icon="" class="svg-icon"></material-icon>

<h6>mixin 自訂圖示 (SVG) 大小 </h6>
<material-icon icon="" class="svg-icon svg-size"></material-icon>
```

樣式程式如下：

```
//chapter16/icon_demo/lib/app_component.scss
// 匯入圖示元件提供的 mixins
@import 'package:angular_components/material_icon/mixins';

.custom-size {
   // 使用 material-icon-size mixin 設定圖示大小
   @include material-icon-size(36px);
}

.svg-icon {
   // 使用 mixin svg-icon 將 SVG 圖型作為圖示
   // 這裡將專案 web 目錄下名為 bilibili.svg 的檔案作為圖示
@include svg-icon(URL('/bilibili.svg'));
}
```

```
.svg-size{
  // 使用 mixin svg-icon-size 設定 SVG 圖示大小
  @include svg-icon-size(16px);
}
```

16.2 滑桿

適用於整數值的材質化滑桿。元件名稱是 MaterialSliderComponent，在範本中使用元素 <material-slider> 創建滑桿元件的實例，可以透過使用滑鼠或鍵盤滑動來控制滑動桿。使用雙值浮點數學運算時，可能導致值不精確，如果向使用者顯示該值，需考慮格式化結果。

1. 輸入屬性

(1) disabled bool：如果禁用滑桿，則其值為 true。

(2) max num：最大進度值。預設值為 100，必須嚴格大於最小進度值。

(3) min num：最小進度值。預設值為 0，必須嚴格小於最大進度值。

(4) step num：步進值。必須為正數。

(5) value num：當前值。

2. 輸出屬性

valueChange Stream<num>：當使用者更改輸入值時，發佈事件。

3. Sass mixin

(1) slider-thumb-color($selector, $color)：用於設定滑動桿的顏色。第一個參數是指向滑桿元素的選擇器，第二個參數是要設定的顏色。

(2) slider-track-color($selector, $left-color, $right-color: $mat-grey)：用於設定滑桿軌道的顏色。第 1 個參數是指向滑桿元素的選擇器。第 2 個參數是滑動桿左側軌道的顏色。第 3 個參數是滑動桿右側軌道的顏色。

創建 Angular 專案 slider_demo，增加依賴項並啟動專案。在根元件中匯入材質化元件庫，將 MaterialSliderComponent 增加到指令清單，將 DomService 和 windowBindings 增加到提供者清單。範例程式如下：

```
//chapter16/slider_demo/lib/app_component.dart
import 'package:angular/angular.dart';
import 'package:angular_components/angular_components.dart';
// 匯入 DomService，用於跨元件同步 DOM 操作
// 例如：在 UI 更新或應用程式事件後檢查更改
import 'package:angular_components/utils/browser/dom_service/angular_2.dart';
// 匯入 windowBindings
// 提供在 Angular 中綁定使用的 Document、HtmlDocument 和 Window
import 'package:angular_components/utils/browser/window/module.dart';
@Component(
   selector: 'my-app',
   styleURLs: ['app_component.css'],
   templateURL: 'app_component.html',
   providers: [DomService,windowBindings],
   directives: [MaterialSliderComponent],
)
class AppComponent{
   // 用於綁定到滑桿元件的 value 屬性
   int value1 = 10;
   int value2 = 30;
   int value3 = 70;
}
```

範本程式如下：

```
//chapter16/slider_demo/lib/app_component.html
<h1> 滑動桿 </h1>
<h6> 預設 </h6>
<material-slider></material-slider>

<h6> 綁定到元件屬性 </h6>
<p> 值：{{value1}}</p>
<material-slider [(value)]="value1"></material-slider>
```

```
<h6>禁用</h6>
<material-slider [(value)]="value2"
                 [disabled]="true"></material-slider>

<h6>mixin 自訂滑動桿顏色</h6>
<material-slider class="thumb-colors"
                 [(value)]="value3"></material-slider>

<h6>mixin 自訂滑軌顏色</h6>
<material-slider class="track-colors"
                 [(value)]="value3"></material-slider>
```

樣式程式如下：

```
//chapter16/slider_demo/lib/app_component.scss
//匯入預先定義材質化樣式
@import 'package:angular_components/css/material/material';
//匯入滑桿元件的 mixins
@import 'package:angular_components/material_slider/mixins';

:host {
   //限定滑桿的寬度和顯示框類型
   material-slider {
      display: inline-block;
      width: 400px;
   }
   //$mat-green, $mat-green-400, $mat-green-100 均是預先定義顏色變數
   //自訂滑動桿的顏色，第一項是選擇器，第二項參數是滑動桿的顏色
   @include slider-thumb-color('material-slider.thumb-colors', $mat-green);
   //自訂軌道的顏色，第一項是選擇器，第二項參數是滑動桿左側軌道的顏色，第三項參
數是滑動桿右側軌道的顏色
   @include slider-track-color('material-slider.track-colors', $mat-green-400,
$mat-green-100);
}
```

16.3 旋轉器

當表示進度和活動在不確定的時間內完成時可用圓形旋轉器顯示，此元件名為 MaterialSpinnerComponent，在範本中使用元素 <material-spinner> 創建滑桿元件的實例。

Sass mixin

material-spinner-thickness($stroke-width)：更改旋轉器的邊框寬度，使其顯得更粗或更細。

修改 material-spinner 的 border-color 屬性可以更改旋轉器的顏色：

```
material-spinner {
   border-color: $mat-red;
}
```

創建 Angular 專案 spinner_demo，將 MaterialSpinnerComponent 增加到指令清單。範例程式如下：

```
//chapter16/spinner_demo/lib/app_component.dart
import 'package:angular/angular.dart';
import 'package:angular_components/angular_components.dart';
@Component(
   selector: 'my-app',
   styleURLs: ['app_component.css'],
   templateURL: 'app_component.html',
   directives: [MaterialSpinnerComponent],
)
class AppComponent {}
```

範本程式如下：

```
//chapter16/slider_demo/lib/app_component.html
<h1> 旋轉器 </h1>
<h6> 預設 </h6>
<material-spinner></material-spinner>
```

```
<h6> 自訂旋轉器顏色 </h6>
<material-spinner class="orange"></material-spinner>
<material-spinner class="green"></material-spinner>

<h6> 配合文字使用 </h6>
<div>
   <material-spinner></material-spinner>
   <span> 處理中 ...</span>
</div>

<h6>mixin 自訂旋轉器的邊框寬度 </h6>
<material-spinner class="custom-width"></material-spinner>
```

樣式程式如下：

```
//chapter16/spinner_demo/lib/app_component.scss
// 匯入旋轉器元件的 mixins
@import 'package:angular_components/material_spinner/mixins';

// 自訂旋轉器的顏色
.green{
   border-color:green;
}
.orange{
   border-color:orange;
}

// 自訂旋轉器的邊框寬度
.custom-width{
   @include material-spinner-thickness(6px);
}
```

16.4 切換按鈕

使用者可以點擊切換按鈕來更改狀態。一般來説只有一個打開或關閉選項時，才使用切換按鈕。元件名稱是 MaterialToggleComponent，在範本中使用元素 <material-toggle> 創建切換按鈕的實例。

1. 輸入屬性

(1) checked bool：切換按鈕的當前狀態。true 為打開，false 為關閉。

(2) disabled bool：啟用或禁用切換按鈕。禁用為 true，啟用為 false。

(3) label String：切換按鈕的標籤。

2. 輸出屬性

checkedChange Stream<bool>：切換按鈕被選中後觸發的事件。

3. Sass mixin

(1) material-toggle-theme($primary-color, $off-btn-color: null)：第一個參數表示切換按鈕打開時按鈕的顏色，第二個參數表示切換按鈕關閉時按鈕的顏色，軌道會根據按鈕顏色自我調整。

(2) flip-toggle-label-position()：在切換框的右側顯示切換標籤。

創建 Angular 專案 toggle_demo，將 MaterialToggleComponent 增加到指令清單。範例程式如下：

```
//chapter16/toggle_demo/lib/app_component.dart
import 'package:angular/angular.dart';
import 'package:angular_components/angular_components.dart';
@Component(
   selector: 'my-app',
   styleURLs: ['app_component.css'],
   templateURL: 'app_component.html',
   directives: [MaterialToggleComponent],
)
class AppComponent {
   //切換按鈕狀態控制變數
   bool toggle1 = true;
   bool bluetooth = false;
   bool finddevice = false;
}
```

範本程式如下：

```
//chapter16/toggle_demo/lib/app_component.html
```

```
<h1> 切換按鈕 </h1>
<h6> 預設 </h6>
<material-toggle></material-toggle>

<h6> 帶有標籤 </h6>
<material-toggle label=" 自動提醒 "></material-toggle>

<h6> 禁用 </h6>
<material-toggle label=" 顯示電量 "
               [disabled]="true"></material-toggle>
<br>
<material-toggle label=" 顯示網速 "
               [checked]="true"
               [disabled]="true"></material-toggle>

<h6> 綁定到元件屬性 </h6>
<p> 狀態：{{toggle1 ? ' 打開 ':' 關閉 '}}</p>
<material-toggle label=" 自動提醒 "
               [(checked)]="toggle1"></material-toggle>

<h6> 切換按鈕互動 </h6>
<material-toggle label=" 藍牙：{{bluetooth ? ' 打開 ':' 關閉 '}}"
               [(checked)]="bluetooth"></material-toggle>
<br>
<material-toggle label=" 發現裝置：{{finddevice ? ' 打開 ':' 關閉 '}}"
               [disabled]="!bluetooth"
               [(checked)]="finddevice"></material-toggle>

<h6> 自訂寬度 </h6>
<material-toggle class="theme-width"
               label=" 自動提醒 "></material-toggle>

<h6> 自訂打開時的顏色 </h6>
<div class="theme-orange">
   <material-toggle label=" 自動提醒 "
               [checked]="true"></material-toggle>
</div>

<h6> 自訂關閉時的顏色 </h6>
```

```
<div class="theme-double">
   <material-toggle label=" 自動提醒 "></material-toggle>
</div>

<h6> 標籤顯示在右邊 </h6>
<div class="theme-fliplabel">
   <material-toggle label=" 自動提醒 "></material-toggle>
</div>
```

樣式程式如下：

```
//chapter16/toggle_demo/lib/app_component.scss
@import 'package:angular_components/css/material/material';
// 匯入切換按鈕提供的 mixins
@import 'package:angular_components/material_toggle/mixins';

// 自訂切換按鈕寬度
.theme-width{
   padding: 12px;
   width: 320px;
}

// 自訂打開時的按鈕顏色
.theme-orange{
   @include material-toggle-theme($primary-color:orange);
}

// 自訂打開和關閉時的按鈕顏色
.theme-double{
   @include material-toggle-theme(
      $primary-color:orange,
      $off-btn-color:gray);
   }

   // 翻轉標籤
   .theme-fliplabel{
      @include flip-toggle-label-position();
}
```

16.5 標籤

16.5.1 固定選單列

固定選單列是具有標籤樣式的按鈕和活動的標籤指示器的選單列元件。固定選單列具有相同大小的標籤按鈕，並且沒有捲動。元件名稱是 FixedMaterialTabStripComponent，在範本中使用元素 <material-tab-strip> 可以創建選單列的實例。

1. 輸入屬性

(1) activeTabIndex int：活動面板的索引，從 0 開始。預設值為 0。

(2) tabIds List<String>：標籤按鈕 ID 的清單。

(3) tabLabels List<String>：標籤按鈕標籤的清單。

2. 輸出屬性

(1) activeTabIndexChange Stream<int>：在觸發 tabChange 事件後發佈的 activeTabIndex 更新串流。

(2) beforeTabChange Stream<TabChangeEvent>：TabChangeEvent 實例的串流，在標籤更改之前發佈。呼叫 TabChangeEvent # preventDefault 將阻止更改標籤。

(3) tabChange Stream<TabChangeEvent>：TabChangeEvent 實例的串流，在標籤已更改時發佈。

3. Sass mixin

(1) tab-panel-tab-strip-width($selector, $tab-width)：設定標籤面板中選單列的寬度。

(2) tab-strip-color($selector, $color, $accent-color)：設定標籤的預設顏色和指示器的顏色。

(3) tab-strip-accent-color($selector, $accent-color)：設定選擇狀態索引籤和指示器的顏色。
(4) tab-strip-tab-color($selector, $color)：設定標籤的預設顏色。
(5) tab-strip-selected-tab-color($selector, $color)：設定選擇狀態索引籤的顏色。
(6) tab-strip-indicator-color($selector, $accent-color)：設定指示器的顏色。
(7) tab-strip-show-bottom-shadow($dp: 2)：設定標籤列底部陰影的寬度。

16.5.2 標籤面板

帶有頂部導覽列的標籤面板，元件名稱是 MaterialTabPanelComponent，在範本中使用元素 <material-tab-pane> 可以創建標籤面板的實例。此元件實例是對 <material-tab-strip> 的擴充，因此 FixedMaterialTabStripComponent 的 Sass mixin 適用於此元件實例。

1. 輸入屬性

(1) activeTabIndex dynamic：活動面板的索引，從 0 開始。預設值為 0。
(2) centerTabs bool：是否將標籤按鈕置中對齊。不然按鈕從左端 (LTR) 對齊。

2. 輸出屬性

(1) beforeTabChange Stream<TabChangeEvent>：TabChangeEvent 實例的串流，在標籤更改之前發佈。呼叫 TabChangeEvent # preventDefault 將阻止更改標籤。
(2) tabChange Stream<TabChangeEvent>：TabChangeEvent 實例的串流，在標籤已更改時發佈。

16.5.3 材質化標籤

材質化風格的標籤，作為 MaterialTabPanelComponent 的一部分顯示或隱藏。元件名稱是 MaterialTabComponent，在範本中使用元素 <material-tab> 可以創建標籤的實例。此元件透過 label 屬性設定按鈕的文字。使用 *deferredContent 範本指令可以延遲實例化標籤的內容。此元件需要與 MaterialTabPanelComponent 配合使用。

輸入屬性

label String：此標籤的標籤。

創建 Angular 專案 tab_demo，將上述元件增加到指令清單。範例程式如下：

```
//chapter16/tab_demo/lib/app_component.dart
import 'package:angular/angular.dart';
import 'package:angular_components/angular_components.dart';
@Component(
   selector: 'my-app',
   styleURLs: ['app_component.css'],
   templateURL: 'app_component.html',
   directives: [
      FixedMaterialTabStripComponent,
      MaterialTabPanelComponent,
      MaterialTabComponent,
      DeferredContentDirective],
   )
   class AppComponent {
   // 標籤預設索引
   var tabIndex = 0;
   // 選項標籤清單
   var tabLabels = <String>['選項1','選項2','選項3'];
   // 用於回應選項更改事件
   void onTabChange(TabChangeEvent event) {
      tabIndex = event.newIndex;
   }
}
```

範本程式如下：

```
//chapter16/tab_demo/lib/app_component.html
<h1> 標籤 </h1>
<h6> 選單列 </h6>
<p> 標籤與內容無直接連結 </p>
<material-tab-strip (tabChange)="onTabChange($event)"
            [tabLabels]="tabLabels">
</material-tab-strip>
<p> 當前活躍標籤索引：{{tabIndex}}</p>

<h6> 自訂選單列寬度 </h6>
<material-tab-strip style="width:300px;"
            (tabChange)="onTabChange($event)"
            [tabLabels]="tabLabels">
</material-tab-strip>
<p> 當前活躍標籤索引：{{tabIndex}}</p>

<h6> 選項面板 </h6>
<p> 標籤與內容相連結 </p>
<material-tab-panel [activeTabIndex]="1">
   <material-tab label=" 選項 1"> 選項 1 的內容 </material-tab>
   <material-tab label=" 選項 2"> 選項 2 的內容 </material-tab>
   <material-tab label=" 選項 3">
      <template deferredContent> 延遲載入內容 </template>
   </material-tab>
</material-tab-panel>

<h6> 選項標籤置中且自訂選單列寬度 </h6>
<material-tab-panel centerStrip [activeTabIndex]="0">
   <material-tab label=" 選項 1"> 選項 1 的內容 </material-tab>
   <material-tab label=" 選項 2"> 選項 2 的內容 </material-tab>
   <material-tab label=" 選項 3">
      <template deferredContent> 延遲載入內容 </template>
   </material-tab>
</material-tab-panel>
```

```
<h6>mixin 自訂選單列寬度、預設標籤顏色、選擇狀態索引籤和指示器顏色、標籤列底部陰影的寬度 </h6>
<material-tab-panel class="tab-panel" [activeTabIndex]="0">
   <material-tab label=" 選項 1"> 選項 1 的內容 </material-tab>
   <material-tab label=" 選項 2"> 選項 2 的內容 </material-tab>
   <material-tab label=" 選項 3">
      <template deferredContent> 延遲載入內容 </template>
   </material-tab>
</material-tab-panel>
```

樣式程式如下：

```
//chapter16/tab_demo/lib/app_component.scss
@import 'package:angular_components/css/material/material';
@import 'package:angular_components/material_tab/mixins';

// 自訂選項面板選單列寬度
@include tab-panel-tab-strip-width('material-tab-panel[centerStrip]', 300px);

// 自訂選項面板選單列寬度
@include tab-panel-tab-strip-width('material-tab-panel.tab-panel', 400px);

// 自訂選項文字顏色
@include tab-strip-tab-color('.tab-panel', blue);

// 自訂當前選項文字顏色
@include tab-strip-accent-color('.tab-panel', $mat-red);

// 自訂選擇項指示器顏色
@include tab-strip-indicator-color('.tab-panel', orange);

// 自訂標籤列底部陰影的寬度
.tab-panel{
   @include tab-strip-show-bottom-shadow($dp: 6);
}
```

16.6 計數卡與計數板

16.6.1 計數卡

計數卡為獨立元件，該元件可在更大的元件中重用或嵌入。元件名稱是 ScorecardComponent，在範本中使用元素 <acx-scorecard> 創建計分卡元件實例。

1. 內容元素

(1) name：標籤區域中的自訂內容。

(2) value：值區域中的自訂內容。

(3) description：描述區域中的自訂內容。為了顯示該部分內容，即使只是將其設定為空字串，也需要設定 description 屬性。

2. 輸入屬性

(1) changeGlyph bool：是否在描述區域顯示變化箭頭，可選。

(2) changeType String：設定計分卡描述的變化類型。用於確定描述的樣式。可能的值為 POSITIVE 表示正，NEGATIVE 表示負，預設值為 NEUTRAL 表示無號。

(3) description String：計分卡上的簡短說明，可選。

(4) extraBig bool：是否對計分卡使用 CSS 類別 big 定義的樣式，可選。

(5) label String：計分卡的標題。

(6) selectable bool：點擊是否可以更改計分卡的選擇狀態。

(7) selected bool：是否選擇了計分卡。

(8) selectedColor Color：選中計分卡時背景要應用的顏色。

(9) suggestionAfter String：描述後的一筆建議文字，可選。

(10) suggestionBefore String：描述前的一筆建議文字，可選。

(11) tooltip String：使用者將滑鼠移過在值上時，該值將顯示在工具提示中。

(12) value String：向使用者顯示的值。

3. 輸出屬性

selectedChange Stream<bool>：selectedChange Stream<bool>。

16.6.2 計數板

計數板管理一行計數卡，元件名稱是 ScoreboardComponent，在明範本中使用元素 <acx-scoreboard> 實例化計分板元件。

1. 屬性

enableUniformWidths bool：計分板上的計分卡是否應具有統一的寬度。

2. 輸入屬性

(1) isVertical bool：計分卡是否垂直顯示。預設為 false。
(2) resetOnCardChanges bool：更改卡片時是否重置卡片選擇。如果增加或刪除卡，並且此卡設定為 true，則將取消選擇所有卡。對於 ScoreboardType.radio，將選擇第一張卡。
(3) scrollable bool：是否允許透過捲動按鈕捲動計分板。捲動屬性可以在應用執行時期動態設定，將根據可捲動狀態增加或刪除視窗大小調整監聽器。
(4) type ScoreboardType：計分板的類型，例如 standard、selectable、radio、toggle。

創建 Angular 專案 scorecard_demo，將上述元件增加到指令清單。元件的使用需要 DomService 和 windowBindings 支援，因此需要在提供者清單中增加它們。範例程式如下：

```
//chapter16/scorecard_demo/lib/app_component.dart
import 'package:angular/angular.dart';
import 'package:angular_components/angular_components.dart';
import 'package:angular_components/utils/browser/dom_service/angular_2.dart';
import 'package:angular_components/utils/browser/window/module.dart';
@Component(
```

```
selector: 'my-app',
styleURLs: ['app_component.css'],
templateURL: 'app_component.html',
providers: [DomService,windowBindings],
directives: [
   ScoreboardComponent,
   ScorecardComponent,
   NgFor],
)
class AppComponent {
   // 預先定義計分板類型變數
   final ScoreboardType selectable = ScoreboardType.selectable;
   final ScoreboardType toggle = ScoreboardType.toggle;
   final ScoreboardType radio = ScoreboardType.radio;
}
```

範本程式如下：

```
//chapter16/scorecard_demo/lib/app_component.html
<h1> 計分卡 </h1>
<h3> 單獨的計分卡 </h3>
<h6> 基礎計分卡 </h6>
<acx-scorecard label=" 餘額 RMB" value="158.22" description="+24.20 (15%)">
</acx-scorecard>

<h6> 正號並提供變化箭頭 </h6>
<acx-scorecard
      label=" 餘額 RMB"
      value="158.22"
      description="+24.20 (15%)"
      [changeGlyph]="true"
      changeType="POSITIVE">
</acx-scorecard>

<h6> 負號並提供變化箭頭 </h6>
<acx-scorecard
      label=" 餘額 RMB"
      value="158.22"
      description="-24.20 (15%)"
      [changeGlyph]="true"
```

```
        changeType="NEGATIVE">
</acx-scorecard>

<h6> 建議 </h6>
        <acx-scorecard
        label=" 餘額 RMB"
        value="158.22"
        description="+24.20 (15%)"
        changeType="POSITIVE"
        suggestionBefore=" 值得關注 "
        suggestionAfter=" 增持 ">
</acx-scorecard>

<h6> 可選擇 </h6>
<acx-scorecard
        label=" 餘額 RMB"
        value="158.22"
        description="-24.20 (15%)"
        [selectable]="true">
</acx-scorecard>

<h6> 提示文字 </h6>
<acx-scorecard
        label=" 餘額 RMB"
        value="158.22"
        description="-24.20 (15%)"
        tooltip=" 多考慮一下 ">
</acx-scorecard>

<h3> 計分板 </h3>
<h6> 基礎 </h6>
<acx-scoreboard>
   <acx-scorecard label=" 餘額 RMB" value="158.22" description="+24.20 (15%)">
   </acx-scorecard>
   <acx-scorecard label=" 餘額 RMB" value="158.22" description="-24.20 (15%)">
   </acx-scorecard>
   <acx-scorecard label=" 餘額 RMB" value="158.22" description="+24.20 (15%)">
   </acx-scorecard>
</acx-scoreboard>
```

```
<h6> 可選擇 </h6>
<acx-scoreboard [type]="selectable">
  <acx-scorecard label=" 餘額 RMB" value="158.22" description="+24.20 (15%)">
  </acx-scorecard>
  <acx-scorecard label=" 餘額 RMB" value="158.22" description="-24.20 (15%)">
  </acx-scorecard>
  <acx-scorecard label=" 餘額 RMB" value="158.22" description="+24.20 (15%)">
  </acx-scorecard>
</acx-scoreboard>

<h6> 可切換 </h6>
<acx-scoreboard [type]="toggle">
  <acx-scorecard label=" 餘額 RMB" value="158.22" description="+24.20 (15%)">
  </acx-scorecard>
  <acx-scorecard label=" 餘額 RMB" value="158.22" description="-24.20 (15%)">
  </acx-scorecard>
  <acx-scorecard label=" 餘額 RMB" value="158.22" description="+24.20 (15%)">
  </acx-scorecard>
</acx-scoreboard>

<h6> 單選 </h6>
<acx-scoreboard [type]="radio">
  <acx-scorecard label=" 餘額 RMB" value="158.22" description="+24.20 (15%)">
  </acx-scorecard>
  <acx-scorecard label=" 餘額 RMB" value="158.22" description="-24.20 (15%)">
  </acx-scorecard>
  <acx-scorecard label=" 餘額 RMB" value="158.22" description="+24.20 (15%)">
  </acx-scorecard>
</acx-scoreboard>

<h6> 垂直 </h6>
<acx-scoreboard [isVertical]="true" [type]="selectable">
  <acx-scorecard label=" 餘額 RMB" value="158.22" description="+24.20 (15%)">
  </acx-scorecard>
  <acx-scorecard label=" 餘額 RMB" value="158.22" description="-24.20 (15%)">
  </acx-scorecard>
  <acx-scorecard label=" 餘額 RMB" value="158.22" description="+24.20 (15%)">
  </acx-scorecard>
</acx-scoreboard>
```

```
<h6> 可捲動 </h6>
<acx-scoreboard scrollable [type]="selectable">
   <acx-scorecard
         *ngFor="let n of [1,2,3,4,5,6,7,8,9]"
         label=" 項目 {{n}}"
         value="{{n}}"
         description="">
   </acx-scorecard>
</acx-scoreboard>

<h6> 自訂內容 </h6>
<acx-scorecard
         label="Estimated earnings"
         value="$158.22"
         description="+$24.20 (15%)"
         changeType="POSITIVE">
   <name> 注入標籤 </name>
   <value> 注入值 </value>
   <description> 注入描述資訊 </description>
   <div> 注入的其他資訊 </div>
</acx-scorecard>
```

16.7 按鈕

16.7.1 按鈕設定

扁平或突起的按鈕，可以選擇帶有波紋效果。元件名稱是 MaterialButton Component，在範本中使用元素 <material-button> 創建按鈕元件的實例。

1. 屬性

(1) icon：刪除按鈕的最小寬度樣式。需要在按鈕中指定確定的圖示，可以使用 <glyph>、<material-icon> 或 <img> 元素。

(2) no-ink：移除按鈕上的波紋效果。

(3) clear-size：將按鈕的最小寬度和邊距置為 0。
(4) dense：將字型大小調整為 13px，將按鈕高度調整為 32px。

2. 輸入屬性

(1) disabled：是否禁用元件。
(2) raised：是否使按鈕具有凸起的陰影。
(3) role：此元件的角色用於 a11y。
(4) tabbable：元件是否可選卡片，適用於標籤。
(5) tabindex：元件標籤索引，如果 tabbable 為 true，disabled 為 false，則使用該值，適用於標籤。

3. 輸出屬性

trigger：透過點擊或按鍵啟動按鈕時觸發。

4. Sass mixin

(1) button-color($selector, $color)：將按鈕字型顏色應用於與選擇器匹配的按鈕。指定的顏色不適用於按鈕，否則它看起來像是啟用的按鈕。
(2) button-disabled-color($selector, $disabled-color)：將按鈕在禁用狀態下的字型顏色應用於與選擇器匹配的按鈕。
(3) button-background-color($selector, $background-color)：將按鈕的背景顏色應用於與選擇器匹配的按鈕。
(4) icon-button-hover-color($selector, $color)：當滑鼠指標浮動在圖示按鈕上時的顏色。
(5) icon-button-color($selector, $color)：圖示按鈕的預設顏色。

預設情況下，不透明度為 25%時，波紋與前景的顏色相同。要自訂顏色，需使用 material-ripple 選擇器。

```
/*
  若無法正常使用，需將其增加到專案的 style.css 檔案中
*/
```

```
#myButton5 material-ripple{
   color:blue;
}
```

實際上波紋的顏色並沒有改變，可能是元件視圖封裝時不夠精確。可以將樣式增加到專案的 style.css 檔案中。

16.7.2 浮動操作按鈕

MaterialFab 是一個浮動操作按鈕。它是一個大圓形的按鈕，使用 mini 屬性可使其變為小的按鈕。它與 MaterialButton 有相同的輸入屬性和輸出屬性，故這裡不再贅述。在 MaterialFab 中指定確定的圖示，可以使用 <glyph>、<material-icon> 或 <img> 元素。元件名稱是 MaterialFabComponent，在範本中使用元素 <material-fab> 創建浮動操作按鈕的實例。它包含 MaterialButton 所有的 mixin，且使用規則是一致的。

創建 Angular 專案 button_demo，將上述元件增加到指令清單。範例程式如下：

```
//chapter16/button_demo/lib/app_component.dart
import 'package:angular/angular.dart';
import 'package:angular_components/angular_components.dart';

@Component(
   selector: 'my-app',
   styleURLs: ['app_component.css'],
   templateURL: 'app_component.html',
   directives: [
      MaterialButtonComponent,
      MaterialIconComponent,
      MaterialFabComponent],
)
class AppComponent{
   //記錄事件觸發次數
   int i=0;
   //回應按鈕 trigger 事件
```

```
    trigger(){
    ++i;
    }
}
```

範本程式如下：

```
//chapter16/button_demo/lib/app_component.html
<h1> 按鈕 </h1>
<h6> 扁平按鈕 </h6>
<material-button> 預設 </material-button>
<material-button no-ink>no-ink</material-button>
<material-button clear-size>clear-size</material-button>
<material-button dense>dense</material-button>
<material-button disabled>disabled</material-button>
<material-button icon>
   <material-icon icon="add_alert"></material-icon>
</material-button>

<h6> 突起按鈕 </h6>
<material-button raised>raised</material-button>
<material-button no-ink raised>no-ink</material-button>
<material-button clear-size raised>clear-size</material-button>
<material-button dense raised>dense</material-button>
<material-button disabled raised>disabled</material-button>
<material-button icon raised>
   <material-icon icon="add_alert"></material-icon>
</material-button>

<h6> 按鈕事件 </h6>
<material-button raised (trigger)="trigger()">trigger</material-button>
<p> 事件觸發次數 {{i}}</p>

<h6>mixin 自訂按鈕 </h6>
<material-button raised class="myButton1"> 藍色背景 </material-button>
<material-button raised class="myButton2"> 藍底白字 </material-button>
<material-button raised class="myButton2" disabled> 禁用按鈕 </material-button>

<h6>mixin 自訂圖示按鈕 </h6>
<material-button icon raised class="myButton3">
```

```
    <material-icon icon="add_alert"></material-icon>
</material-button>
<material-button icon raised class="myButton4">
    <material-icon icon="add_alert"></material-icon>
</material-button>

<h6> 自訂波紋顏色 </h6>
<material-button raised class="myButton5"> 藍色波紋 </material-button>

<h1> 浮動操作按鈕 </h1>
<h6> 扁平按鈕 </h6>
<material-fab>
    <material-icon icon="add"></material-icon>
</material-fab>

<h6> 突起按鈕 </h6>
<material-fab raised>
    <material-icon icon="add"></material-icon>
</material-fab>

<h6>mini 按鈕 </h6>
<material-fab mini>
    <material-icon icon="add"></material-icon>
</material-fab>
<material-fab mini raised>
    <material-icon icon="add"></material-icon>
</material-fab>
```

樣式程式如下：

```
//chapter16/button_demo/lib/app_component.scss
@import 'package:angular_components/material_button/mixins';

// 自訂按鈕背景顏色
@include button-background-color('.myButton1',#4285f4);

// 自訂按鈕背景顏色、文字顏色、禁用狀態下的文字顏色
@include button-background-color('.myButton2',#4285f4);
@include button-color('.myButton2', white);
@include button-disabled-color('.myButton2', white);
```

```
// 自訂滑鼠指標懸浮時圖示按鈕的顏色，和沒有懸浮時的圖示按鈕顏色
@include icon-button-hover-color('.myButton3', #4285f4);
@include icon-button-color('.myButton4', orange);

// 自訂按鈕波紋顏色，預設與按鈕前景顏色一致
// 若無法正常使用，需將其增加到專案的 style.css 檔案中
.myButton5 material-ripple{
   color:blue;
}
```

16.8 進度指示器

進度指示器用於可以確定完成百分比的情況，它讓使用者快速了解一次操作將花費多長時間。元件名稱是 MaterialProgressComponent，在範本中可以使用元素 <material-progress> 創建進度指示器的實例。

1. 輸入屬性

(1) activeProgress int：當前進度值。

(2) indeterminate bool：進度指示器是否是確定性的布林值。預設值為 false。

(3) max int：最大進度值。預設值為 100。

(4) min int：最小進度值。預設值為 0。

(5) secondaryProgress int：次要進度。次要進度以較淺的顏色顯示在主進度的後面。

2. Sass mixin

(1) material-progress-theme($indeterminate-color, $active-color, $secondary-color)：設定不確定進度指示器、主進度指示器、副進度指示器的顏色。

(2) wide-rounded-progress-bar($height)：設定進度指示器軌道的高度。

創建 Angular 專案 progress_demo，將元件 MaterialProgressComponent 和 MaterialButtonComponent 增加到指令清單。範例程式如下：

```
//chapter16/progress_demo/lib/app_component.dart
import 'package:angular/angular.dart';
import 'package:angular_components/angular_components.dart';
@Component(
   selector: 'my-app',
   styleURLs: ['app_component.css'],
   templateURL: 'app_component.html',
   directives: [
      MaterialProgressComponent,
      MaterialButtonComponent],
)
class AppComponent {
      // 設定預設進度值
      int progress = 32;
      // 追加進度
   void addProgress(){
      progress++;
   }
}
```

範本程式如下：

```
//chapter16/progress_demo/lib/app_component.html
<h1> 進度指示器 </h1>
<h6> 主進度指示器 </h6>
<material-progress [activeProgress]="25"></material-progress>

<h6> 不確定進度指示器 </h6>
<material-progress [indeterminate]="true"></material-progress>

<h6> 副進度指示器 </h6>
<material-progress [activeProgress]="25"
                   [secondaryProgress]="75"></material-progress>

<h6> 資料綁定 </h6>
<section class="custom-theme">
```

```
  <material-button raised (trigger)="addProgress()"> 追加進度 </material-button>
  <material-progress [activeProgress]="progress"></material-progress>
</section>

<h6>mixin 自訂進度指示器主題 </h6>
<material-progress class="custom-theme" [indeterminate]="true"></material-
progress>

<h6>mixin 自訂進度指示器軌道高度 </h6>
<material-progress class="custom-height"
                   [activeProgress]="25"
                   [secondaryProgress]="75"></material-progress>
```

樣式程式如下：

```
//chapter16/progress_demo/lib/app_component.scss
@import 'package:angular_components/css/material/material';
// 匯入進度指示器元件定義的 mixins
@import 'package:angular_components/material_progress/mixins';

// 自訂進度指示器主題
.custom-theme {
  @include material-progress-theme(
    // 不確定進度指示器顏色
    $indeterminate-color: $mat-green-100,
    // 主進度指示器顏色
    $active-color: $mat-green-500,
    // 副進度指示器顏色
    $secondary-color: $mat-green-200);
}

// 自訂進度指示器軌道高度
.custom-height{
  @include wide-rounded-progress-bar(6px);
}
```

16.9 選項按鈕

16.9.1 材質化選項按鈕

材質化風格的選項按鈕，元件名稱是 MaterialRadioComponent，在範本中使用元素 <material-radio> 創建選項按鈕的實例。通常與元素 <material-radio-group> 一起使用。一旦選中，則不能透過使用者操作取消選中同一選項按鈕。

1. 屬性

no-ink：設定此屬性可禁用晶片上的波紋效應。

2. 輸入屬性

(1) checked bool：選項按鈕是否被選擇。

(2) disabled bool：選項按鈕是否不應回應事件，並具有不允許進行互動的樣式。

(3) value dynamic：此選項按鈕代表的值，用於帶有選項按鈕群組的選擇模型。

3. 輸出屬性

checkedChange Stream<bool>：選擇狀態更改時發佈事件。

4. Sass mixin

material-radio-color($primary-color,$focus-indicator-color: $primary-color,$modifier:"")：第 1 個參數是按鈕選擇狀態下的顏色，第 2 個參數表示波紋顏色，第 3 個參數是類別選取器。

16.9.2 選項按鈕組

包含多個材料選項按鈕的組，強制在該組中僅選擇一個值。元件名稱是 MaterialRadioGroupComponent，在範本中使用元素 <material-radio-group> 創建按鈕組實例。

可以透過 selected 和 ngModel 獲設定值，但應避免同時使用兩者，因為 ngModel 也透過監聽 onChange 獲設定值，所以這些值可能看起來不同步。

1. 輸入屬性

(1) selected dynamic：當前所選選項按鈕的值，首選 ngModel。

(2) selectionModel SelectionModel<dynamic>：包含值物件的選擇模型。

2. 輸出屬性

selectedChange Stream<dynamic>：選擇更改時發佈事件，首選 ngModelChange。

類別 SelectionModel 用於定義選擇模型，提供管理選定值集合的模式。該類別包含以下建構函數：

(1) SelectionModel.multi({List<T> selectedValues, KeyProvider<T> keyProvider})：創建多選模型。keyProvider 用於相等性檢查。

(2) SelectionModel.single({T selected, KeyProvider<T> keyProvider})：創建單選模型。

Angular 專案 radio_demo，將元件 MaterialRadioComponent 和 MaterialRadioGroupComponent 增加到指令清單。範例程式如下：

```
//chapter16/radio_demo/lib/app_component.dart
import 'package:angular/angular.dart';
import 'package:angular_components/angular_components.dart';
@Component(
   selector: 'my-app',
```

```
    styleURLs: ['app_component.css'],
    templateURL: 'app_component.html',
    directives: [
      MaterialRadioComponent,
      MaterialRadioGroupComponent,
      NgFor],
)
class AppComponent {
    // 設定選項按鈕狀態
    bool radio1 = false;
    bool radio2 = true;
    bool radio3 = false;
    bool radio4 = true;

    // 已選選項
    String selectedOption;

    // 定義單選模型
    final SelectionModel selectionModel = SelectionModel.single();

    // 獲取單選模型的值
    dynamic get selectedValue =>
      selectionModel.selectedValues.isEmpty ? ' 未知 ':
      selectionModel.selectedValues.first;

    // 建構選項
    List<Option> options = [
      Option(' 梨 ','pear'),
      Option(' 香蕉 ','banana'),
      Option(' 桃子 ','peach'),
      Option(' 其他 ','others')
    ];
}
// 定義選項類別
class Option{
    String label;
    String value;
    Option(this.label,this.value);
}
```

範本程式如下：

```
//chapter16/radio_demo/lib/app_component.html
<h1> 選項按鈕 </h1>

<h3> 獨立選項按鈕 </h3>
<h6> 可選擇的選項按鈕 </h6>
<material-radio [(checked)]="radio1"> 可選擇按鈕 </material-radio>

<h6> 預選擇的選項按鈕 </h6>
<material-radio [(checked)]="radio2"> 預選擇按鈕 </material-radio>

<h6> 禁用的選項按鈕 </h6>
<material-radio [disabled]="true" [(checked)]="radio3"> 不可操作的按鈕 </
material-radio>

<h6> 預選擇且禁用的選項按鈕 </h6>
<material-radio [disabled]="true" [(checked)]="radio4"> 預選擇且不可操作的按鈕
</material-radio>

<h3> 分組按鈕 </h3>
<h6> 手動提供選擇項 </h6>
<p> 已選擇項：{{selectedOption}}</p>
<material-radio-group [(selected)]="selectedOption" >
   <material-radio [checked]="true" value="option1"> 選項 1</material-radio>
   <material-radio value="option2"> 選項 2</material-radio>
</material-radio-group>

<h6> 選擇模型 </h6>
<p> 已選擇項：{{selectedValue}}</p>
<material-radio-group [selectionModel]="selectionModel" >
   <material-radio
            *ngFor="let option of options"
            [value]="option.value"
            no-ink>{{option.label}}</material-radio>
</material-radio-group>

<h6>mixin 自訂按鈕在選擇狀態下的顏色、波紋顏色 </h6>
<material-radio class="custom"> 可選擇按鈕 </material-radio>
```

樣式程式如下：

```
//chapter16/radio_demo/lib/app_component.scss
@import 'package:angular_components/material_radio/mixins';

// 自訂按鈕在選擇狀態下的顏色、波紋顏色
@include material-radio-color(green, orange, '.custom');
```

16.10 核取方塊

核取方塊是可以選中或不選中的按鈕。使用者可以點擊核取方塊以選中或取消選中它。一般來説使用核取方塊允許使用者從集合中選擇多個選項。如果只有一個打開或關閉選項，應避免使用單一核取方塊，而應使用切換按鈕。元件名稱是 MaterialCheckboxComponent，在範本中使用元素 <material-checkbox> 創建核取方塊實例。

1. 屬性

no-ink：禁用波紋效果。

2. 輸入屬性

(1) checked bool：使用者設定核取方塊的當前狀態，選中則為 true，否則為 false。

(2) disabled bool：核取方塊處於禁用狀態，不應回應事件，並具有不允許互動的樣式。

(3) indeterminate bool：核取方塊的二選一狀態，非使用者可設定的狀態。在 checked 和 indeterminate 之間，只有一個可以為 true，也可以都為 false。如果是 indeterminate 則為 true，選中或未選中都為 false。

(4) indeterminateToChecked bool：判斷 indeterminate 狀態切換時核取方塊要進入的狀態。為 true 時將被選中，為 false 時將被取消選中。

(5) label String：核取方塊的標籤。

(6) readOnly bool：該核取方塊是否可以透過使用者互動來更改。

(7) themeColor String：核取方塊的顏色和選中時的波紋，預設情況下為 $mat-blue-500。注意，即使未選中核取方塊，themeColor 也會應用到該核取方塊，這與標準材質規範有所不同。除非需要此行為，否則使用 mixin 設定 themeColor。

3. 輸出屬性

(1) change Stream<String>：核取方塊狀態更改時觸發，發送 checkedStr，即 ARIA 狀態。

(2) checkedChange Stream<bool>：選中或取消選中核取方塊時觸發，但設定為 indeterminate 時則不觸發。發送值是 checked 的狀態。

(3) indeterminateChange Stream<bool>：當核取方塊進入和退出 indeterminate 狀態時觸發，但設定為 checked 時則不觸發。

4. Sass mixin

material-checkbox-color($color, $modifier: '')：第一個參數是核取方塊在選擇狀態下的顏色，第二個參數是類別選取器。

Angular 專案 checkbox_demo，將元件 MaterialCheckboxComponent 增加到指令清單。範例程式如下：

```
//chapter16/checkbox_demo/lib/app_component.dart
import 'package:angular/angular.dart';
import 'package:angular_components/angular_components.dart';

@Component(
  selector: 'my-app',
  styleURLs: ['app_component.css'],
  templateURL: 'app_component.html',
  directives: [
    MaterialCheckboxComponent,
    MaterialIconComponent,
    MaterialIconComponent,
```

```
        MaterialToggleComponent],
)
class AppComponent {
    // 初始化各個狀態值
    // 禁用狀態
    bool disabledState = false;
    // 選擇狀態
    bool checkedState = false;
    // 不確定狀態
    bool indeterminateState = false;
    // 不確定行為
    bool indeterminateBehavior = false;
    // 狀態字串
    String statusStr;
}
```

範本程式如下：

```
//chapter16/checkbox_demo/lib/app_component.dart
<h1> 核取方塊 </h1>
<section>
    <h5> 所有狀態的核取方塊 :</h5>
    <div>
        <h6> 沒標籤 </h6>
        <material-checkbox></material-checkbox>

        <h6> 未選中狀態 </h6>
        <material-checkbox label=" 未選中 unchecked"></material-checkbox>

        <h6> 選中狀態 </h6>
        <material-checkbox [checked]="true" label=" 選中 checked"></material-
checkbox>

        <h6> 不確定 </h6>
        <material-checkbox [indeterminate]="true" label=" 不確定狀態
indeterminate"></material-checkbox>

        <h6> 不可用 </h6>
        <material-checkbox [disabled]="true" label=" 不可用 disabled"></material-
checkbox>
```

```
        <h6> 選中且不可用 </h6>
        <material-checkbox [checked]="true" [disabled]="true" label="checked 和
disabled">
        </material-checkbox>

        <h6> 不確定且不可用 </h6>
        <material-checkbox [indeterminate]="true" [disabled]="true"
label="indeterminate 和 disabled">
        </material-checkbox>

        <h6> 選中且不確定 = 不確定 </h6>
        <material-checkbox [indeterminate]="true" [checked]="true"
label="indeterminate 和 checked">
        </material-checkbox>

        <h6> 自訂核取方塊顏色 </h6>
        <material-checkbox
            label=" 紅色核取方塊 "
            themeColor="#FF0000">
        </material-checkbox>

        <h6> 自訂核取方塊內容 ; 頂部對齊 </h6>
        <material-checkbox class="top">
          <div class="custom">
            建議
            <material-icon icon="help" baseline class="help-icon"></material-
icon><br/>
            <textarea cols="40" rows="2"></textarea>
          </div>
        </material-checkbox>
    </div>
</section>

<section>
    <h5> 事件和屬性控制核取方塊的狀態 :</h5>
    <p> 仔細觀察各個狀態的變化 </p>
    <div>
        <material-checkbox
                [disabled]="disabledState"
```

```
            [(checked)]="checkedState"
            [(indeterminate)]="indeterminateState"
            [indeterminateToChecked]="indeterminateBehavior"
            (change)="statusStr=$event"
            label=" 事件和屬性 ">
    </material-checkbox>
  </div>
  <div>
    <material-toggle
            [(checked)]="disabledState"
            label=" 切換到 {{disabledState ? ' 可用 ' : ' 禁用 '}}">
    </material-toggle><br/>
    <material-toggle
            [(checked)]="indeterminateState"
            label=" 切換到 {{indeterminateState ? ' 未設定 indeterminate' :
' 設定 indeterminate'}}">
    </material-toggle><br/>
    <material-toggle
            [(checked)]="indeterminateBehavior"
            label=" 切換到 {{indeterminateBehavior ? ' 從 indeterminate 到
unchecked' : ' 從 indeterminate 到 checked'}}">
    </material-toggle><br/>
  </div>
  <div class="debug-info">
    status = {{statusStr}}<br/>
    checked = {{checkedState}}<br/>
    disabled = {{disabledState}}<br/>
    indeterminate = {{indeterminateState}}<br/>
    indeterminateToChecked = {{indeterminateBehavior}}<br/>
  </div>
</section>

<h6>mixin 自訂選擇狀態下的核取方塊顏色 </h6>
<material-checkbox class="custom"> 自訂顏色 </material-checkbox>
```

樣式程式如下：

```
//chapter16/checkbox_demo/lib/app_component.scss
@import 'package:angular_components/css/material/material';
@import 'package:angular_components/material_checkbox/mixins';
```

```
// 內容與核取方塊頂部對齊
material-checkbox.top {
   align-items: flex-start;
}

.debug-info {
   background: $mat-grey-200;
}

// 自訂核取方塊顏色
@include material-checkbox-color(orange, '.custom');
```

16.11 輸入框

輸入框可以是單行或多行文字欄位，它可以有一個標籤。元件名稱是 MaterialInputComponent，可以在範本中使用元素 <material-input> 創建輸入框的實例。使用時必須在指令清單中宣告 materialInputDirectives 而非 MaterialInputComponent。

1. 屬性

(1) type：輸入的類型，預設是 text。也可以支援 email、password、URL、number、tel 和 search。

(2) multiple：使用者可以輸入透過逗點分隔的多個值。此屬性僅在 type 為 email 時適用，否則忽略。

2. 輸入屬性

(1) label String：此輸入的標籤。如果未在文字標籤中輸入任何內容，則預設顯示文字。使用者輸入文字時消失。

(2) rightAlign bool：輸入內容是否應始終右對齊。預設值為 false。

(3) required bool：是否必須輸入。如果沒有輸入文字，則在第一次失去焦點時顯示驗證錯誤。

(4) requiredErrorMsg String：自訂錯誤訊息，當欄位必填且為空時顯示。

(5) disabled bool：是否禁用輸入。

(6) leadingText String：需要顯示在輸入框前方的文字，例如貨幣符號或類似符號。

(7) leadingGlyph String：在輸入框前方顯示的任何符號，例如連結圖示或類似圖示。

(8) trailingText String：在輸入框的後方顯示的任何文字，例如貨幣符號或類似符號。

(9) trailingGlyph String：在輸入框的後邊顯示的任何符號，例如連結圖示或類似圖示。

(10)hintText dynamic：要在輸入框下方顯示的提示。如果輸入中有錯誤訊息，則不會顯示此文字。

(11)showHintOnlyOnFocus bool：輸入框未聚焦時是否顯示提示文字。

(12)characterCounter int Function(String)：自訂字元計數器函數。接收輸入文字，返回應將文字視為多少個字元。

3. 輸出屬性

(1) blur Stream<FocusEvent>：觸發失去焦點事件時發佈事件。

(2) change Stream<String>：觸發更改事件時發佈事件，例如確認或失去焦點。

(3) focus Stream<FocusEvent>：元素聚焦時觸發的事件。

(4) inputKeyPress Stream<String>：每當輸入文字更改時發佈事件，每次按鍵都會觸發。

擁有 multiline 屬性的輸入框是一個多行文字標籤，它需要提供指令 MaterialMultilineInputComponent。還需要在提供者清單中增加 DomService 和 windowBindings。它還具有兩個特殊屬性：rows 表示多行輸入預設顯示的行數，maxRows 表示顯示的最大行數，超過 maxRows 的所有內容則會出現捲軸。

創 建 Angular 專 案 input_demo， 將 materialInputDirectives 和 MaterialMultilineInputComponent 增加到指令清單。範例程式如下：

```
//chapter16/input_demo/lib/app_component.dart
import 'package:angular/angular.dart';
import 'package:angular_components/angular_components.dart';
import 'package:angular_components/utils/browser/dom_service/angular_2.dart';
import 'package:angular_components/utils/browser/window/module.dart';
import 'package:angular_forms/angular_forms.dart';

@Component(
   selector: 'my-app',
   styleURLs: ['app_component.css'],
   templateURL: 'app_component.html',
   providers: [DomService,windowBindings],
   directives: [
      formDirectives,
      materialInputDirectives,
      MaterialMultilineInputComponent],
)
class AppComponent {
   // 初始化輸入框文字
   String textValue = 'Text value';
}
```

範本程式如下：

```
//chapter16/input_demo/lib/app_component.html
<h6> 單行輸入框 </h6>
<p><material-input></material-input></p>
<p><material-input label=" 標籤 "></material-input></p>
<p><material-input label=" 右對齊 " [rightAlign]="true"></material-input></p>
<p><material-input label=" 必填 " required
                requiredErrorMsg=" 此輸入框必填 "></material-input></p>
<p><material-input [disabled]="true" label=" 此輸入框被禁用 "></material-
input></p>

<h6> 前導和後導 </h6>
<p><material-input label=" 前導文字 " leadingText="$"></material-input></p>
<p><material-input label=" 後導文字 " trailingText=".00"
```

```
            [rightAlign]="true"></material-input></p>
<p><material-input label=" 前導圖示 " leadingGlyph="link"></material-input></p>
<p><material-input label=" 後導圖示 " trailingGlyph="email"></material-input></p>

<h6> 浮動標籤 </h6>
<p><material-input floatingLabel label=" 浮動標籤 "></material-input></p>
<p><material-input floatingLabel label=" 浮動標籤和雙向資料綁定 "
            [ngModel]="textValue"></material-input></p>

<h6> 失去焦點時更新 </h6>
<p><material-input blurUpdate [(ngModel)]="textValue"></material-input></p>
<div> 值 :{{textValue}}</div>

<h6> 更改時更新 </h6>
<p><material-input changeUpdate [(ngModel)]="textValue"></material-input></p>
<div> 值 :{{textValue}}</div>

<h6> 提示文字 </h6>
<p><material-input hintText=" 輸入提示訊息 "></material-input></p>

<p> 僅在聚焦時才顯示提示訊息 </p>
<p><material-input hintText=" 使用 showHintOnlyOnFocus"
            showHintOnlyOnFocus required></material-input></p>

<h6> 字元計數 </h6>
<p><material-input [maxCount]="10" label=" 計數 "></material-input></p>

<h6> 電子郵件 </h6>
<p><material-input floatingLabel label="email" type="email"></material-
input></p>

<h6> 密碼 </h6>
<p><material-input floatingLabel label="password" type="password"></material-
input></p>

<h6>URL 位址 </h6>
<p><material-input floatingLabel label="URL" type="URL"></material-input></p>

<h6> 多行輸入框 </h6>
```

```
<p><material-input multiline floatingLabel
        label=" 預設多行輸入框，隨著字元增多而增加行 "></material-input></p>
<p><material-input multiline floatingLabel rows="2" maxRows="4"
        label=" 顯示 2 行，最多顯示 4 行，隨著字元增多而增加行 "></material-
input></p>
<p><material-input multiline floatingLabel rows="2" [maxCount]="90"
        label=" 最多 90 個字元，否則顯示錯誤訊息 "></material-input></p>
```

16.12 清單

16.12.1 材質化清單

材質化清單是用於與使用者進行互動的一組清單項目的容器元件，它組成了選擇和選單元件的基礎。元件名稱是 MaterialListComponent，在範本中使用元素 <material-list> 可以創建清單的實例。

MaterialListComponent 類別充當清單的根節點，提供樣式和收集清單項目事件的能力。如果需要對清單項目進行分組，則可以使用帶有 group 屬性的元素包裹對應的清單項目。如果需要在組中提供標籤，則可以將 label 屬性置於組內的區塊元素上。

1. 屬性

(1) size String：清單大小，預置大小包括 x-small、small、medium、large 和 x-large，它們的實際寬度分別是 64px*{1.5，3，5，6，7}。其預設大小根據內容自動調整。

(2) min-size String：清單的最小大小，清單寬度至少會達到該指定大小。

2. 輸入屬性

size String：預設寬度 1~5，分別與屬性中的預設寬度對應。預設情況下其值為 0，材質清單將擴充到其父級的完整寬度。

16.12.2 清單項目

材質化清單項目是與使用者互動的區塊級元素，當使用者點擊或按 Enter 鍵時發出並觸發事件。元件名稱是 MaterialListItemComponent，在範本中使用元素 <material-list-item> 創建清單項目的實例。

1. 輸入屬性

(1) disabled bool：禁用觸發器並為清單項目設定禁用樣式。

(2) tabbable bool：元件是否可用作標籤。

(3) tabindex String：元件的標籤索引。與 tabbable 配合使用，當 tabbable 為 true 時有效。

2. 輸出屬性

trigger Stream<UIEvent>：透過點擊，點擊或按鍵啟動按鈕時觸發。

3. 樣式

(1) material-list-item-primary：該樣式常應用於顯示清單項目中的圖示。

(2) material-list-item-secondary：該樣式常用於顯示清單項目的輔助資訊。

創建 Angular 項目 list_demo，將 MaterialListComponent 和 MaterialList ItemComponent 增加到指令清單。範例程式如下：

```
//chapter16/list_demo/lib/app_component.dart
import 'package:angular/angular.dart';
import 'package:angular_components/angular_components.dart';

@Component(
   selector: 'my-app',
   styleURLs: ['app_component.css'],
   templateURL: 'app_component.html',
   directives: [
      MaterialIconComponent,
      MaterialListComponent,
```

```
      MaterialListItemComponent,],
)
class AppComponent {
   // 自訂清單背景顏色
   String bgColor = '#f0c9cf';
   // 回應清單項目的 trigger 事件
   void toggleColor(String color){
      bgColor = color;
   }
}
```

範本程式如下：

```
//chapter16/list_demo/lib/app_component.html
<h1> 清單 </h1>
<h6> 普通清單，預設寬度為父容器的寬度 </h6>
<material-list>
   <material-list-item> 項目 1</material-list-item>
   <material-list-item> 項目 2</material-list-item>
   <material-list-item> 項目 3</material-list-item>
</material-list>

<h6> 寬度為 large 並帶有禁用項目的清單 </h6>
<material-list size="large">
   <material-list-item disabled> 項目 1</material-list-item>
   <material-list-item> 項目 2</material-list-item>
   <material-list-item> 項目 3</material-list-item>
</material-list>

<h6> 分組且帶標籤的清單 </h6>
<material-list size="large">
   <div group>
      <div label> 分組 1</div>
      <material-list-item> 項目 1</material-list-item>
      <material-list-item> 項目 2</material-list-item>
   </div>
   <div group>
      <div label> 分組 2</div>
      <material-list-item> 項目 1</material-list-item>
      <material-list-item> 項目 2</material-list-item>
```

```
  </div>
</material-list>

<h6> 帶圖示和輔助資訊的清單 </h6>
<material-list size="large">
  <material-list-item>
    <material-icon icon="today" class="material-list-item-primary">
    </material-icon>
    項目 1
  </material-list-item>
  <material-list-item disabled>
    項目 2
    <span class="material-list-item-secondary"> 不可選 </span>
  </material-list-item>
  <material-list-item> 項目 3</material-list-item>
  <material-list-item> 項目 4</material-list-item>
</material-list>

<h6> 透過 trigger 事件切換清單背景 </h6>
<material-list size="large" [style.background-color]="bgColor">
  <material-list-item (trigger)="toggleColor('#f07c82')"> 香葉紅 </material-
list-item>
  <material-list-item (trigger)="toggleColor('#eea2a4')"> 牡丹粉紅 </material-
list-item>
  <material-list-item (trigger)="toggleColor('#f03752')"> 海棠紅 </material-
list-item>
</material-list>
```

16.13 片記與片集

16.13.1 片記

片記小元件呈現為帶陰影的圓形框，通常水平排列，它用於呈現簡要資訊。任何物件都可以透過實現 HasUIDisplayName 介面來使用片記。元件名稱是 MaterialChipComponent，在範本中使用元素 <material-chip> 創建片記的實例。

當 removable 屬性為 true 或在片集實例上設定了 selectionModel 時，刪除按鈕才會顯示。

當 hasLeftIcon 為 true 時，應將左圖示內容設定為 MaterialIconComponent 元件實例或 SVG 圖型。

1. 輸入屬性

(1) deleteButtonAriaMessage String：移除按鈕的 Aria 標籤。
(2) hasLeftIcon bool：片記是否應顯示自訂圖示，預設值為 false。
(3) itemRenderer String Function(T)：一個 ItemRenderer 函數，獲取一個物件並返回一個字串。如果 ItemRenderer 不是無狀態的，並且可能為同一輸入項返回不同的值，則需要更新 ItemRenderer 引用，否則該更改將不會得到表現。提供時，將用於生成片記的標籤。
(4) removable bool：片記是否應顯示移除按鈕，預設值為 true。
(5) value dynamic：要繪製的資料模型。在片記的內容中提供標籤，或提供一個 ItemRenderer。

2. 輸出屬性

remove Stream<dynamic>：移除片記時觸發事件，該事件返回片記的值。

16.13.2 片集

片記的集合，將清單中的所有物件以片的形式展示。元件名稱是 MaterialChipsComponent，在範本中使用元素 <material-chips> 創建片集的實例。

輸入屬性

(1) itemRenderer String Function(T)：將清單中的項目呈現為字串的函數。注意：僅當提供 SelectionModel 時，才使用此 ItemRenderer。如果片集是手動繪製的，itemRenderer 屬性也需要手動設定。關於

OnPush 的注意事項：如果 ItemRenderer 不是純函數，並且具有可能以不同方式呈現同一專案的內部狀態，則引用必須更改才能生效。

(2) removable bool：是否可以移除片記。

(3) selectionModel SelectionModel<T>：此元件控制的選擇模型。

創建 Angular 項目 chips_demo，將 MaterialChipsComponent、displayName RendererDirective 和 MaterialChipComponent 增加到指令清單。範例程式如下：

```
//chapter16/chips_demo/lib/app_component.dart
import 'package:angular/angular.dart';
import 'package:angular_components/angular_components.dart';

@Component(
  selector: 'my-app',
  styleURLs: ['app_component.css'],
  templateURL: 'app_component.html',
  directives: [
    MaterialChipsComponent,
    MaterialChipComponent,
    displayNameRendererDirective,
    MaterialIconComponent,
    NgFor],
)
class AppComponent {
  //Label 實現了 HasUIDisplayName 介面
  // 選擇模型的泛型為 <HasUIDisplayName>
  // 需要同指令 displayNameRenderer 一起使用
  //SelectionModel.multi() 方法返回選擇模型 SelectionModel
  //selectedValues 參數用於傳遞一個任意類型的清單
  SelectionModel<HasUIDisplayName> labelSelection = _labelSelectionModel();
  static SelectionModel<HasUIDisplayName> _labelSelectionModel() =>
    SelectionModel.multi(selectedValues: [
      Label(' 書法 '),
      Label(' 國畫 '),
```

```
      Label(' 詩歌 '),
      Label(' 元曲 ')
   ]);

//Subject 未實現 HasUIDisplayName 介面
// 選擇模型的泛型為 <Subject>
// 需要同繪製函數 renderSubjectChip 一起使用
SelectionModel<Subject> subjectSelection =_subjectSelectionModel();
static SelectionModel<Subject> _subjectSelectionModel() =>
   SelectionModel.multi(selectedValues: [
      Subject(1,' 語文 '),
      Subject(1,' 數學 '),
      Subject(1,' 物理 '),
      Subject(1,' 美術 ')
   ]);
// 繪製函數
ItemRenderer<dynamic> renderSubjectChip =
   (dynamic protoChip) {
   return protoChip.name;
};
//Subject 未實現 HasUIDisplayName 介面
// 需要同繪製函數 renderSubjectChip 一起使用
// 透過回應 remove 事件移除 subjects 清單項目
List<Subject> subjects = [
   Subject(1,' 語文 '),
   Subject(1,' 數學 '),
   Subject(1,' 物理 '),
   Subject(1,' 美術 ')
];
//Label 實現了 HasUIDisplayName 介面
// 需要同指令 displayNameRenderer 一起使用
// 透過回應 remove 事件移除 labels 清單項目
List<Label> labels = [
   Label(' 書法 '),
   Label(' 國畫 '),
   Label(' 詩歌 '),
```

```
        Label(' 元曲 ')
    ];
    // 在 labels 清單增加新項目
    void add(String val){
        labels.add(Label(val));
    }
}
// 定義實現 HasUIDisplayName 介面的 Label 類別
class Label implements HasUIDisplayName{
    @override
    //TODO: 實現 uiDisplayName
    String get uiDisplayName => name;
    String name;
    Label(this.name);
}
// 定義一個普通類別 Subject
class Subject{
    int id;
    String name;
    Subject(this.id,this.name);
}
```

範本程式如下：

```
//chapter16/chips_demo/lib/app_component.html
<h1> 片集 </h1>
<h5> 片集 </h5>
<p> 選擇模型 labelSelection</p>
<material-chips [selectionModel]="labelSelection"
                [removable]="true"
                displayNameRenderer>
</material-chips>

<p> 選擇模型 subjectSelection</p>
<material-chips [selectionModel]="subjectSelection"
                [removable]="true"
                [itemRenderer]="renderSubjectChip">
```

```
</material-chips>

<h5> 片記 </h5>
<p> 手動指定內容，移除按鈕不起作用 </p>
<material-chips>
   <material-chip> 預設 </material-chip>
   <material-chip [removable]="false"> 不可移除 </material-chip>
   <material-chip [hasLeftIcon]="true">
      <material-icon left-icon icon="link" size="large"></material-icon>
      帶有左圖示
   </material-chip>
</material-chips>
<h5> 使用資料模型生成片記 </h5>

<p> 資料模型 subjects 清單 </p>
<material-chips>
   <material-chip *ngFor="let subject of subjects"
               [removable]="true"
               [itemRenderer]="renderSubjectChip"
               (remove)="subjects.remove(subject)"
               [value]="subject"></material-chip>
   </material-chips>

<p> 資料模型 labels 清單 </p>
<material-chips>
   <material-chip *ngFor="let c of labels"
               [removable]="true"
               (remove)="labels.remove(c)"
               [value]="c"
               displayNameRenderer></material-chip>
</material-chips>

<input #label type="text" placeholder=" 填入新 label"/>
<button (click)="add(label.value)"> 增加 Label</button>
```

16.14 按鈕組

兩個水平相鄰的按鈕元件，例如：是或否、保存或取消、同意或不同意等。按鈕上的文字可以更改，也可以突出顯示。元件名稱是 MaterialYesNoButtonsComponent，可以使用元素 <material-yes-no-buttons> 創建按鈕組實例。

可以使用 MaterialSaveCancelButtonsDirective 之類的指令提供自訂文字，該指令用 Save 或 Cancel 替換 yes 或 no。

要以相反的順序顯示按鈕，需增加 reverse 屬性。

1. 輸入屬性

(1) disabled bool：是否應禁用按鈕。預設值為 false。

(2) noAutoFocus bool：no 按鈕是否應自動對焦。預設值為 false。

(3) noDisabled bool：是否應禁用 no 按鈕。預設值為 false。

(4) noDisplayed bool：是否顯示 no 按鈕。預設值為 true。

(5) noText String：要在取消按鈕上顯示的文字。舉例來說，關閉、Not now 等。預設值為 No。

(6) pending bool：表示待定狀態，如果值為 true，將隱藏是和否按鈕，並顯示一個旋轉器。這應該用於指示非同步作業，例如：保存或驗證輸入。預設值為 false。

(7) raised bool：是否應該凸出按鈕。預設值為 false。

(8) yesAutoFocus bool：Yes 按鈕是否應該自動對焦。預設值為 false。

(9) yesDisabled bool：是否應禁用 Yes 按鈕。預設值為 false。

(10) yesDisplayed bool：是否顯示 Yes 按鈕。預設值為 true。

(11) yesHighlighted bool：是否應突出顯示 Yes 按鈕。預設值為 false。

(12) yesRaised bool：是否應凸起 Yes 按鈕。預設值為 false。

(13) yesText String：要在 Yes 按鈕上顯示的文字。舉例來說，確定、應用等。預設值為 Yes。

2. 輸出屬性

(1) no Stream<UIEvent>：當 No 按鈕被按下時要呼叫的回呼。發佈的事件是 KeyboardEvent 或 MouseEvent。

(2) yes Stream<UIEvent>：按下 Yes 按鈕時要呼叫的回呼。發佈的事件是 KeyboardEvent 或 MouseEvent。

3. 相關指令

保存取消按鈕組指令 MaterialSaveCancelButtonsDirective，對應元素是 <material-yes-no-buttons[saveCancel]>，它會將 Yes 和 No 替換為 Save 和 Cancel。

提交取消按鈕組指令 MaterialSubmitCancelButtonsDirective，對應元素是 <material-yes-no-buttons[submitCancel]>，它會將 Yes 和 No 更改為 Submit 和 Cancel。

4. Sass mixin

(1) material-yes-button-color($value)：更改 yes 按鈕的顏色。

(2) material-yes-button-text-color($value)：更改 yes 按鈕上的文字顏色。

(3) material-no-button-color($value)：更改 no 按鈕的顏色。

(4) material-no-button-text-color($value)：更改 no 按鈕上的文字顏色。

創建 Angular 項目 yes_no_buttons_demo，增加 MaterialYesNoButtonsComponent、MaterialSaveCancelButtonsDirective、MaterialSubmitCancelButtonsDirective 到指令清單，將 DomService 和 windowBindings 增加到提供者清單。範例程式如下：

```
//chapter16/yes_no_buttons_demo/lib/app_component.dart
import 'package:angular/angular.dart';
import 'package:angular_components/angular_components.dart';
import 'package:angular_components/utils/browser/dom_service/angular_2.dart';
import 'package:angular_components/utils/browser/window/module.dart';
@Component(
   selector: 'my-app',
   styleURLs: ['app_component.css'],
   templateURL: 'app_component.html',
   providers: [DomService,windowBindings],
   directives: [
      MaterialYesNoButtonsComponent,
      MaterialSaveCancelButtonsDirective,
      MaterialSubmitCancelButtonsDirective],
)
class AppComponent {
   // 初始化待定狀態
   bool pending = false;
   // 回應 yes 事件
   void save(){
      pending = true;
      Future.delayed(Duration(seconds: 2),() => pending = false);
   }
}
```

範本程式如下：

```
//chapter16/yes_no_buttons_demo/lib/app_component.html
<h1> 確認和取消按鈕組 </h1>
<h6> 預設 </h6>
<material-yes-no-buttons></material-yes-no-buttons>

<h6> 凸起 </h6>
<material-yes-no-buttons raised></material-yes-no-buttons>

<h6> 僅 Yes 按鈕凸起 </h6>
<material-yes-no-buttons yesRaised></material-yes-no-buttons>
```

```
<h6>Yes 按鈕反白 </h6>
<material-yes-no-buttons yesHighlighted></material-yes-no-buttons>

<h6> 自動聚焦 </h6>
<material-yes-no-buttons yesAutoFocus></material-yes-no-buttons>

<h6> 禁用按鈕 </h6>
<material-yes-no-buttons [disabled]="true"></material-yes-no-buttons>
<material-yes-no-buttons [yesDisabled]="true"></material-yes-no-buttons>
<material-yes-no-buttons [noDisabled]="true"></material-yes-no-buttons>

<h6> 不顯示其中某個按鈕 </h6>
<material-yes-no-buttons [noDisplayed]="false"></material-yes-no-buttons>
<material-yes-no-buttons [yesDisplayed]="false"></material-yes-no-buttons>

<h6> 自訂文字 </h6>
<material-yes-no-buttons noText=" 取消 "
               yesText=" 保存 "></material-yes-no-buttons>
<h6> 翻轉 </h6>
<material-yes-no-buttons reverse
               noText=" 取消 "
               yesText=" 保存 "></material-yes-no-buttons>

<h6> 觸發 yes 事件可觀察到待定狀態 </h6>
<material-yes-no-buttons noText=" 取消 "
               yesText=" 保存 "
               (yes)="save()"
               [pending]="pending"></material-yes-no-buttons>

<h6> 保存和取消 </h6>
<material-yes-no-buttons saveCancel></material-yes-no-buttons>

<h6> 提交和取消 </h6>
<material-yes-no-buttons submitCancel></material-yes-no-buttons>

<h6> 自訂主題 </h6>
```

```
<material-yes-no-buttons raised class="green-yes-no"></material-yes-no-buttons>
<br>
<material-yes-no-buttons class="green-yes-no"></material-yes-no-buttons>
<br>
<material-yes-no-buttons class="blue-text-yes-no"></material-yes-no-buttons>
<br>
<material-yes-no-buttons raised class="yes-no"></material-yes-no-buttons>
```

樣式程式如下：

```
//chapter16/yes_no_buttons_demo/lib/app_component.scss
@import 'package:angular_components/css/material/material';
@import 'package:angular_components/material_yes_no_buttons/mixins';

.green-yes-no{
   // 設定 yes 和 no 按鈕背景顏色，當設定按鈕背景顏色時必須提供 raised 屬性，否則
無效
   @include material-no-button-color(green);
   @include material-yes-button-color(green);
}

.blue-text-yes-no{
   // 設定 yes 和 no 按鈕文字顏色，波紋顏色依賴於文字顏色
   @include material-yes-button-text-color(blue);
   @include material-no-button-text-color(blue);
}

.yes-no{
   //yes 按鈕背景顏色
   @include material-yes-button-color($mat-red-500);
   //yes 按鈕文字顏色
   @include material-yes-button-text-color(white);
   //no 按鈕背景顏色
   @include material-no-button-color($mat-red-500);
   //no 按鈕文字顏色
   @include material-no-button-text-color(white);
}
```

16.15 日期、時間選擇器

16.15.1 日期範圍選擇器

一種採用材料設計風格的日期範圍選擇器。元件名稱是 MaterialDateRangePickerComponent，在範本中使用元素 <material-date-range-picker> 創建日期範圍選擇器的實例。

1. 輸入屬性

(1) minDate Date：無法選擇早於 minDate 的日期。

(2) maxDate Date：無法選擇晚於 maxDate 的日期。

(3) showNextPrevButtons bool：是否顯示「下一個」和「上一個」按鈕。預設值為 true。

(4) range DatepickerComparison：所選日期範圍和比較。此日期選擇器使用 DatepickerComparison 代替普通的 DateRangeComparison 物件，在內部實現增加其他所需的功能，例如「名稱」和「下一個」或「上一個」。

(5) presets List<DatepickerPreset>：使用者可以選擇的預設日期範圍的清單。這些預設項目受 minDate 和 maxDate 的限制，如果它們的終點在 minDate 之前或起點在 maxDate 之後，則將其完全排除。

2. 輸出屬性

rangeChange Stream<DatepickerComparison>：所選日期範圍或比較範圍更改時發佈。

創建 Angular 專案 date_time_picker_demo，在 src 目錄下創建日期範圍選擇器。範例程式如下：

```
//chapter16/date_time_picker_demo/lib/src/date_range_component.dart
import 'package:angular/angular.dart';
import 'package:quiver/time.dart';
```

```
import 'package:angular_components/angular_components.dart';
import 'package:angular_components/utils/browser/window/module.dart';
@Component(
   selector: 'date-range',
   templateURL: 'date_range_component.html',
   providers: [windowBindings, datepickerBindings],
   directives: [
      MaterialDateRangePickerComponent],
)
class DateRangeComponent {
   // 日期選擇器預設清單
   List<DatepickerPreset> presets;

   // 用於綁定日期範圍 range
   DatepickerComparison range;
   DatepickerComparison emptyRange;

   // 日期範圍物件與輸入屬性 [minDate] 和 [maxDate] 配合使用
   DateRange limitRange = DateRange(Date.today().add(months: -3), Date.
today());

   DateRangeComponent() {
   // 創建一個本地時鐘
   var clock = Clock();
   // 預設日期選擇器預設清單接收一個時鐘物件
   presets = defaultPresets(clock);
   // 預設 range 的日期選擇範圍為本周
   range = DatepickerComparison.noComparison(presets
      .singleWhere((preset) => preset.range.title == 'This week')
      .range);
   }
}
```

範本程式如下：

```
//chapter16/date_time_picker_demo/lib/src/date_range_component.html
<h3> 時間範圍選擇器 </h3>
```

```
<h6> 預設 </h6>
<material-date-range-picker>
</material-date-range-picker>

<h6> 限制日期選擇範圍 </h6>
<material-date-range-picker [minDate]="limitRange.start"
                            [maxDate]="limitRange.end"
                            [showNextPrevButtons]="false"
                            [(range)]="emptyRange">
</material-date-range-picker>

<h6> 使用預設清單的日期選擇器 </h6>
<material-date-range-picker
                            [presets]="presets"
                            [(range)]="range"
                            [showNextPrevButtons]="false">
</material-date-range-picker>
```

16.15.2 日期選擇器

材料設計風格的單一日期選擇器。使用者可以輸入自訂日期，或點擊日曆以選擇日期。元件名稱是 MaterialDatepickerComponent，在範本中使用元素 <material-datepicker> 創建日期選擇器實例。

1. 輸入屬性

(1) date Date：所選的日期。常用於綁定到元件屬性。

(2) minDate Date：無法選擇早於 minDate 的日期。

(3) maxDate Date：無法選擇晚於 maxDate 的日期。

2. 輸出屬性

dateChange Stream<Date>：所選日期更改時發佈事件。

16.15.3 時間選擇器

一種採用材料設計風格的單一時間選擇器。元件名稱是 MaterialTime PickerComponent，在範本中使用元素 <material-time-picker> 創建時間選擇器實例。

1. 輸入屬性

(1) time DateTime：所選的時間。

(2) utc bool：是否以 UTC 時區返回時間，預設返回本地時區的時間。

2. 輸出屬性

timeChange Stream<DateTime>：所選時間更改時發佈事件。

16.15.4 日期和時間選擇器

一種採用材料設計風格的單一日期和時間選擇器。元件名稱是 MaterialDateTimePickerComponent，在範本中使用元素 <material-date-time-picker> 創建日期時間選擇器實例。

1. 輸入屬性

(1) dateTime DateTime：所選日期時間。

(2) utc bool：是否使用 UTC 時區中的 dateTime，預設返回本地時區的 dateTime。

2. 輸出屬性

dateTimeChange Stream<DateTime>：選定的 dateTime 更改時發佈事件。

在 src 目錄下創建日期時間選擇器的範例程式如下：

```
//chapter16/date_time_picker_demo/lib/src/date_time_component.dart
import 'package:angular/angular.dart';
```

```
import 'package:angular_components/angular_components.dart';
import 'package:angular_components/utils/browser/window/module.dart';
@Component(
selector: 'date-time',
   templateURL: 'date_time_component.html',
   providers: [windowBindings, datepickerBindings],
   directives: [
      MaterialDatepickerComponent,
      MaterialTimePickerComponent,
      MaterialDateTimePickerComponent,],
)
class DateTimeComponent {
   // 用於綁定到日期選擇器
   Date date = Date.today();

   // 用於綁定到時間選擇器
   DateTime time = DateTime.now();

   // 用於綁定到日期時間選擇器
   DateTime dateTime = DateTime.now();
}
```

範本程式如下：

```
//chapter16/date_time_picker_demo/lib/src/date_time_component.html
<h3> 日期選擇器 </h3>
<material-datepicker [(date)]="date">
</material-datepicker>

<h3> 時間選擇器 </h3>
<material-time-picker [(time)]="time">
</material-time-picker>

<h6> 使用 UTC 時區 </h6>
<material-time-picker [utc]="true"
                      [(time)]="time">
</material-time-picker>
```

```
<h3> 日期時間選擇器 </h3>
<material-date-time-picker [(dateTime)]="dateTime">
</material-date-time-picker>

<h6> 使用 UTC 時區 </h6>
<material-date-time-picker [utc]="true"
                    [(dateTime)]="dateTime">
</material-date-time-picker>
```

在根元件中將 DateRangeComponent 和 DateTimeComponent 增加到指令清單。程式如下：

```
//chapter16/date_time_picker_demo/lib/app_component.dart
import 'package:angular/angular.dart';
import 'src/date_range_component.dart';
import 'src/date_time_component.dart';

@Component(
   selector: 'my-app',
   styleURLs: ['app_component.css'],
   templateURL: 'app_component.html',
   directives: [DateRangeComponent,DateTimeComponent],
)
class AppComponent {
}
```

在範本中增加元素程式如下：

```
//chapter16/date_time_picker_demo/lib/app_component.html
<h1> 日期時間 </h1>
<date-range></date-range>
<date-time></date-time>
```

16.16 步驟指示器

16.16.1 材質化步驟指示器

材質化風格的步驟指示器，它是帶編號的指示器，用於傳達進度或用作導覽工具。元件名稱是 MaterialStepperComponent，在範本中使用元素 <material-stepper> 創建步驟指示器實例。

1. 輸入屬性

(1) keepInactiveStepsInDom bool：如果值為 true，則在非活動狀態下不會從 DOM 移除步驟，而是透過 CSS 隱藏這些步驟。這樣可以在載入複雜的 DOM 之間快速切換步驟。

(2) legalJumps String：合法的跳躍。定義為非繼續或取消按鈕觸發的步進開關。可能的值：none：預設，不允許跳躍；backwards：允許跳躍至已完成的步驟；all：允許任意跳躍，與步驟狀態無關。

(3) noText String：返回上一步按鈕上顯示的文字，預設為 Cancel。

(4) orientation String：步驟的佈局方向。包含兩個值 horizontal 和 vertical，分別代表水平和垂直。

(5) size String：設定大小，用於確定各個步驟標頭元素的大小，包括步驟編號、步驟名稱等。可能的值為 default 和 mini。

(6) stickyHeader bool：指示列出了可用步驟的標頭，是否應該停留在頁面頂部。僅適用於帶有水平標頭的步驟指示器。

(7) yesText String：在下一步按鈕上顯示的文字。預設情況下顯示 Continue，如果原始程式中未提供該輸入屬性，則無法使用。

2. 輸出屬性

activeStepChanged Stream<StepDirective>：當前活動的步驟更改時觸發的事件。

3. Sass mixin

(1) material-stepper-theme($selector:'',$step-color:$mat-blue-500,$disabled-color: $mat-grey-500,$button-color:$mat-blue-500)：第 1 個參數是選擇器，第 2 個參數是步驟的索引顏色，第 3 個參數是處於非活躍狀態的步驟的索引顏色，第 4 個參數是 yes 按鈕的顏色。

(2) material-stepper-step-name-disabled-color($color:$mat-grey-500)：設定不可跳躍步驟名的顏色。

(3) material-stepper-step-name-selectable-color($color:$mat-grey-500)：設定可跳躍步驟名的顏色。

16.16.2 步驟指令

用來標記步驟指示器中的一步。指令名是 StepDirective，在範本中使用帶有 step 屬性的元素創建單一步驟。

1. 輸入屬性

(1) canContinue bool：該步驟是否可以繼續。這可以用來防止繼續執行某個步驟，直到當前步驟的所有部分都滿足驗證要求為止。

(2) cancelHidden bool：在此步驟中是否應隱藏取消按鈕。

(3) complete bool：該步驟是否完成。當進入下一步時設定此值。

(4) completeSummary String：在垂直預設大小的步驟指示器中步驟完成時顯示的摘要文字。對於其他步驟指示器不適用。

(5) hideButtons bool：在此步驟中是否應隱藏按鈕。

(6) name String：顯示為標題。

(7) optional bool：該步驟是否可選。可選步驟帶有一個額外的標籤，表示它們是可選的，應該可以跳過。預設值為 false。

2. 輸出屬性

(1) cancel Stream<AsyncAction<bool>>：點擊取消按鈕時呼叫。如果事件處理常式呼叫 $event.cancel()，則不會取消該步驟。

(2) continue Stream<AsyncAction<bool>>：點擊繼續按鈕時呼叫。如果事件處理常式呼叫 $event.cancel()，則該步驟將不會繼續。

(3) jumpHere Stream<AsyncAction<bool>>：當使用者想要跳至此步驟時呼叫。如果事件處理常式呼叫 $event.cancel()，則該步驟將不會繼續。

創建 Angular 專案 stepper_demo，將 MaterialStepperComponent、StepDirective 和 SummaryDirective 增加到指令清單，在提供者清單中增加 scrollHostProviders、DomService 和 windowBindings。範例程式如下：

```
//chapter16/stepper_demo/lib/app_component.dart
import 'package:angular/angular.dart';
import 'package:angular_components/angular_components.dart';
import 'package:angular_components/model/action/async_action.dart';
import 'package:angular_components/utils/angular/scroll_host/angular_2.dart';
import 'package:angular_components/utils/browser/dom_service/angular_2.dart';
import 'package:angular_components/utils/browser/window/module.dart';
@Component(
  selector: 'my-app',
  styleURLs: ['app_component.css'],
  templateURL: 'app_component.html',
  providers: [scrollHostProviders,DomService,windowBindings],
  directives: [
    MaterialStepperComponent,
    StepDirective,
    SummaryDirective,
    MaterialButtonComponent],
)
class AppComponent{
  // 該方法用於回應 continue 和 cancel 事件
  void validDelayedCheck(AsyncAction<bool> action) {
    // 在這裡沒有進行實際的驗證操作，而是延遲 1s 後允許 continue 和 cancel 操作
繼續執行
```

```
      action.cancelIf(Future.delayed(const Duration(seconds: 1), () {
        // 不取消
        return false;
      }));
    }
}
```

範本程式如下：

```
//chapter16/stepper_demo/lib/app_component.html
<h1> 步驟指示器 </h1>
<h6> 水平步驟指示器，可往回調轉，大小設定值為 default</h6>
<material-stepper legalJumps="backwards"
                  orientation="horizontal"
                  size="default">
  <template step name=" 第 1 步 ">
    <p> 操作內容 </p>
  </template>
  <template step name=" 第 2 步 ">
    <p> 操作內容 </p>
  </template>
  <template step name=" 第 3 步 ">
    <p> 操作內容 </p>
  </template>
  </material-stepper>

<h6> 垂直步驟指示器，可任意調轉，大小設定值為 mini</h6>
<material-stepper legalJumps="all"
                  orientation="vertical"
                  size="mini">
  <template step name=" 第 1 步 ">
    <p> 操作內容 </p>
  </template>
  <template step name=" 第 2 步 ">
    <p> 操作內容 </p>
  </template>
  <template step name=" 第 3 step ">
    <p> 操作內容 </p>
  </template>
```

```
</material-stepper>

<h6> 帶驗證和可選步驟 </h6>
<material-stepper legalJumps="all"
                  orientation="vertical"
                  size="default">
   <template step name=" 第 1 步 "
            (continue)="validDelayedCheck($event)">
      <p> 操作內容 </p>
   </template>
   <template step name=" 第 2 步 " [optional]="true"
            (cancel)="validDelayedCheck($event)">
      <p> 操作內容 </p>
   </template>
   <template step name=" 第 3 步 "
            (continue)="validDelayedCheck($event)">
      <p> 操作內容 </p>
   </template>
   <template step name=" 第 4 步 "
            (continue)="validDelayedCheck($event)"
            (cancel)="validDelayedCheck($event)">
      <p> 操作內容 </p>
   </template>
</material-stepper>

<h6>mixin 步驟的摘要文字和主題及自訂按鈕文字 </h6>
<section class="themed">
   <material-stepper legalJumps="all"
                     orientation="vertical"
                     size="default"
                     yesText=" 繼續 "
                     noText=" 返回 ">
      <template step name=" 第 1 步 ">
         <p> 操作內容 </p>
      </template>
      <template step name=" 第 2 步 " #step2="step">
         <p> 操作內容 </p>
      </template>
      <template [summary]="step2" >
```

```
            第 2 步的摘要文字
        </template>
        <template step name=" 第 3 步 ">
            <p> 操作內容 </p>
        </template>
    </material-stepper>
</section>

<h6>mixin 自訂不可跳躍步驟名的顏色 </h6>
<material-stepper class="custom-disabled"
                      orientation="vertical"
                      size="default"
                      yesText=" 繼續 "
                      noText=" 返回 ">
    <template step name=" 第 1 步 ">
        <p> 操作內容 </p>
    </template>
    <template step name=" 第 2 步 ">
        <p> 操作內容 </p>
    </template>
    <template step name=" 第 3 步 ">
        <p> 操作內容 </p>
    </template>
</material-stepper>

<h6>mixin 自訂可跳躍步驟名的顏色 </h6>
<material-stepper class="custom-selectable"
                  legalJumps="all"
                  orientation="vertical"
                  size="default"
                  yesText=" 繼續 "
                  noText=" 返回 ">
    <template step name=" 第 1 步 ">
        <p> 操作內容 </p>
    </template>
    <template step name=" 第 2 步 ">
        <p> 操作內容 </p>
    </template>
    <template step name=" 第 3 步 ">
```

```
    <p> 操作內容 </p>
  </template>
</material-stepper>
```

樣式程式如下：

```
//chapter16/stepper_demo/lib/app_component.scss
@import 'package:angular_components/css/material/material';
@import 'package:angular_components/material_stepper/mixins';
@import 'package:angular_components/material_yes_no_buttons/mixins';
// 用於設定主題
// 第 1 個參數表示選擇器
// 第 2 個參數表示步驟標頭的顏色
// 第 3 個參數表示 yes 按鈕的顏色
@include material_stepper-theme($selector: '.themed',
                                $step-color: $mat-teal-500,
                                $button-color: $mat-teal-500);
// 自訂不可跳躍步驟名的顏色
.custom-disabled{
  @include material-stepper-step-name-disabled-color($color: green);
}
// 自訂可跳躍步驟名的顏色
.custom-selectable{
  @include material-stepper-step-name-selectable-color($color: orange);
}
```

16.17 對話方塊

遵循材質化風格設計的對話方塊。元件名稱是 MaterialDialogComponent，在範本中使用元素 <material-dialog> 創建對話方塊的實例。

1. 內容元素

(1) [header]：對話方塊的標題內容，需在元素上增加該屬性。

(2) [footer]：對話方塊的頁尾內容，需在元素上增加該屬性。

2. 屬性

(1) headered：在對話方塊標題上增加灰色背景。

(2) info：將對話方塊設定為資訊對話方塊。

3. 輸入屬性

(1) error String：將錯誤顯示在對話方塊的 error 部分。

(2) hideFooter bool：是否隱藏對話方塊頁尾。

(3) hideHeader bool：是否隱藏對話方塊標題。

(4) listenForFullscreenChanges bool：在對話方塊進入或退出全螢幕模式時是否監聽。

(5) shouldShowScrollStrokes bool：當使用者透過捲動存取更多內容時，是否顯示內容部分的上下邊框。預設顯示。

4. 輸出屬性

fullscreenMode Stream<bool>：對話方塊進入或退出全螢幕模式時的事件串流。

5. Sass mixin

material-dialog-fullscreen($width-threshold: 100vw,$height-threshold: 100vh)：控制對話方塊進入全螢幕的條件，第一個參數是可視視窗的寬度，第二個參數是可視視窗的高度。

對話方塊需要與模態元件 ModalComponent 配合使用，模態元件的選擇器是 <modal>，使用該選擇器作為對話方塊的父元素。需要將對話方塊的控制變數雙向資料綁定到模態元件的 visible 變數上。在對話方塊中需要透過手動指定關閉按鈕，按鈕的 trigger 事件修改對話方塊的控制變數。

如果按鈕需要自動聚焦，則需要與指令 AutoFocusDirective 配合使用。如果希望對話方塊能自動關閉，則需要與指令 AutoDismissDirective 配合使用。

創建 Angular 專案 dialog_demo，範例程式如下：

```
//chapter16/dialog_demo/lib/app_component.dart
import 'package:angular/angular.dart';
import 'package:angular_components/angular_components.dart';

@Component(
   selector: 'my-app',
   styleURLs: ['app_component.css'],
   templateURL: 'app_component.html',
   providers: [overlayBindings],
   directives: [
      MaterialDialogComponent,
      MaterialButtonComponent,
      MaterialIconComponent,
      ModalComponent,
      AutoFocusDirective,
      AutoDismissDirective,
   NgIf,],
)
class AppComponent {
   // 基本對話方塊控制變數
   bool basicDialog = false;
   // 捲動對話方塊控制變數
   bool scrollingDialog = false;
   // 資訊對話方塊控制變數
   bool infoDialog = false;
   // 自動關閉對話方塊控制布林量
   bool autoDismissDialog = false;
   // 帶標題背景對話方塊控制布林量
   bool headeredDialog = false;
   // 錯誤對話方塊控制變數
   bool erroDialog = false;
   // 全螢幕對話方塊控制變數
   bool fullscreenDialog = false;
   // 全螢幕狀態
```

```
  bool isInFullscreenMode = false;
  // 錯誤訊息
  String dialogWithErrorErrorMessage;
  // 在錯誤訊息有和無間切換
  void toggleErrorMessage() {
    if (dialogWithErrorErrorMessage == null) {
      dialogWithErrorErrorMessage = ' 自訂錯誤訊息 ';
    } else {
      dialogWithErrorErrorMessage = null;
    }
  }
}
```

範本程式如下：

```
//chapter16/dialog_demo/lib/app_component.html
<h1> 對話方塊 </h1>
<h6> 基本對話方塊 </h6>
<material-button (trigger)="basicDialog = true"
                 [disabled]="basicDialog"
                 raised>
  打開普通對話方塊
</material-button>

<modal [(visible)]="basicDialog">
  <material-dialog>
    <h1 header> 對話方塊標題 </h1>
    <p> 普通對話方塊 </p>
    <div footer>
      <material-button autoFocus clear-size (trigger)="basicDialog = false">
        關閉
      </material-button>
    </div>
  </material-dialog>
</modal>

<material-button (trigger)="scrollingDialog = true"
                 [disabled]="scrollingDialog"
                 raised>
```

```
  打開捲動對話方塊
</material-button>

<modal [(visible)]="scrollingDialog">
  <material-dialog class="scrolling">
    <h1 header> 對話方塊標題 </h1>
    <p class="scrolling-content"> 內容高度超出對話方塊高度 </p>
    <p class="scrolling-content"> 在垂直方向可捲動 </p>
    <p class="scrolling-content"> 更多資訊 </p>
    <div footer>
      <material-button autoFocus clear-size (trigger)="scrollingDialog =
false">
        關閉
      </material-button>
    </div>
  </material-dialog>
</modal>

<h6> 資訊對話方塊 </h6>
<material-button (trigger)="infoDialog = true"
                 [disabled]="infoDialog"
                 raised>
  打開普通對話方塊
</material-button>
<modal [(visible)]="infoDialog">
  <material-dialog info class="info-dialog">
    <div header>
      <material-button icon autoFocus (trigger)="infoDialog = false">
        <material-icon icon="close"></material-icon>
      </material-button>
      <h1> 資訊 </h1>
    </div>
    <p> 資訊對話方塊 </p>
  </material-dialog>
</modal>

<h6> 自動關閉對話方塊 </h6>
<material-button (trigger)="autoDismissDialog = true"
                 [disabled]="autoDismissDialog"
                 raised>
```

```
    打開自動關閉對話方塊
</material-button>
<modal [(visible)]="autoDismissDialog">
    <material-dialog info class="info-dialog"
                    [autoDismissable]="autoDismissDialog"
                    (dismiss)="autoDismissDialog = false">
        <div header>
            <material-button icon autoFocus (trigger)="autoDismissDialog = false">
                <material-icon icon="close"></material-icon>
            </material-button>
            <h1> 資訊 </h1>
        </div>
        <p> 點擊對話方塊以外的地方，對話方塊將自動關閉。</p>
    </material-dialog>
</modal>

<h6> 帶標題背景的對話方塊 </h6>
<material-button (trigger)="headeredDialog = true"
                [disabled]="headeredDialog"
                raised>
    打開帶標題背景的對話方塊
</material-button>
<modal [(visible)]="headeredDialog">
    <material-dialog headered class="info-dialog">
        <div header>
            <h1> 資訊 </h1>
        </div>
        <p> 帶標題背景的對話方塊 </p>
        <div footer>
            <material-button autoFocus clear-size (trigger)="headeredDialog =
false">
                關閉
            </material-button>
        </div>
    </material-dialog>
</modal>

<h6> 帶錯誤訊息的對話方塊 </h6>
<material-button (trigger)="erroDialog = true"
```

```
                  [disabled]="erroDialog"
                  raised>
   打開帶錯誤訊息的對話方塊
</material-button>
<modal [(visible)]="erroDialog">
   <material-dialog headered class="info-dialog"
                   [error]="dialogWithErrorErrorMessage">
      <div header>
         <h1> 資訊 </h1>
      </div>
      <material-button raised (trigger)="toggleErrorMessage()">
         {{dialogWithErrorErrorMessage == null ? ' 顯示 ' : ' 隱藏 '}} 錯誤訊息
      </material-button>
      <div footer>
         <material-button autoFocus clear-size (trigger)="erroDialog =
false">
            關閉
         </material-button>
      </div>
   </material-dialog>
</modal>

<h6> 帶錯誤訊息的對話方塊 </h6>
<material-button (trigger)="fullscreenDialog = true"
                  [disabled]="fullscreenDialog"
                  raised>
   打開帶錯誤訊息的對話方塊
</material-button>
<modal [(visible)]="fullscreenDialog">
   <material-dialog
         headered
         class="fullscreen-dialog"
         [class.fullscreen-mode]="isInFullscreenMode"
         [listenForFullscreenChanges]="true"
         (fullscreenMode)="isInFullscreenMode = $event">
      <div header>
         <h1> 資訊 </h1>
      </div>
      <p *ngIf="isInFullscreenMode">
```

```
            對話方塊當前處於全螢幕模式，放大視窗以退出全螢幕模式。
        </p>
        <p *ngIf="!isInFullscreenMode">
            如果視窗足夠小，則此對話方塊將全螢幕顯示。
        </p>
        <div footer>
            <material-button autoFocus clear-size (trigger)="fullscreenDialog =
false">
                關閉
            </material-button>
        </div>
    </material-dialog>
</modal>
```

樣式程式如下：

```
//chapter16/dialog_demo/lib/app_component.scss
@import 'package:angular_components/css/material/material';
@import 'package:angular_components/material_dialog/mixins';

// 設定對話方塊的高度和寬度，內容超出高度後將可捲動
.scrolling{
    height: 300px;
    width:70%;
}

.scrolling-content{
    height:160px;
}

// 設定資訊對話方塊的寬度
.info-dialog {
    width: 340px;
}

// 透過 mixin 控制對話方塊進入全螢幕的條件
// 第一個參數是高度，第二個參數是寬度，變數 $mat-grid 的值為 8px
.fullscreen-dialog{
    @include material-dialog-fullscreen($mat-grid * 100, $mat-grid * 80);
}
```

16.18 擴充面板

一種材質化風格的擴充面板。元件名稱是 MaterialExpansionPanel，在範本中使用元素 <material-expansionpanel> 創建擴充範本實例。

一個或多個面板組合在一個擴充面板集中。點擊面板時，面板內容會展開。面板由名稱、值、可選的輔助文字和展開的面板內容組成。

當面板內容處於折疊狀態時，具有屬性 value 的內容元素將被用作面板內容的值。

與面板的互動是透過擴充面板集完成的，該集合考慮了集合中其他面板的狀態，並將適當的操作發佈到每個單獨的面板上。

1. 屬性

(1) wide：指示面板展開後的寬度略寬於折疊狀態。

(2) flat：指示面板展開時不應浮動或與其他面板分開。

(3) forceContentWhenClosed：關閉擴充面板時，將擴充面板內容保留在 DOM 中。儘量少使用，此屬性會影響應用性能。

2. 內容引用

focusOnOpen：在內容中使用 #focusOnOpen 標記一個 Focusable 或 DOM 元素，當擴充面板打開時，使該專案被聚焦。

3. 輸入屬性

(1) alwaysHideExpandIcon bool：如果值為 true，則展開圖示永遠不可見。

(2) alwaysShowExpandIcon bool：如果值為 true，則無論是否使用自訂圖示，展開圖示都應始終可見。

(3) cancelDisplayed bool：是否顯示取消按鈕，預設值為 true。

(4) cancelText String：在取消按鈕上顯示的文字。預設值為 Cancel。

(5) closeOnSave bool：如果值為 true，則保存成功後，面板將嘗試關閉。

(6) disableHeaderExpansion bool：如果值為 true，則點擊標題不會展開或折疊面板。

(7) disabled bool：如果值為 true，則面板將保持折疊狀態而無法展開，或如果預設情況下是展開的，則面板將保持展開狀態。

(8) expandIcon String：可選的圖示名稱，用於使用自訂圖示替換展開箭頭。

(9) focusOnOpen dynamic：設定聚焦子元素，當面板打開時，可以聚焦在子元素上。

(10) hideExpandedHeader bool：如果值為 true，當面板展開時，則隱藏顯示面板名稱的標頭。

(11) expanded bool：如果值為 true，則預設情況下會展開面板；如果值為 false，則將關閉面板。

(12) name String：擴充面板的簡稱標籤。

(13) saveDisabled bool：是否禁用保存按鈕。

(14) saveText String：要在保存按鈕上顯示的文字。預設值為 Save。

(15) secondaryText String：一些可選的輔助摘要文字，描述了面板內託管的小元件的狀態。

(16) showSaveCancel bool：是否應顯示保存和取消按鈕，預設值為 true。

4. 輸出屬性

(1) cancel Stream<AsyncAction<bool>>：取消面板時觸發事件。

(2) close Stream<AsyncAction<bool>>：面板嘗試關閉時觸發事件，此操作可能被取消。

(3) expandedChange Stream<bool>：當面板折疊或展開時觸發事件。

(4) expandedChangeByUser Stream<bool>：當使用者折疊或展開面板時觸發事件。

(5) open Stream<AsyncAction<bool>>：面板嘗試打開時觸發事件，此操作可能被取消。

(6) save Stream<AsyncAction<bool>>：保存面板時觸發事件。

擴充面板集合將一組擴充面板組合在一起，一次只能打開一個擴充面板。MaterialExpansionPanelSet 必須將 MaterialExpansionPanel 用作直接子代。元件名稱是 MaterialExpansionPanelSet，在範本中使用元素 <material-expansionpanel-set> 創建擴充範本實例。

創建 Angular 專案 expansionpanel_demo，範例程式如下：

```
//chapter16/expansionpanel_demo/lib/app_component.dart
import 'dart:async';
import 'package:angular/angular.dart';
import 'package:angular_components/angular_components.dart';
import 'package:angular_components/model/action/async_action.dart';

@Component(
  selector: 'my-app',
  styleURLs: ['app_component.css'],
  templateURL: 'app_component.html',
  providers: [overlayBindings],
  directives: [
    AutoFocusDirective,
    MaterialInputComponent,
    MaterialDialogComponent,
    MaterialYesNoButtonsComponent,
    MaterialIconComponent,
    MaterialExpansionPanelSet,
    MaterialExpansionPanel,
    MaterialExpansionPanelAutoDismiss,
    ModalComponent],
)
class AppComponent{
  // 對話方塊顯示控制屬性
```

```
    bool showConfirmation = false;
    // 完成物件，包含完成狀態和資訊
    Completer<bool> dialogFutureCompleter;
    // 顯示對話方塊函數
    void showConfirmationDialog(AsyncAction event) {
        showConfirmation = true;
        // 初始完成物件
        dialogFutureCompleter = Completer();
        // 事件物件根據對話方塊完成狀態控制是否取消
        event.cancelIf(dialogFutureCompleter.future);
    }
    // 關閉對話方塊函數
    void closeDialog(bool proceed) {
        showConfirmation = false;
        if (dialogFutureCompleter != null) {
            // 設定完成狀態
            dialogFutureCompleter.complete(!proceed);
        }
    }
}
```

範本程式如下：

```
//chapter16/expansionpanel_demo/lib/app_component.html
<h1> 擴充面板 </h1>
<h6> 預設面板 </h6>
<material-expansionpanel-set>
    <material-expansionpanel name=" 面板名 "> 預設面板 </material-expansionpanel>
    <material-expansionpanel name=" 面板 2"
                             secondaryText=" 摘要文字 ">
        <p> 面板名和摘要文字 </p>
    </material-expansionpanel>
    <material-expansionpanel name=" 面板 3" >
        <p value> 折疊狀態面板的值 </p>
        <p> 面板 3 設定折疊狀態面板的值 </p>
    </material-expansionpanel>
</material-expansionpanel-set>
```

```
<h6>wide 面板 </h6>
<material-expansionpanel-set>
  <material-expansionpanel wide name=" 隱藏標頭 "
                           [hideExpandedHeader]="true">
    <p> 隱藏標頭 </p>
  </material-expansionpanel>
  <material-expansionpanel wide name=" 自訂按鈕文字 "
                           saveText=" 保存 "
                           cancelText=" 關閉 ">
    <p> 自訂 saveText 和 cancelText</p>
  </material-expansionpanel>
  <material-expansionpanel wide name=" 禁用面板 " disabled>
  </material-expansionpanel>
</material-expansionpanel-set>

<h6>flat 面板 </h6>
<material-expansionpanel-set>
  <material-expansionpanel flat>
    <div name>
      <material-icon icon="notifications" size="medium"></material-icon>
      提示
    </div>
    <p value> 面板可以自訂 name</p>
    <p> 可以將 name 作為元素的屬性，該元素將作為面板的顯示標籤 </p>
  </material-expansionpanel>
  <material-expansionpanel flat name=" 隱藏保持和取消按鈕 "
                           [showSaveCancel]="false">
    <p> 隱藏保持和取消按鈕 </p>
  </material-expansionpanel>
  <material-expansionpanel flat name=" 自訂擴充按鈕 "
                           expandIcon="edit">
    自訂擴充按鈕
  </material-expansionpanel>
</material-expansionpanel-set>

<section>
  <p> 以下面板沒有包含在面板集合中，可以獨立打開 </p>
  <material-expansionpanel name=" 僅能透過擴充圖示打開面板 "
                           [disableHeaderExpansion]="true">
```

```
        點擊面板標頭將不會打開或折疊面板
    </material-expansionpanel>
    <material-expansionpanel name=" 預設展開面板 "
                             [expanded]="true">
        此面板預設展開
    </material-expansionpanel>
    <material-expansionpanel autoDismissable
                             name=" 自動折疊面板 ">
        <p> 當展開面板後，點擊面板之外的地方，面板自動折疊。</p>
    </material-expansionpanel>
    <material-expansionpanel name=" 展開面板自動對焦 ">
        <material-input #focusOnOpen></material-input>
    </material-expansionpanel>
    <material-expansionpanel shouldExpandOnLeft
                             name=" 在面板左側展示擴充圖示 ">
        <p>shouldExpandOnLeft 屬性使得擴充圖示顯示在左側 </p>
    </material-expansionpanel>
    <material-expansionpanel name=" 取消時確認按鈕 "
                             (cancel)="showConfirmationDialog($event)">
        <p> 點擊取消按鈕時，如果取消狀態一直持續沒有完成，則在對話方塊中確定是否
真正取消。</p>
    </material-expansionpanel>
</section>

<!-- 確認對話方塊 -->
<modal [(visible)]="showConfirmation">
    <material-dialog class="confirmation-dialog" >
        <h3> 確定取消嗎 ?</h3>
        <div footer>
            <material-yes-no-buttons raised
                                     yesHighlighted
                                     (yes)="closeDialog(true)"
                                     (no)="closeDialog(false)">
            </material-yes-no-buttons>
        </div>
    </material-dialog>
</modal>
```

16.19 下拉式功能表

材質風格的下拉選擇選單，元件名稱是MaterialDropdownSelectComponent，在範本中使用元素 <material-dropdown-select> 創建下拉式功能表實例。

1. 輸入屬性

(1) options dynamic：設定選擇元件的可用選項。接收 SelectionOptions 或 List。如果傳遞了 List，則將使用 StringSelectionOptions 類別創建選擇選項。如果需要更進階的功能，包括分組選項、自訂過濾或非同步搜索，則可以傳遞 SelectionOptions 的實現。

(2) selection dynamic：設定選擇元件的選擇值或選擇模型。預設使用 SingleSelectionModel，可以自訂 SelectionModel 或使用 MultiSelectionModel。

(3) itemRenderer String Function(T)：將選項物件轉為字串的函數。

2. 輸出屬性

selectionChange Stream<dynamic>：只要更改選擇，就會發出選定的值。對於單選，它將是所選值或為 null。對於多選，它將是所選值的清單或空清單。

創建 Angular 專案 dropdown_select_demo。範例程式如下：

```
//chapter16/dropdown_select_demo/lib/app_component.dart
import 'package:angular/angular.dart';
import 'package:angular_components/angular_components.dart';
import 'package:angular_forms/angular_forms.dart';
@Component(
   selector: 'my-app',
   styleURLs: ['app_component.css'],
   templateURL: 'app_component.html',
   providers: [popupBindings],
   directives: [
```

```
        MaterialDropdownSelectComponent,
        MaterialSelectDropdownItemComponent,
        DropdownSelectValueAccessor,
        MultiDropdownSelectValueAccessor,
        NgModel],
)
class AppComponent{
    // 普通單選值
    String collegeValue;

    // 簡單 List 資料模型
    static List<String> colleges =<String>['文學院','哲學院','數學院','電腦
學院'];

    // 普通多選模型
    final SelectionModel<String> multiSelectModel = SelectionModel<String>.
multi();

    // 生成多選標籤
    String get multiSelectLabel {
        var selectedValues = multiSelectModel.selectedValues;
        if (selectedValues.isEmpty) {
            return '選擇學院';
        } else if (selectedValues.length == 1) {
            return selectedValues.first;
        } else {
            return '${selectedValues.first} + ${selectedValues.length - 1} 更多';
        }
    }

    // 單選與 Angular 表單
    College selectionValue;

    // 複雜模型資料
    static List<College> collegeList = <College>[
        College(1,'文學院'),
        College(2,'哲學院'),
        College(3,'數學院'),
        College(4,'電腦學院')
```

```
    ];

    // 透過 StringSelectionOptions 類別創建選擇選項
    // 其建構函數接收一個 List 物件
    StringSelectionOptions<College> collegeOptions =
    StringSelectionOptions<College>(collegeList);

    // 多選與 Angular 表單
    List<dynamic> selectionValues = [];

    // 顯示項目生成函數
    static final ItemRenderer itemRenderer =
       (item) => (item as HasUIDisplayName).uiDisplayName;

    // 生成多選標籤
    String get selectionValuesLabel {
       final size = selectionValues.length;
       if (size == 0) {
          return '選擇學院';
       } else if (size == 1) {
          return itemRenderer(selectionValues.first);
       } else {
          return '${itemRenderer(selectionValues.first)} + ${size - 1} 更多';
       }
    }
}

// 實現 HasUIDisplayName 介面
// 由該物件組成的選項或清單不需要提供 ItemRenderer 函數
class College implements HasUIDisplayName{
    int id;
    String name;
    @override
    //TODO: 實現 uiDisplayName
    String get uiDisplayName => name;
    College(this.id,this.name);
    @override
    String toString() => uiDisplayName;
}
```

範本程式如下：

```
//chapter16/dropdown_select_demo/lib/app_component.html
<h1> 下拉清單 </h1>
<h6> 普通單選 </h6>
<material-dropdown-select
      buttonText="{{collegeValue == null ? ' 選擇學院 ' : collegeValue}}"
      [(selection)]="collegeValue"
      [options]="colleges">
</material-dropdown-select>

<h6> 普通多選 </h6>
<material-dropdown-select
      [buttonText]="multiSelectLabel"
      [options]="colleges"
      [selection]="multiSelectModel">
</material-dropdown-select>

<h6> 單選和雙向資料綁定 </h6>
<material-dropdown-select
      [buttonText]="selectionValue == null ? ' 選擇學院 ' : selectionValue.name"
      [options]="collegeList"
      [(ngModel)]="selectionValue">
</material-dropdown-select>

<h6> 多選和雙向資料綁定 </h6>
<material-dropdown-select multi
                  [buttonText]="selectionValuesLabel"
                  [options]="collegeOptions"
                  [itemRenderer]="itemRenderer"
                  [(ngModel)]="selectionValues">
</material-dropdown-select>
```

16.20 彈出框

材質化風格的彈出元件，元件名稱是 MaterialPopupComponent，在範本中使用元素 <material-popup> 創建彈出框實例。

注意事項：

(1) 關閉和打開快顯視窗會自動延遲以增加動畫。

(2) 如果內容大小使頁面捲動，則利用 PopupInterface 中定義的 forceSpaceSpaceConstraints 會有所幫助。

(3) 如果內容更改並且需要重新調整位置，需使用在 PopupInterface 中定義的 trackLayoutChanges。

材質快顯視窗還支援延遲載入的內容。

該元件將自身發佈為 DropdownHandle，因此其子代可以透過注入它來控制其可見性。範例程式如下：

```
class MyComponent {
  final DropdownHandle _dropdownHandle;

  MyComponent(this._dropdownHandle);

  void onSomethingThatShouldCloseTheDropdown() {
    _dropdownHandle.close();
  }
}
```

1. 輸入屬性

(1) source PopupSource：設定快顯視窗應該相對創建的來源。

(2) visible bool：設定是否顯示快顯視窗，用於關閉或打開快顯視窗。

(3) enforceSpaceConstraints bool：設定快顯視窗是否應根據相對於視埠的可用空間自動重新定位。

(4) preferredPositions Iterable<Object>：當提供了 enforceSpaceConstraints 屬性時，設定彈出框彈出的位置，它接收一個 RelativePosition 類型的清單，它會依照指定清單依次選擇合適的彈出位置，如果清單中沒有合適的彈出位置，則它將自動選擇彈出位置。

(5) offsetX int：彈出框相對來源在 x 軸上的偏移量，參照點是來源的左上角那個點。

(6) offsetY int：彈出框相對來源在 y 軸上的偏移量，參照點是來源的左上角那個點。

(7) ink bool：將快顯視窗的背景顏色設定為 ink($mat-grey-700)。

(8) matchMinSourceWidth bool：設定快顯視窗是否應將最小寬度設定為來源的寬度。

2. 輸出屬性

visibleChange Stream<bool>：當快顯視窗的 visible 屬性更改時觸發同步事件。

創建 Angular 專案 popup_demo，範例程式如下：

```
//chapter16/popup_demo/lib/app_component.dart
import 'package:angular/angular.dart';
import 'package:angular_components/angular_components.dart';
@Component(
   selector: 'my-app',
   styleURLs: ['app_component.css'],
   templateURL: 'app_component.html',
   providers: [popupBindings],
   exports: [RelativePosition],
   directives: [
      MaterialButtonComponent,
      MaterialPopupComponent,
      PopupSourceDirective,
      MaterialDropdownSelectComponent,],
)
class AppComponent {
```

```
  // 表示彈出框可見性的清單
  List<bool> visible = List.filled(11, false);

  // 單選模型
  SelectionModel<RelativePosition> position =
  RadioGroupSingleSelectionModel(RelativePosition.OffsetBottomRight);

  // 選擇的方向
  RelativePosition get popupPosition => position.selectedValues.first;

  // 手動指定彈出位置
  RelativePosition get customPosition => RelativePosition.OffsetBottomRight;

  // 選項
  final SelectionOptions<RelativePosition> positions =
  SelectionOptions.fromList(positionMap.keys.toList());

  // 標籤生成函數
  ItemRenderer positionLabel =
     (position) => positionMap[position];

  // 按鈕標籤
  String get buttonLabel => positionLabel(popupPosition);
}
// 彈出位置集合
final positionMap = <RelativePosition, String>{
  // 相對來源的 4 個角
  RelativePosition.OffsetBottomRight: '右下角',
  RelativePosition.OffsetBottomLeft: '左下角',
  RelativePosition.OffsetTopRight: '右上角',
  RelativePosition.OffsetTopLeft: '左上角',
  // 行內方向，沿 x 或 y 軸延伸方向
  RelativePosition.InlineBottom: '行內向下',
  RelativePosition.InlineBottomLeft: '行內向左和下',
  RelativePosition.InlineTop: '行內向上',
  RelativePosition.InlineTopLeft: '行內向左和上',
  // 相對來源的正方向
  RelativePosition.AdjacentTop: '上方',
```

```
    RelativePosition.AdjacentRight: '右方',
    RelativePosition.AdjacentLeft: '左方',
    RelativePosition.AdjacentBottom: '下方',
    // 相對來源的兩個方向
    RelativePosition.AdjacentTopLeft: '左邊與來源對齊的上方',
    RelativePosition.AdjacentTopRight: '右邊與來源對齊的上方',
    RelativePosition.AdjacentLeftTop: '上邊與來源對齊的左方',
    RelativePosition.AdjacentRightTop: '上邊與來源對齊的右方',
    RelativePosition.AdjacentRightBottom: '下邊與來源對齊的右方',
    RelativePosition.AdjacentBottomRight: '右邊與來源對齊的下方',
    RelativePosition.AdjacentBottomLeft: '左邊與來源對齊的下方',
    RelativePosition.AdjacentLeftBottom: '下邊與來源對齊的左方',
};
```

範本程式如下：

```
//chapter16/popup_demo/lib/app_component.html
<h1>彈出框</h1>
<section>
    <p>由系統根據可視視窗大小自動選擇彈出方向</p>
    <material-button
            raised
            popupSource
            #source="popupSource"
            (trigger)="visible[0]= !visible[0]">
        {{visible[0]? '關閉' : '打開'}} 彈出框
    </material-button>
    <material-popup [source]="source" [(visible)]="visible[0]">
        <div style="height: 100px;width:200px;">彈出框內容</div>
    </material-popup>
</section>

<section>
    <p>帶 ink 背景並手動指定彈出方向，在可視視窗能完全容納彈出框時按指定方向彈
出。否則由它自動選擇彈出方向</p>
    <material-button
            raised
            popupSource
            #source1="popupSource"
```

```
        (trigger)="visible[2]= !visible[2]">
      {{visible[2]? '關閉' : '打開'}} 彈出框
    </material-button>

    <material-popup [source]="source1" [(visible)]="visible[2]"
                    [enforceSpaceConstraints]="true"
                    ink
                    [preferredPositions]="[customPosition,popupPosition]">
      <div style="height: 100px;width:200px;"> 彈出框內容 </div>
    </material-popup>
</section>

<section>
    <p> 請選擇方向，然後點擊打開彈出框 </p>
<material-dropdown-select
        [selection]="position"
        [options]="positions"
        [itemRenderer]="positionLabel"
        [buttonText]="buttonLabel">
</material-dropdown-select>
<div style="height:60px;"></div>
<material-button
        raised
        popupSource
        #customsrc="popupSource"
        class="trigger"
        (trigger)="visible[1]= !visible[1]">
    {{visible[1]? '關閉' : '打開'}} 彈出框
</material-button>
<material-popup [source]="customsrc"
                [(visible)]="visible[1]"
                [enforceSpaceConstraints]="true"
                [preferredPositions]="[popupPosition]">
    <div style="height: 200px; padding: 24px;">
      <p> 彈出框內容 </p>
      <material-button
            raised
            (trigger)="visible[1]= false">
          關閉
```

```
      </material-button>
    </div>
</material-popup>
</section>

<section>
    <p>offsetX 為 30px，offsetY 為 60px</p>
    <material-button
          raised
          popupSource
          #source2="popupSource"
          (trigger)="visible[3]= !visible[3]">
      {{visible[3]? '關閉' : '打開'}} 彈出框
    </material-button>

    <material-popup [source]="source2" [(visible)]="visible[3]"
                    [offsetX]="30"
                    [offsetY]="60">
      <div style="height: 100px;width:200px;">彈出框內容</div>
    </material-popup>
</section>

<section>
    <p>最小寬度等於來源的寬度，透過匯出的 RelativePosition 物件指定彈出方向</p>
    <material-button
          raised
          popupSource
          #source3="popupSource"
          (trigger)="visible[4]= !visible[4]">
      {{visible[4]? '關閉' : '打開'}} 彈出框，最小寬度等於來源的寬度
    </material-button>
    <material-popup [source]="source3" [(visible)]="visible[4]"
                    [enforceSpaceConstraints]="true"
                    [preferredPositions]="[RelativePosition.AdjacentTop]"
                    [matchMinSourceWidth]="true">
      <p>彈出框內容</p>
    </material-popup>
</section>
```

16.21 選項選單

選項選單常用於在多個選項中選取一個或多個項目的使用場景。

16.21.1 選項容器

MaterialSelectComponent 是用於從集合中選擇專案的容器，並標記所選的選項，在範本中使用元素 <material-select> 創建選項容器實例。

可以透過 SelectionOptions 實例或在範本中強制寫入指定選項，透過範本或對照選擇模型將選項標記為已選擇項。

輸入屬性

(1) disabled bool：選擇是否應顯示為禁用。預設值為 false。
(2) itemRenderer String Function(T)：將選擇選項指定的值轉為字串，並呈現。
(3) options SelectionOptions<T>：SelectionOptions 實例提供的呈現選項。
(4) selection SelectionModel<T>：此容器的選擇模型，可以是單選或多選模型。
(5) width dynamic：呈現清單的寬度，可用值從 1 到 5。

16.21.2 選擇項目

MaterialSelectItemComponent 是可以被選擇的一種特殊的清單項，在範本中使用元素 <material-select-item> 創建選擇項目實例。

1. 輸入屬性

(1) isHidden bool：是否隱藏該選項，預設值為 false。
(2) itemRenderer String Function(T)：將選項呈現為 String 的函數。
(3) selected bool：手動將選項標記為已選。

(4) selection SelectionModel<T>：選擇模型，可以是單選或多選模型。

(5) useCheckMarks bool：核取記號，如果值為 true，將替代核取方塊來指示多選中的已選選項。

(6) value T：當前選擇項表示的值。如果物件實現了 HasUIDisplayName，則它將使用 uiDisplayName 欄位作為項目的標籤進行繪製。不然必須提供 itemRenderer 屬性，標籤才由該元件生成。

2. 輸出屬性

trigger Stream<UIEvent>：透過點擊，點擊或按鍵啟動按鈕時觸發。

創建 Angular 項目 select_demo，範例程式如下：

```
//chapter16/select_demo/lib/app_component.dart
import 'package:angular/angular.dart';
import 'package:angular_components/angular_components.dart';
@Component(
   selector: 'my-app',
   styleURLs: ['app_component.css'],
   templateURL: 'app_component.html',
   directives: [
      MaterialSelectComponent,
      MaterialSelectItemComponent,
      NgFor,
      displayNameRendererDirective],
)
class AppComponent {
   String season;

   //用作選項的清單
   static List<Season> seasonList = [
      Season('spring','春'),
      Season('summer','夏'),
      Season('autumn','秋'),
      Season('winter','冬'),
   ];

   //單選模型，接收的是 String
```

```
    final SelectionModel<String> singleSelection =
    SelectionModel.single();

    // 回應選項的 trigger 事件，將單選模型的值設定值給變數 season
    void selected(){
      season = singleSelection.selectedValues.first;
    }

    // 建構選擇項，用於 options 參數
    final SelectionOptions<Season> seansonOptions =
    SelectionOptions.fromList(seasonList);

    // 單選模型，接收單一 Season 物件，用於 selection 參數
    final SelectionModel<Season> singleSeasonSelection =
    SelectionModel.single();

    // 多選模型，可接收多個 Season 物件，用於 selection 參數
    final SelectionModel<Season> multiSelection =
    SelectionModel.multi();

    // 獲取儲存在多選模型中的已選項
    List<Season> get selectedList => multiSelection.selectedValues.toList();
}

// 實現 HasUIDisplayName 介面的 Season 類別
class Season implements HasUIDisplayName{
    String en;
    String cn;
    @override
    // 實現 uiDisplayName
    String get uiDisplayName => cn;
    Season(this.en,this.cn);
}
```

範本程式如下：

```
//chapter16/select_demo/lib/app_component.html
<h1> 選項 </h1>
<h6> 手動指定選項 </h6>
<material-select [width]="2">
```

```
   <material-select-item (trigger)="season = 'spring'"
                  [selected]="season == 'spring'"> 春 </material-select-item>
   <material-select-item (trigger)="season = 'summer'"
                  [selected]="season == 'summer'"> 夏 </material-select-item>
   <material-select-item (trigger)="season = 'autumn'"
                  [selected]="season == 'autumn'"> 秋 </material-select-item>
   <material-select-item (trigger)="season = 'winter'"
                  [selected]="season == 'winter'"> 冬 </material-select-item>
</material-select>

<h6> 使用 NgFor 生成選項 </h6>
<material-select [width]="2">
   <material-select-item *ngFor="let s of seasonList"
                  (trigger)="season = s.en"
                  [selected]="season == s.en">{{s.cn}}</material-select-item>
</material-select>

<h6> 使用單選模型，並手動指定選項 </h6>
<p> 透過單選模型設定值 season 的值：{{season}}</p>
<material-select [width]="2" [selection]="singleSelection">
   <material-select-item value="spring" (trigger)="selected()"> 春 </material-
select-item>
   <material-select-item value="summer" (trigger)="selected()"> 夏 </material-
select-item>
   <material-select-item value="autumn" (trigger)="selected()"> 秋 </material-
select-item>
   <material-select-item value="winter" (trigger)="selected()"> 冬 </material-
select-item>
</material-select>

<h6> 使用單選模型，並提供選項 </h6>
<material-select [selection]="singleSeasonSelection"
                 [options]="seansonOptions"
                 displayNameRenderer
                 width="2"></material-select>

<h6> 使用多選模型，並提供選項 </h6>
<p> 透過多選模型設定值已選項：<span *ngFor="let s of selectedList">{{s.cn }}</
span></p>
```

```
<material-select [selection]="multiSelection"
                 [options]="seansonOptions"
                 displayNameRenderer
                 width="2"></material-select>
```

16.22 工具提示

16.22.1 工具提示指令

工具提示可以附加到任何元素上，背景是水墨。指令名是 MaterialTooltip Directive，作用於範本中帶有 materialTooltip 屬性的元素。

輸入屬性

(1) alignPositionX String：快顯視窗在水平方向上的對齊方式。

(2) alignPositionY String：快顯視窗在垂直方向上的對齊方式。

(3) tooltipPositions List<RelativePosition>：工具提示應嘗試顯示的位置。

(4) materialTooltip String：要在工具提示中顯示的文字。

16.22.2 工具提示卡片

工具提示卡片用於展示目標元素相關的資訊，在提示訊息較多時使用。元件名稱是 MaterialPaperTooltipComponent，在範本中使用元素 <material-tooltip-card> 創建工具提示卡片實例。

其目標可以是任何元素，此元件需與指令 ClickableTooltipTargetDirective 結合使用，且需將 focusContents 設定為 true。

如果工具提示內容是另一個元件，則需使用指令 DeferredContentDirective 載入元件。

支援 header 和 footer 元素作為內容，其他內容將採用工具提示正文樣式。

輸入屬性

(1) focusContents bool：打開時工具提示內容是否應自動聚焦。
(2) offsetX int：工具提示最終定位位置的 x 偏移量。
(3) offsetY int：工具提示最終定位位置的 y 偏移量。
(4) preferredPositions List<RelativePosition>：嘗試顯示工具提示的相對位置。
(5) for TooltipTarget：此工具提示所針對的元素。

16.22.3 工具提示目標指令

工具提示目標指令將元素標記為工具提示的目標，它會立即打開工具提示，且工具提示不會聚焦。指令名是 ClickableTooltipTargetDirective，作用於帶有 clickableTooltipTarget 屬性的元素，匯出為 tooltipTarget。該指令與工具提示元件一起使用，例如 MaterialPaperTooltipComponent，它可以完全控制簡單工具提示的內容。

1. 輸入屬性

(1) alignPositionX String：快顯視窗在水平方向上的對齊方式。
(2) alignPositionY String：快顯視窗在垂直方向上的對齊方式。

2. 輸出屬性

tooltipActivate Stream<bool>：啟動工具提示時觸發的事件。

16.22.4 圖示提示

顯示工具提示的圖示。元件名稱是 MaterialIconTooltipComponent，在範本中使用元素 <material-icon-tooltip> 創建圖示提示實例。

與在 MaterialIconComponent 上顯示 MaterialTooltipCard 大致相同，除此之外它在點擊時也會顯示工具提示，與沒有點擊觸發器的 MaterialTooltip Target 相反。

1. 屬性

(1) icon：圖示的名稱。

(2) size：圖示的大小。

(3) type：圖示的類型。可能的值：help 顯示 help_outline 圖示。info 顯示 info_outline 圖示。error 顯示 error_outline 圖示。

2. 輸入屬性

(1) offsetX int：工具提示最終定位位置的 x 偏移量。

(2) offsetY int：工具提示最終定位位置的 y 偏移量。

(3) preferredPositions List<RelativePosition>：嘗試顯示工具提示的相對位置。

創建 Angular 專案 tooltip_demo，範例程式如下：

```
//chapter16/tooltip_demo/lib/app_component.dart
import 'package:angular/angular.dart';
import 'package:angular_components/angular_components.dart';
@Component(
   selector: 'my-app',
   styleURLs: ['app_component.css'],
   templateURL: 'app_component.html',
   providers: [popupBindings,materialTooltipBindings],
   directives: [
      MaterialTooltipDirective,
      MaterialPaperTooltipComponent,
      ClickableTooltipTargetDirective,
      MaterialIconTooltipComponent,
      MaterialIconComponent],
)
class AppComponent {}
```

範本程式如下：

```
//chapter16/tooltip_demo/lib/app_component.html
<h1> 工具提示 </h1>
<h6> 工具提示指令 </h6>
<material-icon
         icon="print"
         materialTooltip=" 列印 ">
</material-icon>
<p materialTooltip=" 段落 "> 段落元素 </p>

<h6> 工具提示卡片 </h6>
<p>
可點擊的工具提示
   <material-icon
            icon="help_outline"
            clickableTooltipTarget
            size="medium"
            #clickableRef="tooltipTarget">
   </material-icon>
</p>
<material-tooltip-card [for]="clickableRef" focusContents>
   <header> 標頭 </header>
   <p> 提示內容 </p>
   <footer> 註腳 </footer>
</material-tooltip-card>

<span clickableTooltipTarget #spanRef="tooltipTarget"> 懸浮標籤 </span>
<material-tooltip-card [for]="spanRef" focusContents>
   <header> 標頭 </header>
   <p> 普通文字 </p>
   <footer> 註腳 </footer>
</material-tooltip-card>

<h6> 圖示工具提示 </h6>
<p> 附帶圖示工具提示的段落
   <material-icon-tooltip icon="help">
      說明資訊
   </material-icon-tooltip>
</p>
```

16.23 佈局元件

應用佈局是由樣式、指令和元件組成的系統，它根據材質化設計規範實現了應用欄、抽屜和導覽樣式。創建 Angular 專案 layout_demo。

樣式由 package:angular_components/app_layout/layout.scss.css 檔案提供。只需將該檔案增加到 @Component 註釋的 styleURLs 清單中。最好的做法是放在所有樣式檔案前，這樣可以根據需要進行微調。範例程式如下：

```
//chapter16/layout_demo/lib/app_component.dart
import 'package:angular/angular.dart';

@Component(
   selector: 'my-app',
   styleURLs: ['package:angular_components/app_layout/layout.scss.css','app_
component.css'],
   templateURL: 'app_component.html',
)
class AppComponent {}
```

16.23.1 應用欄

應用欄是透過現有 HTML 元素與 CSS 樣式配合使用的，導覽列樣式如表 16-1 所示。

表 16-1 導覽列樣式

class	說明
material-header	將元素作為標頭的容器，即頁眉
shadow	將陰影應用於標頭元素
dense-header	使得標頭內元素更加緊湊
material-header-row	標頭中的行，只能包含一行
material-drawer-button	標頭行中左側的按鈕，用於呼叫出抽屜
material-header-title	標頭的標題

class	說明
material-spacer	放在標題和導覽元素中間，用於填充它們之間的空間
material-navigation	導覽元素，僅與錨標籤 <a> 一起使用

標頭就是常說的導覽列，範例程式如下：

```
//chapter16/layout_demo/lib/app_component.html
<header class="material-header shadow dense-header">
   <div class="material-header-row">
      <material-button icon
                      class="material-drawer-button">
         <material-icon icon="menu"></material-icon>
      </material-button>
      <span class="material-header-title"> 應用欄 </span>
      <div class="material-spacer"></div>
      <nav class="material-navigation">
         <a> 連結 1</a>
      </nav>
      <nav class="material-navigation">
         <a > 連結 2</a>
      </nav>
      <nav class="material-navigation">
         <a > 連結 3</a>
      </nav>
   </div>
</header>
```

執行結果如圖 16-1 所示。

▲ 圖 16-1 導覽列

16.23.2 抽屜

共包含 3 種類型的抽屜：永久、持久和臨時抽屜。所有抽屜都由元素 <material-drawer> 實例化，這些抽屜的實現方式略有差異，以保證各自的最佳性能和使用場景。對於抽屜之外的內容，可由 <material-content> 元素或帶有 CSS 類別 material-content 的元素包裹。

1. 永久抽屜

抽屜會固定在頁面中，不能被關閉。使用時將 permanent 屬性增加到 <material-drawer> 元素上即可。範例程式如下：

```
//chapter16/layout_demo/lib/src/permanent_component.dart
import 'package:angular/angular.dart';
import 'package:angular_components/angular_components.dart';
@Component(
   selector: 'permanent-drawer',
   styleURLs: ['package:angular_components/app_layout/layout.scss.css'],
   templateURL: 'permanent_component.html',
)
class PermanentComponent {

}
```

範本程式如下：

```
//chapter16/layout_demo/lib/src/permanent_component.html
<material-drawer permanent>
   <div>
      <!-- 抽屜內容 -->
      固定抽屜
   </div>
</material-drawer>
<material-content>
   <header class="material-header shadow">
      <div class="material-header-row">
         標頭
      </div>
   </header>
   <div>
      內容
   </div>
</material-content>
```

2. 持久抽屜

持久抽屜可以透過觸發器關閉或打開，例如：按鈕。將 persistent 屬性增加到 <material-drawer> 元素，然後將指令 MaterialPersistentDrawerDirective 增加到元件的指令清單，以實例化持久抽屜。該指令將自身匯出為 drawer，可以透過範本引用變數控制，例如：使用該指令的 toggle() 方法關閉或打開抽屜。抽屜支援結構指令 deferredContent，當抽屜關閉時該指令允許動態增加或刪除抽屜中的內容。抽屜中的內容通常是由導覽組成的，實際上可以增加任何東西。範例程式如下：

```
//chapter16/layout_demo/lib/src/persistent_component.dart
import 'package:angular/angular.dart';
import 'package:angular_components/angular_components.dart';
@Component(
   selector: 'persistent-drawer',
   styleURLs: ['package:angular_components/app_layout/layout.scss.css'],
   templateURL: 'persistent_component.html',
   directives: [
      MaterialButtonComponent,
      MaterialIconComponent,
      MaterialPersistentDrawerDirective,
      DeferredContentDirective,
   ]
)
class PersistentComponent {

}
```

範本程式如下：

```
//chapter16/layout_demo/lib/src/persistent_component.html
<material-drawer persistent #drawer="drawer">
   <div *deferredContent>
      <!-- 抽屜內容 -->
      持久抽屜
   </div>
</material-drawer>
<material-content>
   <header class="material-header shadow">
```

```
        <div class="material-header-row">
            <!-- 透過抽屜的 toggle() 方法關閉或打開抽屜 -->
            <material-button icon
                    class="material-drawer-button" (trigger)="drawer.toggle()">
                <material-icon icon="menu"></material-icon>
            </material-button>
            <!-- 其他標頭資訊 -->
        </div>
    </header>
    <div>
        可透過點擊導覽列選單按鈕打開或關閉抽屜
    </div>
</material-content>
```

3. 臨時抽屜

臨時抽屜是位於所有頁面內容上方的抽屜。使用時將 MaterialTemporary DrawerComponent 增加到指令清單，並將 temporary 屬性應用於 <material-drawer> 元素。

臨時抽屜有一個可選的屬性 overlay，當抽屜打開時，在非抽屜內容上方顯示半透明的隱藏。範例程式如下：

```
//chapter16/layout_demo/lib/src/temporary_component.dart
import 'package:angular/angular.dart';
import 'package:angular_components/angular_components.dart';
@Component(
    selector: 'temporary-drawer',
    styleURLs: ['package:angular_components/app_layout/layout.scss.css'],
    templateURL: 'temporary_component.html',
    directives: [
        MaterialButtonComponent,
        MaterialIconComponent,
        MaterialTemporaryDrawerComponent
    ]
)
class TemporaryComponent {

}
```

範本程式如下：

```
//chapter16/layout_demo/lib/src/temporary_component.html
<material-drawer temporary #drawer="drawer" overlay>
   <div>
      <!-- 抽屜內容 -->
      臨時抽屜
   </div>
</material-drawer>
<material-content>
   <header class="material-header shadow">
      <div class="material-header-row">
         <!-- 透過 toggle 關閉或打開抽屜 -->
         <material-button icon
                  class="material-drawer-button" (trigger)="drawer.toggle()">
            <material-icon icon="menu"></material-icon>
         </material-button>
         <!-- 其他標頭資訊 -->
      </div>
   </header>
   <div>
      內容
   </div>
</material-content>
```

所有抽屜都具有 HTML 屬性 end，該屬性將抽屜放置在另一側。如果原來抽屜在左側則放在右側，如果原來在右側則放在左側。範例程式如下：

```
<material-drawer temporary end>
</material-drawer>
```

4. 應用欄與抽屜互動

應用欄常與抽屜協作工作，以滿足應用程式整體佈局。應用程式欄可以位於 <material-content> 元素的內部，也可以位於外部。如果位於 <material-content> 的內部，它將與內容佈局在一起。如果位於 <material-content> 的外部，永久和持久抽屜及內容將位於應用欄下方。導覽列位於抽屜及內容上方的範例程式如下：

```
<header class="material-header shadow dense-header">
  <div class="material-header-row">
    <material-button icon
            class="material-drawer-button" (trigger)="drawer.toggle()">
      <material-icon icon="menu"></material-icon>
    </material-button>
    <span class="material-header-title">應用欄</span>
    <div class="material-spacer"></div>
    <nav class="material-navigation">
      <a>連結 1</a>
    </nav>
  </div>
</header>

<material-drawer persistent #drawer="drawer" overlay>
  <!-- 抽屜內容 -->
</material-drawer>
<material-content>
  <!-- 內容 -->
</material-content>
```

5. 導覽樣式

佈局樣式表中還提供了抽屜內的導覽元素樣式，導覽元素由清單元件和特殊的 CSS 類別組成。將清單元件 MaterialListComponent 作為抽屜的子元素，在清單元件中透過組元素將內容分組，組元素是在元素上使用 group 屬性指定的。

CSS 類別 mat-drawer-spacer 是可選的，當應用欄位於元素 <material-content> 內部時，抽屜內容將留出與應用欄等高的空間，使用 dense-header 屬性的導覽列不適用。

使用元件 MaterialListItemComponents 作為抽屜中的項目，項目通常放在組元素中，如果每個組需要標籤，則可以在區塊元素上使用 label 屬性。範例程式如下：

```
<material-drawer permanent>
   <material-list>
      <!-- 空出與應用欄等高的空間 -->
      <div group class="mat-drawer-spacer"></div>
      <!-- 不帶標籤的組元素 -->
      <div group>
         <material-list-item>
            <material-icon icon="inbox"></material-icon>Inbox
         </material-list-item>
         <material-list-item>
            <material-icon icon="star"></material-icon>Star
         </material-list-item>
      </div>
      <!-- 帶標籤的組元素 -->
      <div group>
         <div label>Tags</div>
         <material-list-item>
            <material-icon icon="star"></material-icon>Favorites
         </material-list-item>
      </div>
   </material-list>
</material-drawer>
```

6. 堆疊式抽屜

持久抽屜和臨時抽屜具有輸入屬性 visible 和輸出屬性 visibleChange。因此可以使用雙向資料綁定控制抽屜的打開與關閉，其效果與呼叫 toggle() 方法一致。

堆疊式抽屜可以在抽屜中放置另一個抽屜，並且可以一直巢狀結構下去。為了好的互動體驗，建議最多使用兩個抽屜。堆疊式抽屜繼承了臨時抽屜，因此也具有輸入屬性 visible 和輸出屬性 visibleChange。堆疊式抽屜只能透過雙向資料綁定到 visible 屬性來控制抽屜的打開與關閉。

將 MaterialStackableDrawerComponent 增加到元件的指令清單，並定義兩個布林屬性，用於初始兩個抽屜的 visible 屬性。範例程式如下：

```
//chapter16/layout_demo/lib/src/stackable_component.dart
import 'package:angular/angular.dart';
import 'package:angular_components/angular_components.dart';

@Component(
  selector: 'stackable-drawer',
  styleURLs: ['package:angular_components/app_layout/layout.scss.css'],
  templateURL: 'stackable_component.html',
  directives: [
    MaterialButtonComponent,
    MaterialIconComponent,
    MaterialStackableDrawerComponent,
    MaterialListComponent,
    MaterialListItemComponent
  ]
)
class StackableComponent {
  // 用於控制抽屜 1 的打開或關閉
  bool drawer1Visable = false;
  // 用於控制抽屜 2 的打開或關閉
  bool drawer2Visable = false;
}
```

範本程式如下：

```
//chapter16/layout_demo/lib/src/stackable_component.html
<header class="material-header shadow">
  <div class="material-header-row">
    <material-button icon
        class="material-drawer-button" (trigger)="drawer1Visable = true">
      <material-icon icon="menu"></material-icon>
    </material-button>
    <span class="material-header-title"> 應用欄 </span>
    <div class="material-spacer"></div>
    <nav class="material-navigation">
      <a> 連結 1</a>
    </nav>
    <nav class="material-navigation">
      <a> 連結 2</a>
```

```
    </nav>
    <nav class="material-navigation">
      <a > 連結 3</a>
    </nav>
  </div>
</header>
<material-drawer stackable [(visible)]="drawer1Visable">
  <!-- 抽屜內容 -->
  <material-list>
    <!-- 空出與應用欄等高的空間 -->
    <div group class="mat-drawer-spacer"></div>
    <!-- 不帶標籤的組元素 -->
    <div group>
      <material-list-item>
        <material-icon icon="inbox"></material-icon>Inbox
      </material-list-item>
      <material-list-item>
        <material-icon icon="star"></material-icon>Star
      </material-list-item>
      </div>
      <!-- 帶標籤的組元素 -->
      <div group>
        <div label>Tags</div>
        <material-list-item>
          <material-icon icon="star"></material-icon>Favorites
        </material-list-item>
    </div>
  </material-list>
  <div> 第一個抽屜 </div>
  <material-button (trigger)="drawer2Visable = true"> 顯示第二個抽屜
  </material-button>
  <material-drawer stackable [(visible)]="drawer2Visable">
    <!-- 抽屜內容 -->
    <div> 第二個抽屜 </div>
  </material-drawer>
</material-drawer>
<material-content>
  <!-- 內容 -->
</material-content>
```

將上述所有抽屜元件增加到根元件的指令清單，更新根元件。範例程式如下：

```
//chapter16/layout_demo/lib/app_component.dart
import 'package:angular/angular.dart';
import 'package:angular_components/angular_components.dart';

import 'src/permanent_component.dart';
import 'src/persistent_component.dart';
import 'src/stackable_component.dart';
import 'src/temporary_component.dart';

@Component(
  selector: 'my-app',
  styleURLs: ['package:angular_components/app_layout/layout.scss.css','app_
component.css'],
  templateURL: 'app_component.html',
  directives: [
    PermanentComponent,
    PersistentComponent,
    TemporaryComponent,
    StackableComponent,
    MaterialButtonComponent,
    MaterialIconComponent,
    NgSwitch,
    NgSwitchWhen,
    NgSwitchDefault
  ]
)
class AppComponent {
  // 快取當前抽屜名
  String currentDrawer = '';
  // 切換抽屜範例
  switchDrawer(String drawer){
    currentDrawer = drawer;
  }
}
```

範本程式如下：

```
//chapter16/layout_demo/lib/app_component.html
<div [ngSwitch]="currentDrawer">
   <permanent-drawer *ngSwitchCase="'permanent'"></permanent-drawer>
   <persistent-drawer *ngSwitchCase="'persistent'"></persistent-drawer>
   <temporary-drawer *ngSwitchCase="'temporary'"></temporary-drawer>
   <stackable-drawer *ngSwitchCase="'stackable'"></stackable-drawer>
   <div *ngSwitchDefault>
      <header class="material-header shadow dense-header">
      <div class="material-header-row">
         <material-button icon
                        class="material-drawer-button">
         <material-icon icon="menu"></material-icon>
         </material-button>
         <span class="material-header-title"> 應用欄 </span>
         <div class="material-spacer"></div>
         <nav class="material-navigation">
            <a> 連結 1</a>
         </nav>
         <nav class="material-navigation">
            <a> 連結 2</a>
         </nav>
         <nav class="material-navigation">
            <a> 連結 3</a>
         </nav>
      </div>
      </header>
   </div>
</div>
<div style="width: 600px;margin:10px auto;">
   <material-button raised (trigger)="switchDrawer('permanent')"> 切換到永久抽屜
</material-button>
   <material-button raised (trigger)="switchDrawer('persistent')"> 切換到持久抽屜
</material-button>
   <material-button raised (trigger)="switchDrawer('temporary')"> 切換到臨時抽屜
</material-button>
```

```
    <material-button raised (trigger)="switchDrawer('stackable')">切換到堆疊式
抽屜
</material-button>
    <material-button raised (trigger)="switchDrawer('')">切換到應用欄
</material-button>
</div>
```

刷新瀏覽器，點擊不同按鈕切換到不同的應用程式佈局，以觀察不同抽屜的使用範例。

Chapter 17

專案實戰 Deadline

前面學習了 Dart 基礎、服務端程式設計和 Web 框架 Angular 的相關知識，本章透過專案實戰將這些內容貫穿起來。

實戰專案 Deadline 用於規劃大小交易，並記錄完成時間。

17.1 MySQL 資料庫

資料庫負責對資料進行管理、維護和使用。主流的資料庫有 Oracle、SQL Server、DB 2、Sysbase 和 MySQL 等，本節介紹 MySQL 資料庫的安裝與使用。

17.1.1 資料庫安裝

MySQL 資料庫由 Oracle 公司負責提供技術支援和維護，並提供多個版本供選擇，其中社區版 MySQL Community Edition 適合於個人開發者和中小型公司，本書也採用該版本用於專案實戰中。

社區版提供多個平台版本，以滿足在 Windows、Linux 和 macOS 等作業系統上安裝和執行。本書以 Windows 作業系統為例，其他作業系統安裝步驟類似，下載網址是 https://dev.MySQL.com/downloads/Windows/installer/8.0.html，如圖 17-1 所示。

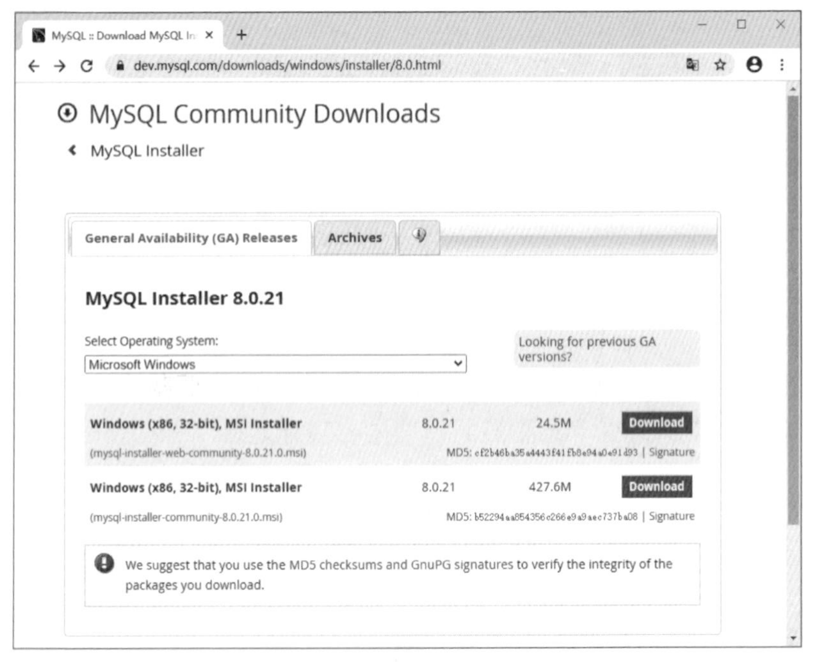

▲ 圖 17-1　社區版 MySQL 下載

此版本下載的安裝檔案是 mysql-installer-community-8.0.21.0.msi，雙擊該檔案啟動安裝過程，這裡對重要步驟說明。

1. 安裝類型

如圖 17-2 所示是安裝類型選擇對話方塊。在此對話方塊中包含 5 種可選擇的安裝類型：Developer Default、Server only、Client only、Full 和 Custom。本書選擇 Server only 並點擊 Next 按鈕。

進入步驟 Installation，點擊 Execute 按鈕。待狀態處於 Complete 時，繼續點擊 Next 按鈕。進入步驟 Product Configuration，繼續點擊 Next 按鈕。

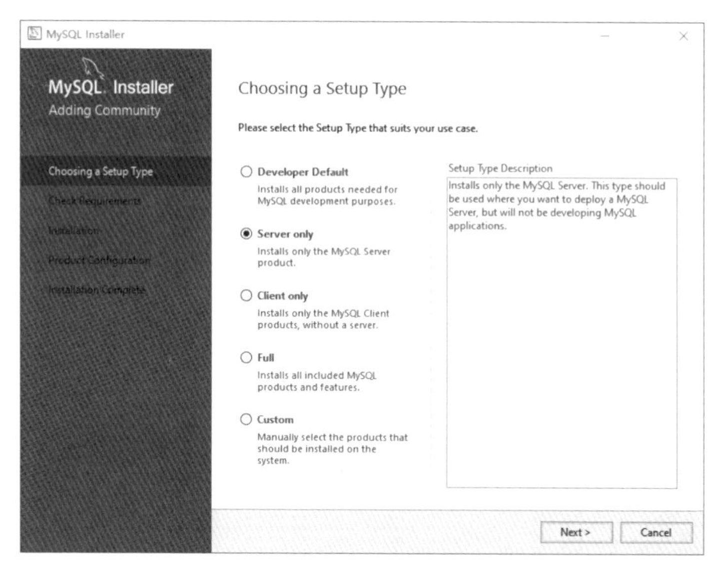

▲ 圖 17-2 安裝類型選擇對話方塊

2. 設定

所需檔案安裝完成後，會進入 MySQL 的設定過程。如圖 17-3 所示是資料庫類型選擇對話方塊，Standalone 是單一伺服器，InnoDB Cluster 是資料庫叢集。

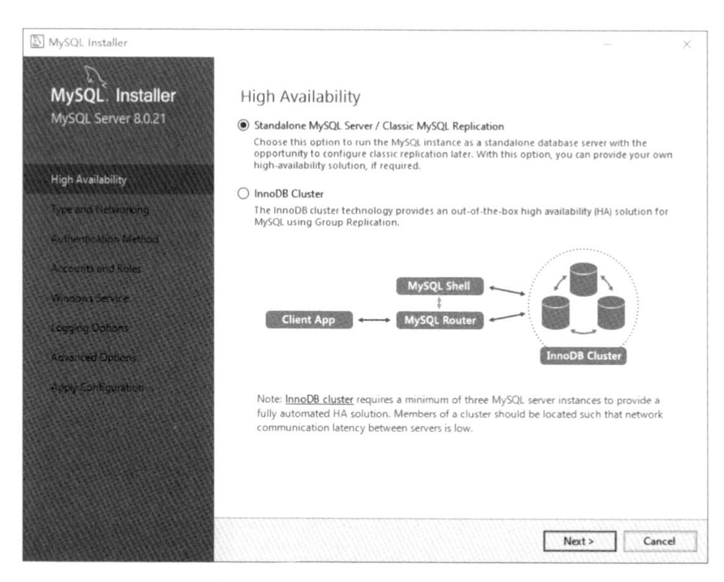

▲ 圖 17-3 資料庫類型選擇對話方塊

選擇 Standalone 即可，點擊 Next 按鈕進入如圖 17-4 所示的伺服器設定類型對話方塊。

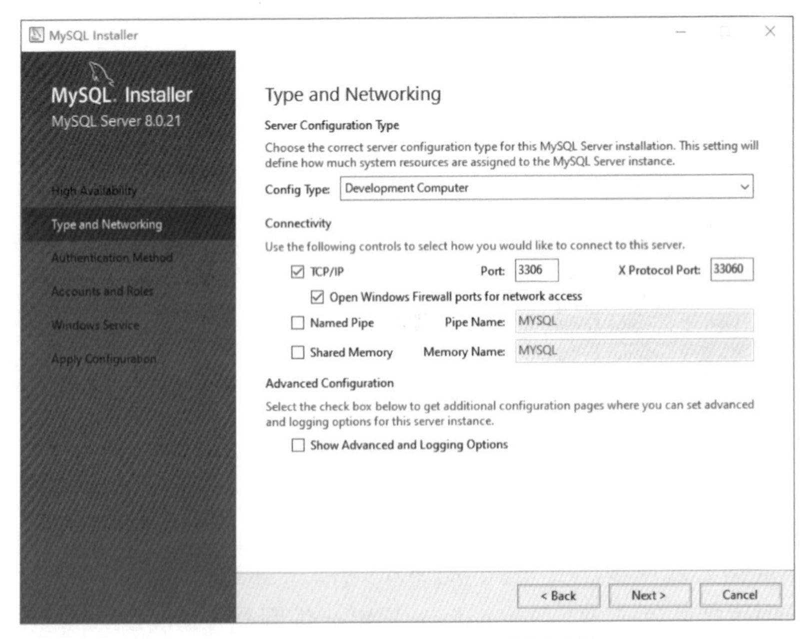

▲ 圖 17-4 伺服器設定類型對話方塊

在此對話方塊可以選擇設定類型、通訊協定和通訊埠等。點擊 Config Type 下拉清單可以選擇以下設定類型：

(1) Development Computer：該選項代表個人用桌面工作站，該設定將使用最少的系統資源執行 MySQL 伺服器。
(2) Server Computer：該選項代表伺服器，該設定為 MySQL 伺服器分配適當的系統資源。
(3) Dedicated Computer：該選項表示伺服器只允許 MySQL 服務，該設定為 MySQL 伺服器分配所有可用的系統資源。

這裡選擇 Development Computer，點擊 Next 按鈕進入如圖 17-5 所示的授權方式對話方塊。它包含以下兩種授權方式：

(1) Use Strong Password Encryption for Authentication：使用強式密碼加密授權。

(2) Use Legacy Authentication Method：傳統授權方法，保留對 5.x 版本的相容性。

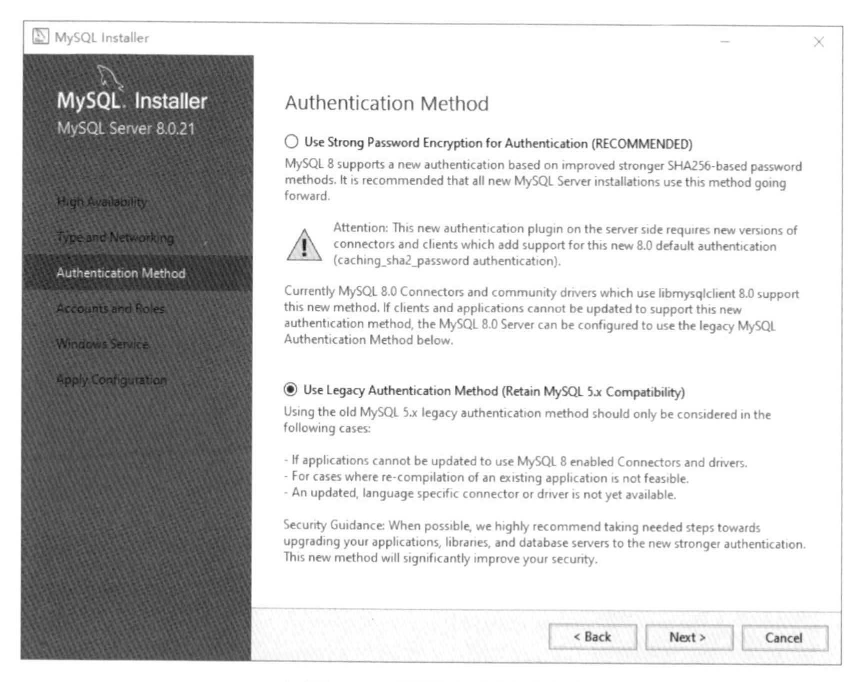

▲ 圖 17-5 授權方式對話方塊

這裡選擇傳統授權方法，點擊 Next 按鈕進入如圖 17-6 所示的帳號和角色設定對話方塊。在該對話方塊中設定帳號 root 的密碼，也可以增加其他帳號。

這裡設定好帳號 root 的密碼後，點擊 Next 按鈕進入如圖 17-7 所示的設定 Windows 服務對話方塊。在此對話方塊中可以將 MySQL 資料庫設定成為一個 Windows 服務，Windows 服務可以在後台隨著 Windows 系統的啟動而啟動。當前版本預設的服務名稱是 MySQL80。

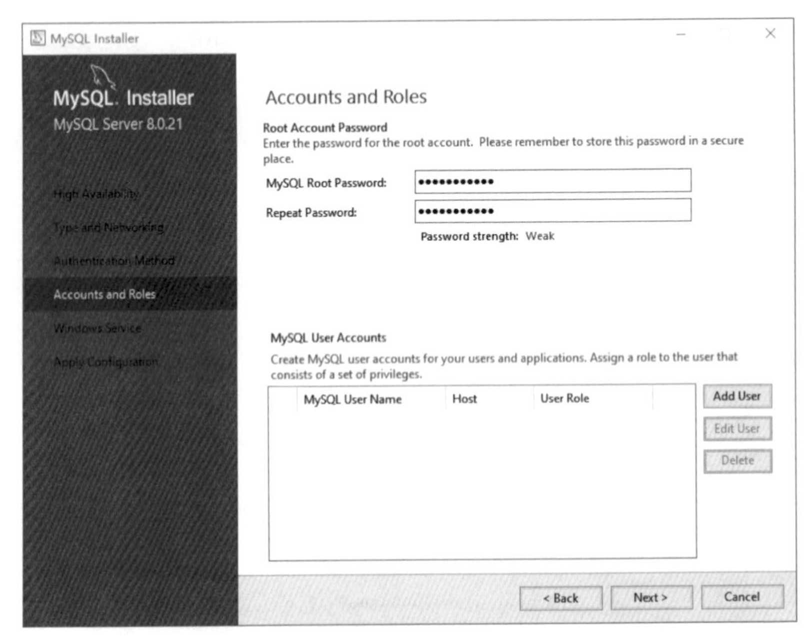

▲ 圖 17-6 帳號和角色設定對話方塊

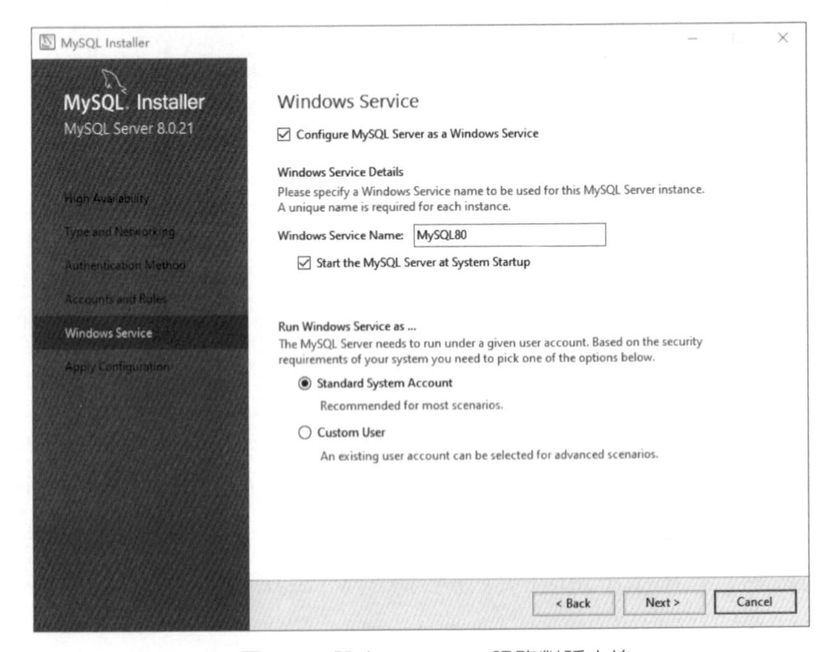

▲ 圖 17-7 設定 Windows 服務對話方塊

完成如圖 17-7 所示的設定後，就不需要進行手動設定了，只需點擊 Next 按鈕即可。

17.1.2 資料庫連接

MySQL 是用戶端 / 伺服器結構的，所以應用程式必須連接到伺服器才能使用其服務功能。下面介紹使用 MySQL 自己的用戶端和第三方用戶端連接到伺服器。

1. 附帶用戶端

MySQL for Windows 版本提供一個選單項目可以快速連接伺服器，打開步驟：按右鍵螢幕左下角的 Windows 圖示，在應用清單中找到 MySQL 8.0 Command Line Client，點擊後會打開一個終端視窗，如圖 17-8 所示。

▲ 圖 17-8 MySQL 命令列用戶端

這個工具就是 MySQL 命令列用戶端工具，可以使用 MySQL 命令列用戶端工具連接到 MySQL 伺服器，要求輸入 root 密碼。輸入密碼後按 Enter 鍵，如果密碼正確，則連接到 MySQL 伺服器，如圖 17-9 所示。

▲ 圖 17-9 命令列用戶端連接到伺服器

2. Navicat

Navicat forMySQL 是管理和開發 MySQL 的理想解決方案，它是一套單一的應用程式。這套全面的前端工具為資料庫管理、開發和維護提供了一款直觀而強大的圖形介面。

下載網址是 http://www.navicat.com.cn/download/navicat-for-MySQL，它提供了 Windows、Linux、macOS 3 個平台的用戶端供選擇。雙擊下載的檔案即可安裝應用，安裝過程很簡單，一直點擊「下一步」按鈕即可。

打開 Navicat 用戶端，在介面中點擊「連接」按鈕，在下拉清單中選擇 MySQL，如圖 17-10 所示。點擊 MySQL 選項後會彈出新建連接對話方塊，在對話方塊中輸入自訂連接名和帳號 root 的密碼，最後點擊「確定」按鈕即可，如圖 17-11 所示。

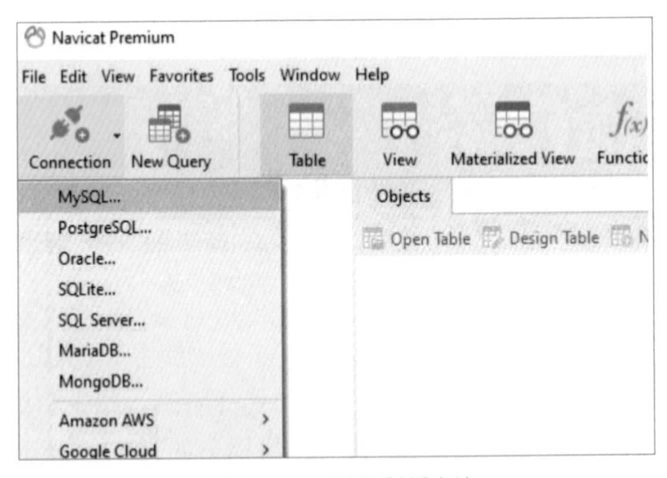

▲ 圖 17-10 連接對話方塊

MySQL80

localhost

3306

root

●●●●●●●●●●●●

▲ 圖 17-11 新建 MySQL 連接對話方塊

建立新連接後，雙擊連接名以建立連接，按右鍵連接名在選單中選擇新建資料庫，打開新建資料庫對話方塊，如圖 17-12 所示。在資料庫名稱處輸入 deadline，字元集是 utf8mb4，點擊「確定」按鈕完成資料庫的建立。

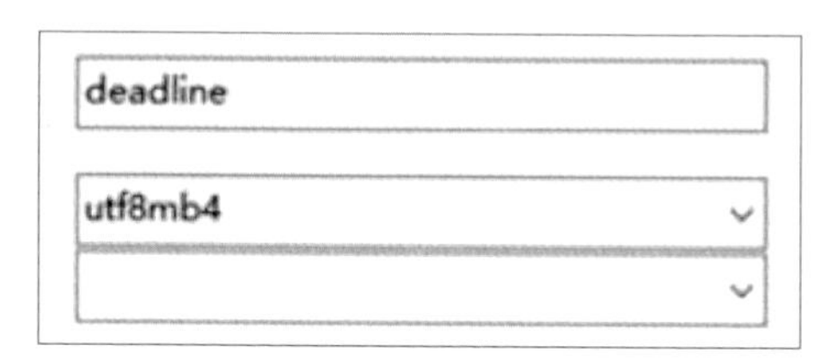

▲ 圖 17-12　新建資料庫對話方塊

雙擊資料庫 deadline，在展開的表上按右鍵，在選項選單中選擇新建表 plan，資料結構如圖 17-13 所示。

id	int		☑	☐	🔑1
title	varchar	255	☑	☐	
remarks	varchar	255	☐	☐	
plantime	timestamp		☑	☐	
endtime	timestamp		☐	☐	
theme	varchar	10	☐	☐	
complete	tinyint	1	☐	☐	

▲ 圖 17-13　新建表 plan

計畫清單表結構如表 17-1 所示。

表 17-1　計畫表

欄位名稱	資料類型	長度	主鍵	備註
id	int	0	是	計畫 id
title	varchar	255	否	標題
remarks	varchar	255	否	備註
plantime	timestamp	0	否	計畫完成時間
endtime	timestamp	0	否	實際完成時間
theme	varchar	10	否	主題
complete	tinyint	1	否	完成狀態

17.2 資料庫連接套件

套件 MySQL1 提供了應用程式與 MySQL 資料庫通訊的能力，可以透過 SQL 敘述查詢、增加、刪除和更新資料。

創建 shelf 應用程式，命名為 deadline_server。增加依賴並執行 pub get 命令：

```
//chapter17/deadline_server/pubspec.yaml
name: deadline_server
description: A web server built using the shelf package.
#version: 1.0.0
#homepage: https://www.example.com

environment:
  #sdk 版本依賴資訊
  sdk: '>=2.7.0 <3.0.0'

dependencies:
  #shelf 框架套件
  shelf: ^0.7.5
  #shelf 框架路由套件
  shelf_router: ^0.7.2
  #shelf 框架異常處理套件
  shelf_exception_handler: ^0.2.0
  #JSON 和 Dart 物件相互轉換套件
  json_string: ^2.0.1
  #MySQL 資料庫驅動
  MySQL1: ^0.17.1

dev_dependencies:
  # 路由生成器
  shelf_router_generator: ^0.7.2+2
  build_runner: ^1.3.1
```

在專案的 test 目錄下創建 MySQL_test.dart 檔案，將在該檔案中演示如何連接資料庫、執行 SQL 敘述，以及關閉連接。

17.2.1 連接設定

套件 MySQL1 中的類別 ConnectionSettings 用於設定連接資訊，常用屬性和方法如下：

(1) host：資料庫所在主機位址，預設為 localhost。
(2) port：資料庫對外開放的通訊埠，MySQL 資料庫預設是 3306 通訊埠。
(3) user：可以存取資料庫的用戶名。
(4) password：用戶名對應的密碼。
(5) db：資料庫名稱。
(6) ConnectionSettings({String host: 'localhost', int port: 3306, String user, String password, String db, bool useCompression: false, bool useSSL: false, int maxPacketSize: 16 * 1024 * 1024, Duration timeout: const Duration(seconds: 30), int characterSet: CharacterSet.UTF8MB4})：建構函數，透過指定參數創建 ConnectionSettings 實例。

範例程式如下：

```
// 資料庫連接設定
   var settings = ConnectionSettings(
   host: 'localhost',
   port: 3306,
   user: 'root',
   password: 'rootflzx3QC',
   db: 'deadline'
);
```

17.2.2 連接與執行

套件 MySQL1 中的類別 MySQLConnection 用於與資料庫連接。使用 connect 類別方法打開連接，完成後，必須呼叫 close 方法關閉連接。以下是一些常用方法：

(1) connect(ConnectionSettings c)：類別方法。接收 ConnectionSettings 類型的資料庫設定資訊連接到對應資料庫，返回 MySQLConnection 的實例物件並包裝在 Future 中。

(2) close()：關閉資料庫連接。

(3) query(String sql, [List<Object> values])：在資料庫上執行 sql 查詢，並將 values 拼湊到對應 sql 參數上。返回由 Future 封裝的 Results 物件。

(4) queryMulti(String sql, Iterable<List<Object>> values)：為 values 中每組對應 sql 參數執行一次 sql 查詢。返回由 Future 封裝的 List<Results> 物件。

範例程式如下：

```
// 返回資料庫連線物件
var conn = await MySQLConnection.connect(settings);
// 儲存當前時間
var datetime = DateTime.now().toUtc();
// 插入資料
var result1 = await conn.query('insert into plan(title,remarks,plantime)
values(?,?,?)',
      ['新計畫','備註資訊',datetime]);
   // 批次插入資料
var result2 = await conn.queryMulti('insert into plan(title,remarks,plantime)
values(?,?,?)',
   [
      ['新計畫 1','備註資訊 1',datetime],
      ['新計畫 2','備註資訊 2',datetime],
      ['新計畫 3','備註資訊 3',datetime]
   ]
);
```

17.2.3 結果集

Results 是執行 query 或 queryMulti 方法返回的結果行的可疊代物件。常用屬性和方法如下：

(1) affectedRows：返回受影響的行數。

(2) fields：返回欄位的清單。
(3) insertId：對於插入敘述如果表中有一個自動增加長的列，則會返回該值。否則為空。
(4) isEmpty：如果返回結果中沒有元素則返回 true。
(5) length：返回查詢結果中元素的數量。
(6) single：檢查此可疊代物件是否只有一個元素，然後返回該元素。如果為空或具有多個元素，則拋出 StateError。
(7) forEach(void f(Row element))：按疊代順序將函數 f 應用於此集合的每個元素。

Row 表示一行資料，可以透過索引或名字來檢索欄位。按名字檢索欄位時，欄位必須是有效的 Dart 識別符號，且不能是 List 物件的欄位。常用屬性和方法如下：

(1) fields：返回欄位和值組成的 Map 集合。
(2) values：返回由值組成的 List 集合。

範例程式如下：

```
//result1 是 Results 類型的資料
print(' 受影響的行 :${result1.affectedRows}');
print(' 自動增加長列返回的 id:${result1.insertId}');

// 查詢資料
var result3 = await conn.query('select * from plan');

print(' 返回資料中包含表的欄位 :${result3.fields}');
print(' 返回資料中包含的筆數 :${result3.length}');

//result3 是 List<Results> 類型的資料
result3.forEach((row){
   // 列印由欄位和值組成的 Map
   print(row.fields);
   // 列印值組成的 List
   print(row.values);
});
```

完整資料庫測試程式如下：

```
//chapter17/deadline_server/bin/MySQL1_test.dart
import 'package:MySQL1/MySQL1.dart';

void main() async{
   // 資料庫連接設定
   var settings = ConnectionSettings(
      host: 'localhost',
      port: 3306,
      user: 'root',
      password: 'rootflzx3QC',
      db: 'deadline'
   );
   // 返回資料庫連線物件
   var conn = await MySQLConnection.connect(settings);
   // 儲存當前時間
   var datetime = DateTime.now().toUtc();
   // 插入資料
   var result1 = await conn.query('insert into plan(title,remarks,plantime)
values(?,?,?)',
      [' 新計畫 ',' 備註資訊 ',datetime]);
   // 批次插入資料
   var result2 = await conn.queryMulti('insert into
plan(title,remarks,plantime) values(?,?,?)',
      [
         [' 新計畫 1',' 備註資訊 1',datetime],
         [' 新計畫 2',' 備註資訊 2',datetime],
         [' 新計畫 3',' 備註資訊 3',datetime]
      ]
   );

   //result1 是 Results 類型的資料
   print(' 受影響的行 :${result1.affectedRows}');
   print(' 自動增加長列返回的 id:${result1.insertId}');

   // 查詢資料
   var result3 = await conn.query('select * from plan');
```

```
    print('返回資料中包含表的欄位：${result3.fields}');
    print('返回資料中包含的筆數：${result3.length}');

    //result3 是 List<Results> 類型的資料
    result3.forEach((row){
      //列印由欄位和值組成的 Map
      print(row.fields);
      //列印值組成的 List
      print(row.values);
    });

}
```

17.2.4 工具類別

現在編寫一個工具類別，用於返回 MySQLConnection 的實例，這樣就不需要每次連接資料庫都創建設定資訊。範例程式如下：

```
//chapter17/deadline_server/bin/utils/dbconn.dart
import 'package:MySQL1/MySQL1.dart';

class Db{
  //返回資料庫連線物件 MySQLConnection
  static Future<MySQLConnection> conn()async{
    //資料庫連接設定
    var settings = ConnectionSettings(
      host: 'localhost',
      port: 3306,
      user: 'root',
      password: 'rootflzx3QC',
      db: 'deadline'
    );
    return await MySQLConnection.connect(settings);
  }
}
```

因為 conn 方法是類別方法，使用時可以直接透過類別 Db 呼叫。

17.3 編寫服務端

使用 shelf 套件創建專案，並命名為 deadline_server。

17.3.1 實體類別

首先定義資料模型類別 Plan，該類別的屬性與表 plan 的欄位一一映射。範例程式如下：

```
//chapter17/deadline_server/bin/entity.dart
import 'package:json_string/json_string.dart';

class Plan with Jsonable{
   final int id;
   //計畫的標題
   String title;
   //計畫的備註資訊
   String remarks;
   //計畫完成時間
   DateTime plantime;
   //實際完成時間，預設
   DateTime endtime;
   //計畫的標記主題
   String theme;
   //計畫完成狀態，預設未完成，0表示未完成，1表示完成
   int complete;
   Plan({this.id,this.title,this.remarks,this.plantime,this.endtime,this.
theme,this.complete = 0});
   //將Dart物件轉為JSON資料
   @override
   Map<String, dynamic> toJson() {
      return {
         'id':'$id',
         'title':title,
         'remarks':remarks,
         'plantime':plantime,
         'endtime':endtime,
```

```
            'theme':theme,
            'complete':'$complete',
        };
    }
    //將 JSON 資料轉為 Dart 物件
    static Plan fromJson(Map<String,dynamic> json){
        return Plan(
            id:json['id'],
            title:json['title'],
            remarks:json['remarks'],
            plantime:DateTime.parse(json['plantime']).toUtc(),
            endtime:json['endtime']== null ? null:DateTime.
parse(json['endtime']).toUtc(),
            theme:json['theme'],
            complete:json['complete']
        );
    }
}
```

該類別使用了 json_string 套件提供的 Mixin 類別 Jsonable，用於將 Dart 類別轉為 Json 資料。提供的 fromJson 方法用於將 Json 資料轉化為 Plan 物件的實例。

17.3.2 服務類別

提供對計畫進行增、刪、改、查的服務類別。範例程式如下：

```
//chapter17/deadline_server/bin/service.dart
import 'entity.dart';
import 'utils/dbconn.dart';

class PlanService {
    //創建新計畫
    Future<int> create(Plan plan) async {
        var conn = await Db.conn();
        try {
            var result = await conn.query(
                'insert into plan(title,remarks,plantime,theme) '
```

```
            'values(?,?,?,?)',
            [plan.title, plan.remarks, plan.plantime, plan.theme]);
        // 創建成功，返回插入 id
        if (result.affectedRows > 0) {
            return result.insertId;
        }
        // 創建失敗，返回 0
        return 0;
    } catch (e) {
        rethrow;
    } finally {
        await conn.close();
    }
}

// 獲取所有計劃
Future<List<Map>> getALL() async {
    var conn = await Db.conn();
    var list = <Map>[];
    try {
        var result = await conn.query('select * from plan');
        result.forEach((row) {
            // 修正 MySQL1 套件對於 UTC 時間的解析 Bug
            row.fields.update('plantime', (item) {
                if (item == null) return null;
                return DateTime.parse(item.toLocal().toIso8601String() + 'Z');
            });
            row.fields.update('endtime', (item) {
                if (item == null) return null;
                return DateTime.parse(item.toLocal().toIso8601String() + 'Z');
            });
            // 將單行資料增加到 List 集合
            list.add(row.fields);
        });
        // 返回所有資料
        return list;
    } catch (e) {
        rethrow;
    } finally {
```

```
      await conn.close();
    }
  }

  // 獲取指定 id 的計畫資訊
  Future<Map<String, dynamic>> getById(int id) async {
    var conn = await Db.conn();
    try {
      var result = await conn.query('select * from plan where id=?', [id]);
      // 如果結果不為空，則返回單筆資料
      if (result.isNotEmpty) return result.single.fields;
      // 否則返回 null
      return null;
    } catch (e) {
      rethrow;
    } finally {
      await conn.close();
    }
  }

  // 更新指定 id 的計畫資訊
  Future<bool> update(Plan plan) async {
    var conn = await Db.conn();
    try {
      var result = await conn.query(
        'update plan set title=?,remarks=?,plantime=?,theme=? where id=?',
        [plan.title, plan.remarks, plan.plantime, plan.theme, plan.id]);
      // 如果更新成功，則返回 true
      if (result.affectedRows > 0) return true;
      // 如果更新失敗，則返回 false
      return false;
    } catch (e) {
      rethrow;
    } finally {
      await conn.close();
    }
  }

  // 更新指定 id 的計畫資訊的狀態
```

```
    Future<bool> updateStatus(int id, int complete) async {
      var conn = await Db.conn();
      try {
        var endtime;
        if (complete == 1) {
          // 完成狀態時確定實際完成時間
          endtime = DateTime.now().toUtc();
        }
        var result = await conn.query(
          'update plan set endtime=?,complete=? where id=?',
          [endtime, complete, id]);
        // 如果更新成功，則返回 true
        if (result.affectedRows > 0) return true;
        // 如果更新失敗，則返回 false
        return false;
      } catch (e) {
        rethrow;
      } finally {
        await conn.close();
      }
    }

    // 更新指定 id 刪除計畫
    Future<bool> delete(int id) async {
      var conn = await Db.conn();
      try {
        var result = await conn.query(
          'delete from plan where id=?',[id]);
        // 如果刪除成功，則返回 true
        if (result.affectedRows > 0) return true;
        // 如果刪除失敗，則返回 false
        return false;
      } catch (e) {
        rethrow;
      } finally {
        await conn.close();
      }
    }
}
```

17.3.3 時間轉換類別

在實際專案中需要對日期類型的資料轉碼為字串。程式如下：

```
//chapter17/deadline_server/bin/utils/to_encodable.dart
dynamic myEncode(dynamic item){
   if(item is DateTime){
      return item.toLocal().toIso8601String();
   }
   return item;
}
```

17.3.4 路由器

採用路由註釋編寫路由器，用於回應用戶端請求。範例程式如下：

```
//chapter17/deadline_server/bin/routers.dart
import 'dart:convert';
import 'package:shelf/shelf.dart';
import 'package:shelf_router/shelf_router.dart';
import 'package:json_string/json_string.dart';
import 'service.dart';
import 'entity.dart';
import 'utils/to_encodable.dart';
part 'routers.g.dart';
// 定義路由器
class Dl{
   // 實例化服務 PlanService
   final PlanService _planService = PlanService();

   //GET 請求，獲取所有計劃
   @Route.get('/plans')
   Future<Response> _getAll(Request request) async{
      var dls = await _planService.getALL();
      return Response.ok(jsonEncode(dls,toEncodable: myEncode));
   }

   //GET 請求，獲取指定 id 的計畫資訊
```

```
@Route.get('/plan/<id>')
Future<Response> _getById(Request request) async{
  var id = int.parse(params(request,'id'));
  var dl = await _planService.getById(id);
  return Response.ok(jsonEncode(dl,toEncodable: myEncode));
}

//POST 請求，創建新計畫
@Route.post('/plan')
Future<Response> _create(Request request) async{
  var plan1,plan2;
  await utf8.decoder.bind(request.read()).join().then((content){
    // 將 Json 內容解碼為 Employee 實例
    plan1 = JsonString(content).decodeAsObject(Plan.fromJson);
  });
  // 創建計畫並快取插入 id
  var id = await _planService.create(plan1);
  // 如果創建成功則透過插入 id 獲取創建的計畫
  if(id != null && id != 0){
    plan2 = await _planService.getById(id);
  }
  return Response.ok(jsonEncode(plan2,toEncodable: myEncode));
}

//PUT 請求，更新指定 id 的計畫資訊
@Route.put('/plan/<id>')
Future<Response> _update(Request request) async{
  var id = int.parse(params(request,'id'));
  var plan1,plan2;
  await utf8.decoder.bind(request.read()).join().then((content){
    // 將 Json 內容解碼為 Employee 實例
    plan1 = JsonString(content).decodeAsObject(Plan.fromJson);
  });
  // 更新計畫並返回結果
  var result = await _planService.update(plan1);
  // 如果更新成功則透過 id 獲取更新後的計畫
  if(id != null && result){
    plan2 = await _planService.getById(id);
  }
```

```
    return Response.ok(jsonEncode(plan2,toEncodable: myEncode));
}

//PUT 請求，更新指定 id 的計畫的完成狀態
@Route.put('/plan/<id>/<status>')
Future<Response> _updateStatus(Request request) async{
    // 解析請求中的參數 id
    var id = int.parse(params(request,'id'));
    // 解析請求中的參數 status
    var status = int.parse(params(request,'status'));
    // 更新狀態
    var result = await _planService.updateStatus(id, status);
    var plan;
    // 如果更新成功則透過 id 獲取更新後的計畫
    if(result){
        plan = await _planService.getById(id);
    }
    return Response.ok(jsonEncode(plan,toEncodable: myEncode));
}

//DELETE 請求，更新指定 id 的計畫資訊
@Route.delete('/plan/<id>')
Future<Response> _delete(Request request) async{
    var id = int.parse(params(request,'id'));
    //Map 物件，用於儲存回應資訊
    Map res = Map();
    if(id != null){
        res.addAll({'id':id});
        // 刪除計畫並返回結果
        var result = await _planService.delete(id);
        // 儲存刪除結果
        res.addAll({'delete':result});
    }
    return Response.ok(jsonEncode(res));
}

// 攔截所有指向未知路由的請求
@Route.all('/<ignored|.*>')
Future<Response> _notFound(Request request) async{
```

```
    return Response.notFound('頁面未找到');
  }
  // 返回本路由器的處理函數
  Handler get handler => _$DlRouter(this).handler;
}
```

17.3.5 跨域中介軟體

在實際專案中服務端通常需要回應本地之外的請求，因此需要為服務端提供回應跨域請求的能力。創建中介軟體，用於設定跨域請求。範例程式如下：

```
//chapter17/deadline_server/bin/utils/middle_cors.dart
import 'package:shelf/shelf.dart';

// 創建中介軟體並提供 corsHeaders 設定
final cors = createCorsHeadersMiddleware(
  corsHeaders:{
    'Access-Control-Allow-Origin': '*',
    'Access-Control-Expose-Headers': 'Authorization, Content-Type',
    'Access-Control-Allow-Headers': 'Authorization, Origin, X-Requested-
With, Content-Type, Accept',
    'Access-Control-Allow-Methods': 'GET, POST, PUT, PATCH, DELETE'
  }
);

Middleware createCorsHeadersMiddleware({Map<String, String> corsHeaders}) {
  // 未提供 corsHeaders 時，使用預設設定
  corsHeaders ??= {'Access-Control-Allow-Origin': '*'};

  // 請求處理常式
  Response handleOptionsRequest(Request request) {
    if (request.method == 'OPTIONS') {
      return Response.ok(null, headers: corsHeaders);
    } else {
      return null;
    }
  }
```

```
  // 回應處理常式，將 corsHeaders 的設定應用於回應
  Response addCorsHeaders(Response response) => response.change(headers:
corsHeaders);

  // 創建中介軟體並返回
  return createMiddleware(requestHandler: handleOptionsRequest,
responseHandler: addCorsHeaders);
}
```

17.3.6 介面卡

本專案採用 shelf 框架提供的介面卡來啟動服務。範例程式如下：

```
//chapter17/deadline_server/bin/server.dart
import 'package:shelf/shelf.dart';
import 'package:shelf_exception_handler/shelf_exception_handler.dart';
import 'package:shelf/shelf_io.dart' as io;
import 'utils/middle_cors.dart';
import 'routers.dart';

const _hostname = 'localhost';

void main() async {
  // 返回路由器的處理常式
  var routerHandler = Dl().handler;
  // 連接中介軟體和處理常式
  var handler = const Pipeline()
    .addMiddleware(exceptionHandler())
    .addMiddleware(logRequests())
    .addMiddleware(cors)
    .addHandler(routerHandler);
  // 啟動服務
  var server = await io.serve(handler, _hostname,1024);
  print('服務位址 http://${server.address.host}:${server.port}');
}
```

打開編輯器中的命令列工具，執行以下命令：

```
pub run build_runner build
```

命令執行成功後，路由生成器會生成一個新檔案。生成的程式如下：

```
//chapter17/deadline_server/bin/routers.g.dart
//GENERATED CODE - DO NOT MODIFY BY HAND

part of 'routers.dart';

//**************************************************************
//ShelfRouterGenerator
//**************************************************************

Router _$DlRouter(Dl service) {
   final router = Router();
   router.add('GET', r'/plans', service._getAll);
   router.add('GET', r'/plan/<id>', service._getById);
   router.add('POST', r'/plan', service._create);
   router.add('PUT', r'/plan/<id>', service._update);
   router.add('PUT', r'/plan/<id>/<status>', service._updateStatus);
   router.add('DELETE', r'/plan/<id>', service._delete);
   router.all(r'/<ignored|.*>', service._notFound);
   return router;
}
```

然後執行 server.dart 檔案即可啟動服務。

17.4 編寫用戶端

創建 Angular 應用程式，命名為 deadline_web。將服務端的 entity.dart 檔案複製到專案的 lib/src/plan 目錄下。

17.4.1 管道

編寫範本中會用到的管道類別 TimeIntervalPipe 和 CompletePipe，前者用於計算時間間隔，後者根據完成狀態篩選計畫清單。程式如下：

```
//chapter17/deadline_web/lib/src/plan/date_pipe.dart
import 'package:angular/angular.dart';
import 'entity.dart';
//時間間隔計算管道
@Pipe('timeInterval', pure: true)
class TimeIntervalPipe implements PipeTransform{
  String transform(DateTime now,DateTime plantime) {
    if (plantime == null) return null;
    var planMills = plantime.millisecondsSinceEpoch;
    var nowMills = now.millisecondsSinceEpoch;
    //計畫期限內
    if(nowMills < planMills){
      var mills = planMills - nowMills;
      return '${_str(mills)}';
    }
    //逾時
    if(nowMills > planMills){
      var mills = nowMills - planMills;
      return '${_str(mills)}';
    }
  }
  //計算、天、時、分、秒
  String _str(int mills){
    var str = '';
    //相距天數
    var days = mills~/(3600*24*1000);
    //小時
    var hours = (mills%(3600*24*1000))~/(3600*1000);
    //分鐘
    var minutes =(mills%(3600*24*1000))%(3600*1000)~/(60*1000);
    //秒
    var seconds = (mills%(3600*24*1000))%(3600*1000)%(60*1000)~/1000;

    if(days>0) str += '$days 天 ';
    if(hours>0) str += '$hours 時 ';
    if(minutes>0) str += '$minutes 分 ';
    if(seconds>=0) str += '$seconds 秒 ';

    return str;
```

```
    }
}
// 根據 complete 值過濾集合
@Pipe('complete', pure: true)
class CompletePipe implements PipeTransform{
    List<Plan> transform(List<Plan> value,int complete){
        return value.where((plan) => plan.complete == complete).toList();
    }
}
```

17.4.2 服務

編寫向伺服器發出請求的服務類別，本專案採用 http 套件提供的方法向伺服器發出請求，並處理回應。程式如下：

```
//chapter17/deadline_web/lib/src/plan/plan_service.dart
import 'dart:async';
import 'dart:convert';
import 'package:angular/core.dart';
import 'package:http/http.dart' as http;
import 'package:json_string/json_string.dart';
import 'entity.dart';

// 請求伺服器位址和通訊埠
const _URL = 'http://localhost:1024';
// 定義服務
@Injectable()
class PlanService{

    // 從伺服器獲取所有計劃
    Future<List<Plan>> getAll() async{
        var plans;
        await http.get('$_URL/plans').then((response){
            if(response.statusCode == 200){
                // 解析伺服器回應資訊並轉為 Plan 物件集合
                plans = JsonString(response.body).decodeAsObjectList(Plan.fromJson);
            }
        });
```

```
    return plans;
}

// 根據指定 id 從伺服器獲取單一計畫
Future<Plan> getById(int id)async{
    var plan;
    await http.get('$_URL/plan/$id').then((response){
        if(response.statusCode == 200){
            // 解析伺服器回應資訊並轉為 Plan 物件
            plan = JsonString(response.body).decodeAsObject(Plan.fromJson);
        }
    });
    return plan;
}

// 提交創建請求
Future<Plan> post(Plan pl)async{
    var plan;
    await http.post('$_URL/plan',body:jsonEncode(pl)).then((response){
        if(response.statusCode == 200){
            // 解析伺服器回應資訊並轉為 Plan 物件
        plan = JsonString(response.body).decodeAsObject(Plan.fromJson);
        }
    });
    return plan;
}

// 更新計畫
Future<Plan> put(Plan pl)async{
    var plan;
    await http.put('$_URL/plan/${pl.id}',body: jsonEncode(pl)).then((response){
        if(response.statusCode == 200){
            plan = JsonString(response.body).decodeAsObject(Plan.fromJson);
        }
    });
    return plan;
}

// 更新計畫的完成狀態
```

```
  Future<Plan> putStatus(int id,int status)async{
    var plan;
    await http.put('$_URL/plan/$id/$status').then((response){
      if(response.statusCode == 200){
        plan = JsonString(response.body).decodeAsObject(Plan.fromJson);
      }
    });
    return plan;
  }

  // 刪除計畫
  Future<bool> delete(int id)async{
    var result = false;
    await http.delete('$_URL/plan/$id').then((response){
      if(response.statusCode == 200){
        // 解析伺服器回應資訊並轉為 Map 集合
        var map = JsonString(response.body).decodedValueAsMap;
        if(map != null) {
          // 獲取刪除結果
          result = map['delete'];
        }
      }
    });
    return result;
  }
}
```

17.4.3 增加計畫元件

創建元件 PlanAddComponent，用於增加計畫。程式如下：

```
//chapter17/deadline_web/lib/src/plan/plan_add_component.dart
import 'dart:async';
import 'package:angular/angular.dart';
import 'package:angular_components/angular_components.dart';
import 'package:angular_components/utils/browser/window/module.dart';
import 'package:angular_router/angular_router.dart';
import 'package:angular_forms/angular_forms.dart';
```

```
import 'plan_service.dart';
import 'entity.dart';
import 'package:intl/intl.dart';
import '../route_paths.dart';

@Component(
  selector: 'plan-add',
  templateURL: 'plan_add_component.html',
  directives: [
    coreDirectives,
    formDirectives,
    NgFor,
    NgIf,
    materialInputDirectives,
    MaterialDateTimePickerComponent,
    MaterialMultilineInputComponent,
    MaterialYesNoButtonsComponent,
    MaterialSubmitCancelButtonsDirective
  ],
  providers: [windowBindings,datepickerBindings,ClassProvider(PlanService)],
)


class PlanAddComponent{
  final PlanService _planService;
  final Router _router;
  Plan plan_add = Plan();
  // 設定使用者可選擇的最小時間點
  DateTime minDateTime = DateTime.now();

  // 定義日期格式
  DateFormat dateFormat = DateFormat("yy 年 MM 月 dd 日 ");
  DateFormat timeFormat = DateFormat("HH 時 mm 分 ");

  // 控制 yes 按鈕待定狀態
  bool pending = false;

  // 注入服務 _planService 和路由 _router
  PlanAddComponent(this._planService,this._router){
```

```
        // 初始化計畫時間
        plan_add.plantime = DateTime.now().add(Duration(hours: 1));
    }

    // 回應 yes 按鈕的點擊事件
    void add() async{
        pending = true;
        var dl = await _planService.post(plan_add);
        if(dl == null){
            getWindow().alert(' 增加失敗 !');
        }else{
            getWindow().alert(' 增加成功 !');
        }
        pending = false;
    }

    // 回應 no 按鈕的點擊事件，前往計畫清單
    Future<NavigationResult> gotoList() =>
        _router.navigate(RoutePaths.plans.toURL());
}
```

增加計畫元件的範本程式如下：

```
//chapter17/deadline_web/lib/src/plan/plan_add_component.html
<form style="width:360px;border:1px solid #eee;margin:0 auto;padding:30px;">
    <div>
        <material-input label=" 事件 *"
                        floatingLabel
                        required
                        requiredErrorMsg=" 此輸入框必填 "
                        blurUpdate
                        [(ngModel)]="plan_add.title"></material-input>
    </div>

    <div>
        <material-input multiline
                        floatingLabel
                        blurUpdate
                        rows="2"
                        maxRows="4"
```

```
                        label=" 備註資訊 "
                        [(ngModel)]="plan_add.remarks"></material-input>
   </div>
   <div>
      <text> 截止時間 </text>
      <material-date-time-picker [(dateTime)]="plan_add.plantime"
                        [outputDateFormat]="dateFormat"
                        [outputTimeFormat]="timeFormat"
                        [minDateTime]="minDateTime"
                        required>
      </material-date-time-picker>
   </div>
   <div>
      <text> 標注顏色 </text>
      <input type="color" [(ngModel)]="plan_add.theme">
   </div>
<material-yes-no-buttons yesAutoFocus
                        submitCancel
                        raised
                        reverse
                        noText=" 返回清單 "
                        [pending]="pending"
                        yesText=" 增加 "
                        (yes)="add()"
                        (no)="gotoList()"
                        [yesDisabled]="plan_add.title==null || plan_add.
title==''"></material-yes-no-buttons>
</form>
```

17.4.4 編輯計畫元件

增加元件 PlanEditComponent，用於更新計畫。程式如下：

```
//chapter17/deadline_web/lib/src/plan/plan_edit_component.dart
import 'package:angular/angular.dart';
import 'package:angular_components/angular_components.dart';
import 'package:angular_components/utils/browser/window/module.dart';
import 'package:angular_forms/angular_forms.dart';
```

```
import 'package:angular_router/angular_router.dart';
import 'package:intl/intl.dart';

import 'plan_service.dart';
import 'entity.dart';
import '../route_paths.dart';

@Component(
  selector: 'plan-edit',
  templateURL: 'plan_edit_component.html',
  directives: [
    coreDirectives,
    formDirectives,
    NgFor,
    NgIf,
    materialInputDirectives,
    MaterialDateTimePickerComponent,
    MaterialMultilineInputComponent,
    MaterialYesNoButtonsComponent,
    MaterialSubmitCancelButtonsDirective
  ],
  providers: [windowBindings,datepickerBindings,ClassProvider(PlanService)],
)
// 實現 OnActivate 介面
class PlanEditComponent implements OnActivate{
  final PlanService _planService;
  final Router _router;
  Plan plan_edit = Plan();

  // 設定使用者可選擇的最小時間點
  DateTime minDateTime = DateTime.now();

  // 定義日期顯示格式
  DateFormat dateFormat = DateFormat("yy 年 MM 月 dd 日 ");
  DateFormat timeFormat = DateFormat("HH 時 mm 分 ");

  // 控制 yes 按鈕待定狀態
  bool pending = false;
```

```
    // 注入服務 _planService 和路由 _router
    PlanEditComponent(this._planService,this._router);

    // 實現 onActivate 路由生命週期函數
    @override
    void onActivate(RouterState previous, RouterState current) async{
      // 解析當前路由的參數 id
      var id = current.parameters['id'];
      // 獲取需要編輯的計畫
      if(id != null){
      plan_edit = await _planService.getById(int.parse(id));
      }
    }

    // 回應 yes 按鈕的點擊事件，更新計畫
    void update() async{
      pending = true;
      var plan = await _planService.put(plan_edit);
      if(plan != null){
        plan_edit = plan;
        getWindow().alert(' 更新成功 !');
      }else{
        getWindow().alert(' 更新異常 !');
      }
        pending = false;
    }

    // 回應 no 按鈕的點擊事件，前往計畫清單
    Future<NavigationResult> gotoList() =>
      _router.navigate(RoutePaths.plans.toURL());
}
```

編輯計畫的元件範本程式如下：

```
//chapter17/deadline_web/lib/src/plan/plan_edit_component.html
<form *ngIf="plan_edit != null" style="width:360px;border:1px solid
#eee;margin:0 auto;padding:30px;">
    <div>
      <material-input label=" 事件 *"
```

```
                    floatingLabel
                    required
                    requiredErrorMsg=" 此輸入框必填 "
                    blurUpdate
                    [(ngModel)]="plan_edit.title"></material-input>
</div>

<div>
   <material-input multiline
                    floatingLabel
                    blurUpdate
                    rows="2"
                    maxRows="4"
                    label=" 備註資訊 "
                    [(ngModel)]="plan_edit.remarks"></material-input>
</div>

<div>
   <text> 截止時間 </text>
   <material-date-time-picker [(dateTime)]="plan_edit.plantime"
                    [outputDateFormat]="dateFormat"
                    [outputTimeFormat]="timeFormat"
                    [minDateTime]="minDateTime"
                    required>
   </material-date-time-picker>
</div>
<div>
   <text> 標注顏色 </text>
   <input type="color" [(ngModel)]="plan_edit.theme">
</div>
<material-yes-no-buttons yesAutoFocus
                    submitCancel
                    raised
                    reverse
                    noText=" 返回清單 "
                    yesText=" 更新 "
                    [pending]="pending"
                    (yes)="update()"
                    (no)="gotoList()"
```

```
                    [yesDisabled]="plan_edit.title==null || plan_edit.
title==''"></material-yes-no-buttons>
</form>
```

17.4.5 計畫清單元件

增加元件 PlanListComponent，用於展示所有的計畫。程式如下：

```
//chapter17/deadline_web/lib/src/plan/plan_list_component.dart
import 'dart:async';
import 'package:angular/angular.dart';
import 'package:angular_components/angular_components.dart';
import 'package:angular_components/utils/browser/window/module.dart';
import 'package:angular_router/angular_router.dart';

import 'plan_service.dart';
import 'entity.dart';
import 'date_pipe.dart';
import '../route_paths.dart';

@Component(
  selector: 'plan-list',
  styleURLs: ['plan_list_component.css'],
  templateURL: 'plan_list_component.html',
  directives: [
    MaterialListComponent,
    MaterialListItemComponent,
    MaterialCheckboxComponent,
    MaterialIconComponent,
    MaterialButtonComponent,
    MaterialTooltipDirective,
    coreDirectives,
    NgFor,
    NgIf,
  ],
  providers: [popupBindings, windowBindings,ClassProvider(PlanService)],
```

```
  pipes: [TimeIntervalPipe, CompletePipe, AsyncPipe, DatePipe],
)
class PlanListComponent implements OnInit {
  final PlanService _planService;
  // 儲存完成狀態的計畫
  List<Plan> plans_complete = <Plan>[];
  // 儲存未完成狀態的計畫
  List<Plan> plans_uncomplete = <Plan>[];
  DateTime time_now = DateTime.now();
  Timer _timer;
  final Router _router;

  PlanListComponent(this._planService, this._router) {
    // 計時器，每隔 1s 更新一次 time_now 的值
    _timer = Timer.periodic(Duration(seconds: 1), (timer) {
      time_now = DateTime.now();
    });
  }

  @override
  Future<Null> ngOnInit() async {
    // 獲取所有計劃
    var plans_get = await _planService.getAll();
    if (plans_get != null) {
      // 根據計畫完成狀態分組
      plans_get.forEach((item) {
        if (item.complete == 1) {
          plans_complete.add(item);
        } else {
          plans_uncomplete.add(item);
        }
      });
      // 根據計畫事件排序
      plans_uncomplete.sort(sortByPlantime);
      plans_complete.sort(sortByPlantime);
    }
  }
```

```
// 更新指定 id 的計畫
void edd(int id) async {
   var index = plans_uncomplete.indexWhere((d) => d.id == id);
   if (index != null) {
      var dl2 = await _planService.put(plans_uncomplete[index]);
      plans_uncomplete[index]= dl2;
   }
}

// 刪除指定 id 的計畫
void delete(int id) async {
   var index = plans_uncomplete.indexWhere((d) => d.id == id);
   var result = await _planService.delete(id);
   if(result){
      plans_uncomplete.removeAt(index);
      getWindow().alert(' 刪除成功 !');
   }
}

// 更新指定 id 計畫的狀態
void eddStatus(int id, int status) async {
   var index = plans_uncomplete.indexWhere((d) => d.id == id);
   var dl = await _planService.putStatus(id, status);
   if (dl != null && dl.complete == 1) {
      plans_complete.add(dl);
      plans_uncomplete.removeAt(index);
      plans_complete.sort(sortByPlantime);
   }
}

// 回應核取方塊 checkedChange 事件
void onChecked(int id, bool isChecked) {
   eddStatus(id, isChecked ? 1 : 0);
}

// 根據計畫完成時間排序的函數
```

```
    int sortByPlantime(a, b) {
        return a.plantime.millisecondsSinceEpoch
        .compareTo(b.plantime.millisecondsSinceEpoch);
    }

    // 跳躍
    void onSelect(int id) {
        _gotoDetail(id);
    }

    // 跳躍到指定 id 計畫詳細頁
    Future<NavigationResult> _gotoDetail(int id) =>
        _router.navigate(RoutePaths.edit.toURL(parameters: {'id': '$id'}));
}
```

範本程式如下：

```
//chapter17/deadline_web/lib/src/plan/plan_list_component.html
<div> 待完成 </div>
<material-list>
    <material-list-item *ngFor="let plan of plans_uncomplete; let i = index">
        <div class="item-dl" [style.border-left-color]="plan.theme">
            <div>
                <material-checkbox [disabled]="false"
                            [checked]="false"
                            (checkedChange)="onChecked(plan.id,$event)"
                            materialTooltip=" 標記為完成 "></material-checkbox>
                <span class="action">
                    <material-icon *ngIf="" icon="{{time_now.millisecondsSinceEpoch
> plan.plantime.millisecondsSinceEpoch ? 'more_time' : 'timelapse'}}"
                            materialTooltip="{{time_now.millisecondsSinceEpoch >
plan.plantime.millisecondsSinceEpoch ? ' 逾時 ' : ' 倒計時 '}}"></material-icon>
                    <material-button icon (trigger)="delete(plan.id)"
materialTooltip=" 刪除 ">
                        <material-icon icon="delete"></material-icon>
                    </material-button>
                    <material-button icon (trigger)="onSelect(plan.id)"
```

```
materialTooltip=" 編輯 ">
                  <material-icon icon="edit"></material-icon>
               </material-button>
            </span>
         </div>
         <div>
            <span class="title">{{plan.title}}</span>
            <span class="current">
               {{time_now | timeInterval:plan.plantime}}
            </span>
         </div>
         <p class="remarks">{{plan.remarks}}</p>
         <div class="time">
            <span class="plantime">{{plan.plantime | date:'yyyy.MM.dd
HH:mm'}}</span>
         </div>
      </div>
   </material-list-item>
</material-list>
<div> 已完成 </div>
<material-list>
   <material-list-item *ngFor="let plan of plans_complete">
      <div class="item-dl" [style.border-left-color]="plan.theme">
         <material-checkbox [disabled]="true"
                           [checked]="true"></material-checkbox>
         <div>
            <span class="title">{{plan.title}}</span>
            <span class="current"> 已完成 </span>
         </div>
         <p class="remarks">{{plan.remarks}}</p>
         <div class="time">
            <span class="plantime">{{plan.plantime | date:'yyyy.MM.dd
HH:mm'}}</span>
            <span class="endtime">{{plan.endtime | date:' - yyyy.MM.dd
HH:mm'}}</span>
         </div>
```

```
        </div>
    </material-list-item>
</material-list>
```

樣式程式如下：

```
//chapter17/deadline_web/lib/src/plan/plan_list_component.css
material-list{
    width:720px;
}

material-list-item{
    padding:0;
}

.item-dl{
    width:720px;
    padding:6px 10px;
    border-left: 3px solid #5dbe8a;
    border-bottom: 1px solid #eee;
}
.title{
    font-weight: bold;
}
.remarks{
    margin: 1px 0;
    font-size:small;
    font-weight: 300;
}
.time{
    font-size:12px;
    line-height:14px;
    font-weight: 100;
    font-style: italic;
}

.current{
    float:right;
    align-items: center;
```

```
    display: flex;
}

.action{
    float:right;
    align-items: center;
    display: flex;
}
```

增加未知頁回應元件。程式如下：

```
//chapter17/deadline_web/lib/src/not_found_component.dart
import 'package:angular/angular.dart';

@Component(
    selector: 'not-found',
    template: '<h2> 頁面未找到 </h2>',
)
class NotFoundComponent {}
```

17.4.6 路由

增加路由路徑定義檔案。程式如下：

```
//chapter17/deadline_web/lib/src/route_paths.dart
import 'package:angular_router/angular_router.dart';

class RoutePaths{
    // 增加計畫的路由路徑
    static final add = RoutePath(path: 'plan/add');
    // 編輯計畫的路由路徑
    static final edit = RoutePath(path: 'plan/edit/:id');
    // 計畫清單的路由路徑
    static final plans = RoutePath(path: 'plans');
}
```

增加路由定義檔案。程式如下：

```
//chapter17/deadline_web/lib/src/routes.dart
import 'package:angular_router/angular_router.dart';

import 'plan/plan_list_component.template.dart' as plan_list_template;
import 'plan/plan_add_component.template.dart' as plan_add_template;
import 'plan/plan_edit_component.template.dart' as plan_edit_template;
import 'not_found_component.template.dart' as not_found_template;

import 'route_paths.dart';
export 'route_paths.dart';

class Routes{
  // 計畫清單的路由定義，連接增加計畫到元件
  static final plans = RouteDefinition(
    routePath: RoutePaths.plans,
    component: plan_list_template.PlanListComponentNgFactory,
  );
  // 增加計畫的路由定義
  static final add = RouteDefinition(
    routePath: RoutePaths.add,
    component: plan_add_template.PlanAddComponentNgFactory,
  );
  // 編輯計畫的路由定義
    static final edit = RouteDefinition(
    routePath: RoutePaths.edit,
    component: plan_edit_template.PlanEditComponentNgFactory,
  );

  static final all = <RouteDefinition>[
    plans,
    add,
    edit,
    // 攔截首頁並跳躍到路由計畫清單
    RouteDefinition.redirect(
      path: '',
      redirectTo: RoutePaths.plans.toURL(),
    ),
    // 攔截未知頁
```

```
      RouteDefinition(
         path: '.*',
         component: not_found_template.NotFoundComponentNgFactory,
      ),
   ];
}
```

17.4.7 佈局

專案通常會利用根元件進行佈局，並且會將其作為路由元件。程式如下：

```
//chapter17/deadline_web/lib/app_component.dart
import 'package:angular/angular.dart';
import 'package:angular_router/angular_router.dart';
import 'package:angular_components/angular_components.dart';
import 'src/routes.dart';

@Component(
   selector: 'my-app',
   styleURLs:
   ['package:angular_components/app_layout/layout.scss.css','app_component.
css'],
   templateURL: 'app_component.html',
   providers: [routerProviders],
   directives: [
      routerDirectives,
      MaterialPersistentDrawerDirective,
      MaterialToggleComponent,
      MaterialIconComponent,
      MaterialButtonComponent,
      MaterialListComponent,
      MaterialListItemComponent,
      ],
   exports: [RoutePaths,Routes],
)
class AppComponent{}
```

根元件範本程式如下：

```
//chapter17/deadline_web/lib/app_component.html
<header class="material-header shadow dense-header">
   <div class="material-header-row">
      <material-button icon
               class="material-drawer-button" (trigger)="drawer.toggle()">
         <material-icon icon="menu"></material-icon>
      </material-button>
      <span class="material-header-title"> 應用欄 </span>
   </div>
</header>
<material-drawer persistent #drawer="drawer">
   <!-- 抽屜內容 -->
   <material-list>
      <!-- 不帶標籤的組元素 -->
      <div group>
         <material-list-item>
            <a [routerLink]="RoutePaths.plans.toURL()"><material-icon
icon="list"></material-icon> 返回所有計劃 </a>
         </material-list-item>
<material-list-item>
         <a [routerLink]="RoutePaths.add.toURL()"><material-icon icon="add"></
material-icon> 增加 </a>
         </material-list-item>
      </div>
      <!-- 帶標籤的組元素 -->
      <div group>
         <div label>Tags</div>
         <material-list-item>
            <material-icon icon="star"></material-icon>Favorites
         </material-list-item>
      </div>
   </material-list>
</material-drawer>
<material-content style="width:720px;margin:0 auto;padding:30px;">
   <!-- 內容 -->
```

```
    <router-outlet [routes]="Routes.all"></router-outlet>
</material-content>
```

樣式程式如下：

```
//chapter17/deadline_web/lib/app_component.css
material-list-item a{
   width: 100%;
   align-items: center;
   display: flex;
   text-decoration:none;
   color:rgba(0,0,0,0.87);
}
material-list-item a:hover{
   text-decoration: none;
}
```

執行專案，將在瀏覽器中展示計畫清單，可以透過導覽前往增加和編輯計畫的頁面。